21 Sept 79

BASIC APPLIED STATISTICS

Steve and Anne

With compliments

John

One for the family!!

STATISTICS: Textbooks and Monographs

A SERIES EDITED BY

D. B. OWEN, Coordinating Editor
Department of Statistics
Southern Methodist University
Dallas, Texas

PAUL D. MINTON
Virginia Commonwealth University
Richmond, Virginia

JOHN W. PRATT
Harvard University
Boston, Massachusetts

Volume 1: The Generalized Jackknife Statistic, *H. L. Gray and W. R. Schucany*

Volume 2: Multivariate Analysis, *Anant M. Kshirsagar*

Volume 3: Statistics and Society, *Walter T. Federer*

Volume 4: Multivariate Analysis: A Selected and Abstracted Bibliography, 1957-1972, *Kocherlakota Subrahmaniam and Kathleen Subrahmaniam* (out of print)

Volume 5: Design of Experiments: A Realistic Approach, *Virgil L. Anderson and Robert A. McLean*

Volume 6: Statistical and Mathematical Aspects of Pollution Problems, *John W. Pratt*

Volume 7: Introduction to Probability and Statistics (in two parts) Part I: Probability; Part II: Statistics, *Narayan C. Giri*

Volume 8: Statistical Theory of the Analysis of Experimental Designs, *J. Ogawa*

Volume 9: Statistical Techniques in Simulation (in two parts), *Jack P. C. Kleijnen*

Volume 10: Data Quality Control and Editing, *Joseph I. Naus*

Volume 11: Cost of Living Index Numbers: Practice, Precision, and Theory, *Kali S. Banerjee*

Volume 12: Weighing Designs: For Chemistry, Medicine, Economics, Operations Research, Statistics, *Kali S. Banerjee*

Volume 13: The Search for Oil: Some Statistical Methods and Techniques, *edited by D. B. Owen*

Volume 14: Sample Size Choice: Charts for Experiments with Linear Models, *Robert E. Odeh and Martin Fox*

Volume 15: Statistical Methods for Engineers and Scientists, *Robert M. Bethea, Benjamin S. Duran, and Thomas L. Boullion*

Volume 16: Statistical Quality Control Methods, *Irving W. Burr*

Volume 17: On the History of Statistics and Probability, *edited by D. B. Owen*

Volume 18: Econometrics, *Peter Schmidt*

Volume 19: Sufficient Statistics: Selected Contributions, *Vasant S. Huzurbazar (edited by Anant M. Kshirsagar)*

Volume 20: Handbook of Statistical Distributions, *Jagdish K. Patel, C. H. Kapadia, and D. B. Owen*

Volume 21: Case Studies in Sample Design, *A. C. Rosander*

Volume 22: Pocket Book of Statistical Tables, *compiled by R. E. Odeh, D. B. Owen, Z. W. Birnbaum, and L. Fisher*

Volume 23: The Information in Contingency Tables, *D. V. Gokhale and Solomon Kullback*

Volume 24: Statistical Analysis of Reliability and Life-Testing Models: Theory and Methods, *Lee J. Bain*

Volume 25: Elementary Statistical Quality Control, *Irving W. Burr*

Volume 26: An Introduction to Probability and Statistics Using BASIC, *Richard A. Groeneveld*

Volume 27: Basic Applied Statistics, *B. L. Raktoe and J. J. Hubert*

OTHER VOLUMES IN PREPARATION

BASIC APPLIED STATISTICS

B. L. Raktoe and J. J. Hubert

Department of Mathematics and Statistics
University of Guelph
Guelph, Ontario

MARCEL DEKKER, INC. New York and Basel

Library of Congress Cataloging in Publication Data
Raktoe, B L
Basic applied statistics

(Statistics, textbooks and monographs ; v. 27)
Bibliography: p.
Includes index.
1. Statistics. 2. Mathematical statistics.
I. Hubert, John Joseph, (Date) joint author.
II. Title

QA276.12.R34 519.5 79-727
ISBN 0-8247-6537-0

MARCEL DEKKER, INC.
270 Madison Avenue, New York, New York 10016

Current printing (last digit):
10 9 8 7 6 5 4 3 2 1

PRINTED IN THE UNITED STATES OF AMERICA

To

Shanti

Sant

Jody

Jason

PREFACE

This book is based on lecture notes and has been used successfully in an introductory course in statistics for students in the social sciences. The main features of the book are:

1. It has been written for students without grade 12 mathematics, and the development of the topics uses only elementary algebra.

2. Throughout the book we have adopted the rigorous approach of defining the concepts formally and stating the results as theorems (without proofs). This produces a coherent and concise treatment of the subject matter.

3. There is an abundance of real world examples from the social sciences.

4. At the end of each chapter there are many exercises, for a total of over 350. There are demonstrative experiments for which the students, individually or as a group, can discover other properties that are not discussed in the chapter. We have provided the answers to most of the exercises.

5. For the examples and exercises we have appealed to over 60 recently published papers, for which the references are included in the bibliography.

6. There is an emphasis on sampling distributions and their applications. All probabilistic interpretations of calculated results are from the repetitive sampling viewpoint.

7. We have included chapters on experimental design and sampling from finite populations, topics typically missing in similar texts.

The first four chapters deal with noninferential or descriptive statistics. Chapters 5 to 9 develop the notions of probability and probability distributions and provide sampling distributions of Z, t, χ^2, and F. Chapters 10 to 13 then utilize these distributions in problems of estimation and testing. The fundamentals of regression and correlation analysis are given in Chapter 14. The last two chapters, 15 and 16, contain the fundamentals of design and analysis of experiments and sampling from finite populations, respectively. The topics in these two chapters are developed in a spirit which differs from most introductory texts. These sixteen chapters can be taught in one semester consisting of thirty-nine to forty-two lecture hours.

The authors are grateful to the Literary Executor of the late Sir Ronald A. Fisher, F.R.S.; to Dr. Frank Yates, F.R.S.; and to Longman Group Ltd., London, for permission to reprint in abridged form Tables IV and VII from their book, *Statistical Tables for Biological, Agricultural and Medical Research* (sixth edition, 1974); to the *Biometrika* Trustees, for permission to reproduce tables of percentage points of the t and F distributions; to McGraw-Hill Book Co., for permission to reproduce a table on binomial probabilities; and to the Chemical Rubber Co., for permission to reproduce an abridged table of random numbers.

Our sincere thanks go to June Hubert for her expert typing and suggestions for improvement in style.

B. L. Raktoe

J. J. Hubert

CONTENTS

1

INTRODUCTION AND SOME FUNDAMENTAL CONCEPTS

Statistics as a scientific discipline is a recognized branch of the mathematical sciences. As such it uses many of the other branches as a tool, especially the theory of probability. On the other hand, statistics itself is a tool for other disciplines. If its methods are applied to biology, one uses the terms biological statistics or biometrics. Similarly, the usefulness of statistics led to the development of many new disciplines, such as econometrics, sociometrics, bibliometrics, and technometrics.

Whenever data are collected with the purpose of studying a phenomenon, techniques are needed to properly collect, analyze, infer from, and present the data. This leads us to the following definition of statistics:

Statistics is a branch of the mathematical sciences concerned with

1. Development of methods to collect data
2. Development of methods to analyze data
3. Development of methods to infer from data
4. Development of methods to present data

These aspects are known as design of surveys and experiments, statistical analysis, statistical inference, and statistical presentation, respectively. Before going into these aspects, we should note that in any scientific investigation there must be a well-formulated problem outlining the phenomenon to be studied. If a survey is called for, then the plan to obtain the data is a *survey design*. On the other hand, if the scientist is planning to carry out a controlled experiment, then the investigation requires an *experimental design*. Any process of observation or measurement is referred to as an *experiment* or a *survey*. The result which one obtains from such a process (a simple "yes" or "no" answer, an instrument reading, the result of certain calculations, etc.) is called an *outcome*. The methods of summarizing the data depend on how the data were collected, i.e., on the type of survey that was conducted or the type of experimental design used. Sometimes averages are sufficient. In more complex investigations the statistical analysis can be quite complicated. In most practical situations a complete enumeration of every item in a prescribed population is not feasible due to time limitations or budget restrictions. The solution is obtained by extracting a *sample*, so that after the analysis an inference is made from sample quantities to population quantities. Since repeated samples do not produce the same result, a reliability coefficient is needed to reflect the performance of the actual sample drawn from the population. It is here that the theory of probability comes in, as will be seen later. Let us now define certain terms which will be used over and over again, because without precise definitions it is impossible to do anything in statistics.

DEFINITION 1.1 *A population is the collection of all individuals on which one or more variables are to be observed, counted, or measured.*

Populations are also referred to as *universes*, and individuals as *elements* or *units*. A population, then, need not consist of people. All the milk-producing cows in Canada as of December 31, 1975 form a population. Similarly, all the housewives in Toronto, Canada, form a population with, for example, number of marriages being the variable of interest.

Populations can be *finite* or *infinite*. All populations in nature are finite. The population of integers is infinite. Populations can be hypothetical or theoretical; for example, all the repeated IQ measurements we could have made of a high school student lead to a population consisting of the student being identified by the first test, the second test, etc.

If it is clear from the context, then the population consisting of individuals is replaced by the collection of values of the variable of interest. Thus the collection of grade 8 students in Wellington County, Ontario on December 31, 1977 is represented by the collection of average grades if this is the variable of interest. In this case we would speak of the population of average grades.

In the definition of population we have used the term "variable." The following definitions clarify the notion of a variable and also provide the various types which are encountered in actual surveys and experiments.

DEFINITION 1.2 *A variable is a quantity that varies from individual to individual in a population.*

The individual outcomes taken on by a variable need not be numerical values. Also, a variable whose outcomes do not vary from individual to individual is known as a *constant*. Such "constant populations" are of no statistical interest, because observing one individual is sufficient to describe the population. Fortunately (for statisticians, that is) nature abhors constancy, and it is nature's variability which makes statistics such a rich descipline. It is clear that there would be neither the discipline of statistics nor statisticians without variability, because there would exist no "statistical" problem in a "constant population."

Variables whose individual outcomes are not numerical are known as *qualitative* variables, in contrast to *quantitative* variables, whose outcomes are always real numbers. Thus for the population of schizophrenic medical patients, the variable which classifies a patient as "paranoid" or "nonparanoid" is a qualitative variable. On the other hand, the IQ scores based on the "Stanford Achievement Test," or the weight in kilograms of individuals, are quantitative variables, because they take on real numbers for every individual.

A much more important distinction between quantitative variables is given in the following definitions.

DEFINITION 1.3 *A discrete variable is a quantitative variable whose individual values are real numbers separated by definite distances.*

DEFINITION 1.4 *A continuous variable is a quantitative variable whose individual values can be any of the possible values from an interval of real numbers.*

The number of questions answered correctly in an examination consisting of 10 true-false questions and given to pupils in a well-defined population is a discrete variable because any individual can have 0, 1, 2, ..., 10 of the questions correct. On the other hand, variables such as height, weight, and chronological age are examples of continuous variables.

By the way, populations whose elements can take on only two outcomes with respect to a variable are known as "dichotomous" populations. An example of a dichotomous variable is the "yes-no" variable that would occur as a response to the question whether one is voting or not voting for a candidate in a given election.

No doubt, the reader can provide examples of "trichotomous" populations. If meaningful integer scores can be attached to a qualitative variable, it then becomes a discrete variable. And, if values from an interval of real values can be assigned to a qualitative variable, it becomes a continuous variable. For example, the qualitative variable, sex, is presently treated by many psychologists as a continuous variable by assigning an individual a value in the interval 0 to 1, where 0 means the highest degree of femininity, and 1 means the highest degree of masculinity.

Quantitative variables are usually measured in units such as kilograms, centimeters, or degrees Celsius; or they consist of counts, scores, or numerical ratings.

The degree to which variables in general may be quantified with respect to scales of measurements has led to their division into the four following types.

DEFINITION 1.5 *A variable with a nominal scale assumes levels which are specified name categories.*

This definition implies that all qualitative variables have a nominal scale, since these are classifiers. As an example, consider the variable "religion" with the name categories Protestant, Catholic, Jewish, and others.

DEFINITION 1.6 *A variable with an ordinal scale is a nominal scale variable which has the added property that its levels can be ranked in some order.*

The ranking implies a "less than or more than" meaning to the categories and is indicated by either symbols or numerical values. No specific distances are connotated by the rankings; i.e., units are of no concern. For example, a psychologist may rank three groups of 5-year-old children in terms of the variable "rapidity of learning a task" as:

Rapidity of Learning of 5-Year-Old Children	
Fast learner	or 3
Average learner	or 2
Slow learner	or 1

DEFINITION 1.7 *A variable with an interval scale is an ordinal scale variable such that distances between the levels can be assigned.*

IQ test scores of seventh grade school children form a good example of a variable on an interval scale.

IQs of Seventh Grade Children

Student	Score
1	137
2	136
3	135
4	121
5	120
6	119

Note that the difference (i.e., distance) between, say, 137 and 136, is the same as between, say, 120 and 119. An inherent difficulty with an interval scale is that ratios have no meaning; e.g., a student whose IQ score is 1.5 times higher than that of another student is not 1.5 times as intelligent as that other student, because of the absence of an origin. That is, zero IQ score does not mean no intelligence.

DEFINITION 1.8 *A variable with a ratio scale is an interval scale variable whose levels can be expressed as ratios.*

This definition implies that the scale has an origin, so that ratios are meaningful. Height, age, and income are examples of variables which are on a ratio scale. A person who earns $20,000 yearly earns twice as much as a person who earns $10,000 yearly.

Let us now define the concepts of a census and a sample.

DEFINITION 1.9 *A census is an enumeration of every individual in a population with respect to one or more variables.*

A common example of a census or complete enumeration is the property tax census, which is carried out in every municipality of Canada. The main variable of interest is the value of the property, because this determines the amount of municipal tax to be paid by the property owner.

For most practical situations complete enumerations are time consuming and costly, so that an economical and reliable method is necessary to obtain information on a specified population. For example, it would be impossible to check every drop of water in a city water supply for purity. The concept which helps us out is that of a sample:

DEFINITION 1.10 *A sample is a partial enumeration of a population with respect to one or more variables.*

This definition implies that the number of elements in a sample can include one or more elements but fewer than the total number in the population being sampled. The number of elements in a sample is referred to as the *sample size*, and the letter n is traditionally used for this. If N is used to indicate the size of the population, then one may write $1 \le n < N$, where $\le$ means "less than or equal to" and $<$ means "less than." The process or procedure of extracting a sample from a population is called the *sampling technique*. It is the device of sampling which makes statistics a powerful science, because the idea is to infer to the population at large via a sample, which contains only part of the information. This is important enough to warrant the diagram shown in Figure 1.1.

On the left-hand side of Figure 1.1 we have a symbolic representation of a population. Each cross depicts an element and the top arrow depicts the extraction of a sample, which is the small blob on the right-hand side. If the sample is "properly" extracted and it has "captured" the population, then our inference back to the population, as shown by the lower arrow, will be quite "good." It is obvious that repeated samples produce different results. Thus we will need a measure of how reliably or how precisely a particular sample has performed. The amazing thing is (as will be seen later) that we can compute the reliability or precision in many settings from one sample. To have a deeper understanding of what is going on when sampling a population, we need the following concept:

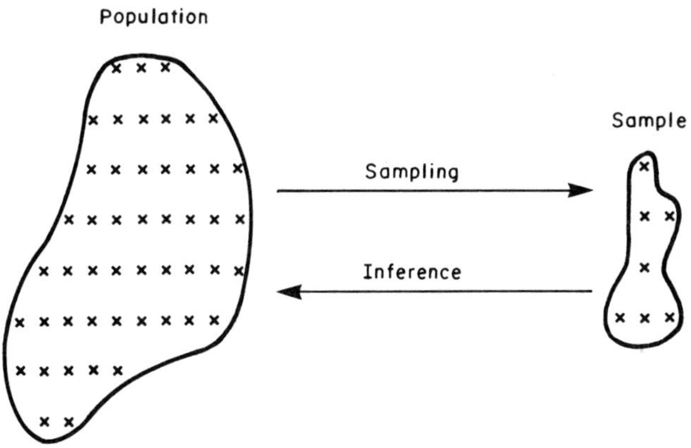

Figure 1.1 Sampling a population.

DEFINITION 1.11 *A parameter is a quantity derived from all the individual values of a variable in a population.*

In typical investigations parameters are not known initially, and it is the task to "determine" them either via a complete enumeration or a sample. Thus a parameter can only be known exactly if we use a complete enumeration, while a sample will provide an estimate of a parameter. Suppose that the parameter of interest is the average or mean IQ of all 1977 grade 12 male students in Canada. By carrying out a complete enumeration of all high schools and obtaining the IQ of each grade 12 male student, one can find the parameter by averaging all the individual IQs. Therefore a quicker way to obtain an idea of the mean IQ is to carry out a sample survey and then estimate the parameter by calculating the average of the sample values. This leads to

DEFINITION 1.12 *A sample statistic or simply a statistic is any quantity calculated from the individual values in the sample.*

This definition reserves the word "statistics" for sample quantities, while in the literature we encounter various other meanings, such as the outcomes of football games, the number of deaths on holiday weekends, or the measurements of a girl in a beauty contest. Of course, "Statistics" with a capital S refers to the discipline, which we have defined earlier. Thus in the example above, if a sample of 100 students was selected the *statistic* to *estimate* the population mean would be the sample mean of 100 IQ scores.

EXERCISES

1.1 True or false:
   (a) Parameters are sample quantities.
   (b) A sample of observations is a collection of individual observations from a population selected by a specified procedure.
   (c) All populations in nature are finite.
   (d) A census is a partial enumeration of a population.

(e) Statistics provide estimates of parameters.
(f) Parameters are known when we use a sample rather than a complete enumeration.
(g) Measurements of continuous variables cannot be represented by decimal fractions.
(h) Integers can sometimes be used to represent discrete quantities.
(i) A sampling technique is a method of extracting a sample from a population.
(j) An ordinal scale number tells us the position of a quantity in the ranking.
(k) Qualitative variables have outcomes which are always real numbers.
(l) The ratio scale differentiates and orders objects according to a standard unit and with reference to an absolute zero point.
(m) Arithmetic operations can be performed on numbers from interval and nominal scales, but they cannot be performed on numbers from ratio and ordinal scales.

1.2 Which of the four types of scales does each of the following involve? Is each variable continuous or discrete?
(a) A person's place in a seniority system
(b) Life expectancy
(c) Barometric pressure
(d) Postal code
(e) First, second, and third place in a contest to determine who has the longest beard
(f) The volume of a soup can
(g) The year in a date (1886, 1966, etc.)
(h) The number of windows of a building in a given city

1.3 For the following, state the main variable or variables of interest and indicate the type.
(a) A test in social studies was given to 185 students and marks were recorded.

(b) A study was performed on a sample of individuals 16 years and over to detect any association between education level and marriage adjustment.

(c) A standardized intelligence test is administered to 82 children to detect any difference in intelligence from the general population.

(d) At a fitness club, members' weights and measurements were recorded before and after a 6-week period to decide whether there was an improvement in body fitness.

(e) The average amount of electricity consumed per household per month in an urban city.

(f) The sex of a student entering university.

(g) In a chemistry laboratory, the freezing and boiling points of water were recorded in degrees Fahrenheit, and various temperatures were recorded above and below zero, the reference point.

1.4 Give examples of populations in sociology, psychology, economics, politics, geography, and history. You must indicate a variable of interest for each example.

1.5 Give an example of a government sponsored census and state the reasons why it was carried out.

1.6 Find examples (in books or journals of your own interest) for qualitative, quantitative, discrete, continuous, nominal, ordinal, interval, and ratio scale variables.

1.7 Give an example of a hypothetical or theoretical population with a continuous variable of interest.

1.8 For a population of your own choice describe a parameter and indicate how this can be estimated.

1.9 On what scale is the variable "net income of professors"?

Net Income of Professors

Far above average
Above average
Average
Below average
Far below average

1.10 The following data concern the variable "crimes of violence" which was measured by the number of offences committed against individuals in Alberta during the years 1962 to 1966.

1962	4144
1963	5028
1964	5170
1965	5589
1966	7141

What is the scale of this variable? Explain your answer.

1.11 The following data are from Clark and Johnston (1974). What types of variables are represented?

Psychometric Data of Five Subjects

Case	Date of birth	PPVT	PMTQ
1	Sept. 21, 1950	83	76
2	Apr. 30, 1950	63	48
3	Mar. 28, 1950	58	36
4	Sept. 25, 1950	70	83
5	Mar. 18, 1950	88	91

PPVT = Peabody Picture Vocabulary Test

PMTQ = Porteus Maze Test Quotient

2

DATA PRESENTATION

When a survey or an experiment is conducted, there usually arises a need to reduce the raw data to the form of tables, graphs or charts, so that a better overview can be obtained of the phenomenon under study. This is especially the case if there are a large number of units being observed, measured, classified, or counted. Often a simple table will do the trick. In other situations the use of a frequency histogram is appropriate. The objective of this chapter is to discuss *some* of the methods which have been adopted in published literature to present data.

2.1 TABLES

DEFINITION 2.1 *A statistical table is an array consisting of rows and/or columns with a clear heading depicting the quantities appearing in each row and/or column.*

If there are no columns, such a table is called a *one-way table*. On the other hand, if both rows and columns are meaningful, then the array is referred to as a *two-way* table. These notions can be generalized to *higher-way* tables, which are usually called *k-way* or *k-dimensional* tables. The more dimensions a table has the more difficult it becomes to understand the data. Table 2.1 is a very simple example of a one-way table. It refers to the variable "size of the student population in the University of Guelph taking an introductory statistics course." A simple table like this seems to show that the

Table 2.1 Size of the Student Body Enrolled in Statistics 100 in the University of Guelph for a 2-Year Period

Semester	No. of Students
Fall 1970	80
Winter 1971	61
Fall 1971	92
Winter 1972	69

winter population for that course is smaller than the fall population. Also, from the table it appears that there is an increase through time, which might be explained by increased popularity of the course or the overall increase in student enrollment in the University of Guelph. Table 2.1 is a one-way table because we are interested in one factor only, namely, time, with the four semesters being the four levels.

If we were interested in taking into account another factor, for example, sex, then a two-way table would do the job. Such a "refinement" is illustrated in one of the next type of presentations.

2.2 GRAPHICAL PRESENTATIONS

DEFINITION 2.2 *A bar graph is a graph consisting of horizontal or vertical rectangles or bars, the height of each bar depicting a magnitude of a variable.*

The bar graph is especially useful in bringing out the *relative magnitudes* of a variable at various levels of a factor. If there are too many levels under consideration, the bar graph becomes crowded. The spacing of the bars depends on the phenomenon being presented. There is no general working rule as to spacing of the bars. In practice, one often encounters a space of at least half the base of a bar.

If a bar graph is such that each bar is divided in parts to depict the magnitude of a variable with respect to the levels of another factor, then we speak of a *component bar graph*. Usually, too many components reduce the value of the presentation because of difficulty in visualization.

A side-by-side placement of the components in a bar graph results in a *grouped bar graph*. Examples of component and grouped bar graphs are shown in Figures 2.1 and 2.2, respectively.

Note that the sizes according to sex are brought out more sharply in the grouped bar graph than in the component bar graph. This is an effect of the side-by-side placement of the bars.

A very popular way to depict parts contributing to a total is presented by the following definition:

DEFINITION 2.3 *A circle chart or pie chart is a circle partitioned into segments such that the angle of each segment is proportional to the contribution of each part in the total value of a variable, where the total corresponds to 360°.*

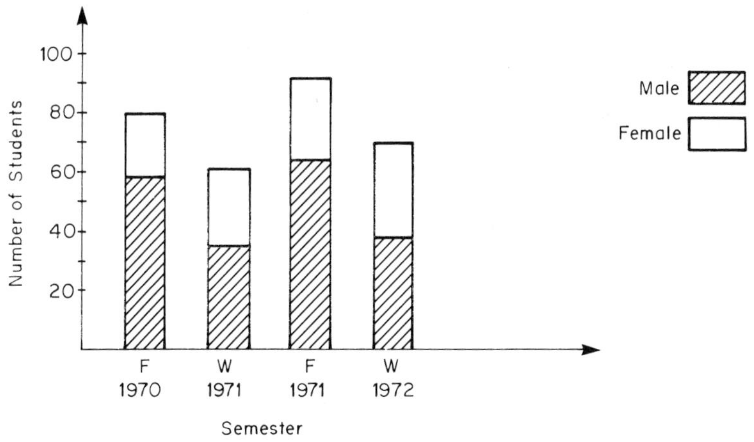

Figure 2.1 Size of the student body of Statistics 100 at the University of Guelph, according to semester and sex, in a component bar graph.

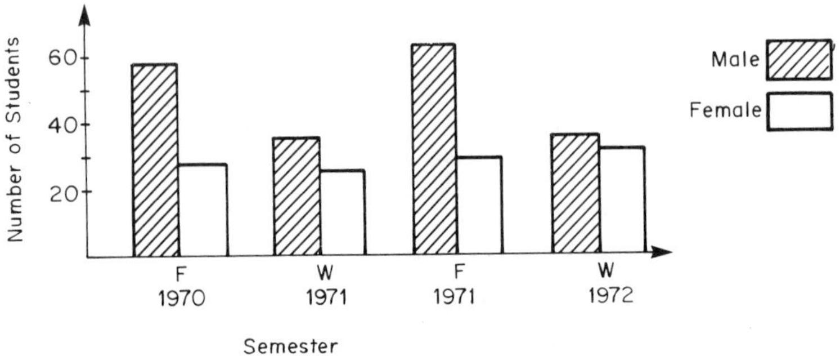

Figure 2.2 Size of the student body of Statistics 100 at the University of Guelph, in a grouped bar graph.

If the total is partitioned into too many parts, the resulting circle chart will be crowded and the relative magnitudes will not be clearly observable. To illustrate the construction of a pie chart, let T be the total magnitude of a variable and A, B, C be the magnitudes of the parts; i.e., $T = A + B + C$. Since T corresponds to 360°, then geometrically the angle of the segment corresponding to A is given by $(A/T) \cdot 360°$. Similarly the angles for B and C are obtained as $(B/T) \cdot 360°$ and $(C/T) \cdot 360°$.

Thus if we take the fall semester of 1970 in Figure 2.1, the angles corresponding to 58 males and 22 females are obtained as: $(58/80) \cdot 360° = 261°$ and $(22/80) \cdot 360° = 99°$, respectively. Note that the second number could in this case be obtained by subtraction, i.e., $360° - 261° = 99°$. The circle chart for the fall semester of 1970 is presented in Figure 2.3.

An extension of the circle chart is the "cylinder chart"; the reader is invited to find the discussion of this elsewhere.

In the next definition we introduce a graph which has become extremely popular, but can be ambiguous because the parts of the graph are hard to make proportional to the magnitudes represented.

DEFINITION 2.4 *A pictorial chart or ideograph is a graph where magnitudes of a variable are depicted by pictures of objects, people, etc.*

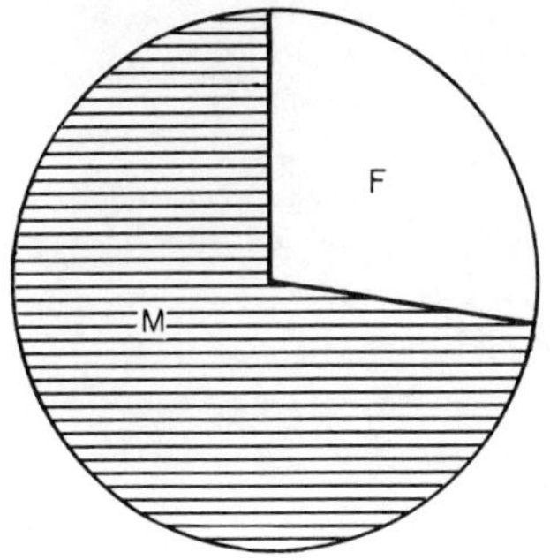

Figure 2.3 A circle chart of the student body, according to sex, of the course Statistics 100 at the University of Guelph for the fall semester of 1970.

This type of chart is illustrated in Figure 2.4, which shows that parts of a figure can lead to misrepresentations of magnitudes.

DEFINITION 2.5 *A broken line graph is a graph connecting magnitudes of a variable.*

Such graphs occur in situations where data are analyzed through time. The purpose of so-called *time-series* data is to bring out the changes and point out trends if they exist. Figure 2.5 is an illustration in point.

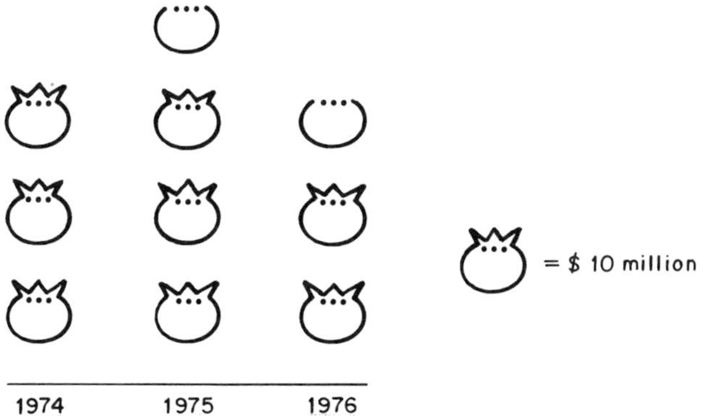

Figure 2.4 Savings account deposits in a local bank, for the years 1974, 1975, and 1976.

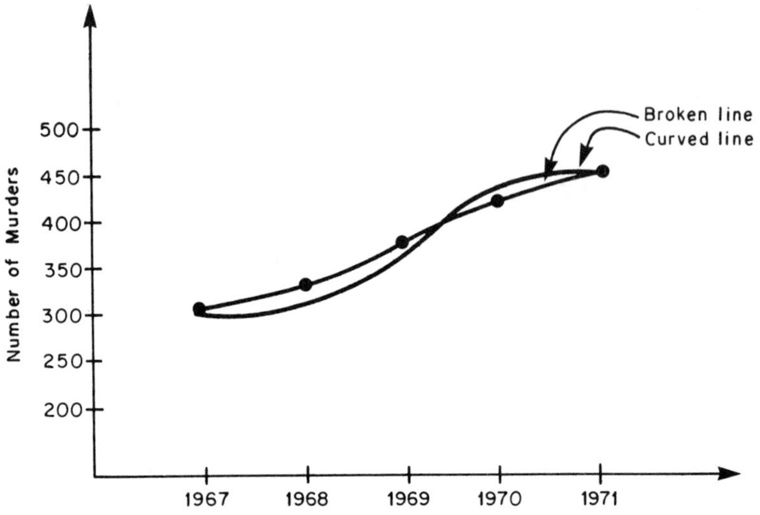

Figure 2.5 Homicides in Canada for 1967-1971 (from Michalos, 1974).

If a continuous curve is used to depict the magnitudes, then the resulting graph is called a *curve line graph*. In Figure 2.5 the curve line graph is a "freehand" curve. If such a curve is fitted to the data using a technique, then the freehand curve is a rough approximation if the technique gives a "better" fit. More uses of curves will be encountered later. Also, note that a broken line graph or curve line graph can always be obtained from a bar graph by connecting the magnitudes at the midpoints. Finally, changes in a broken line graph can be easily overemphasized by expanding the scale on the vertical axis of the graph.

Let us now proceed with the presentation of some celebrated methods of quantitative data. Suppose the following represents a set of measurements (in centimeters) on the Müller-Lyer (1889) illusion variable from 13 subjects: 4, 2, 2, 1, 3, 4, 2, 6, 6, 6, 7, 8, 4. A measurement here is a rounded off number in the sense that 4 cm means a measurement *greater than or equal to* 3.5 cm and *less than* 4.5 cm.

The other measurements have a similar meaning. The numbers 3.5 and 4.5 are known in the literature as *exact limits for* 4 cm.

One way to group these data in a table is shown in Table 2.2.

Before proceeding further we state the following preliminary definitions, which will be needed to define *frequency distributions*. Such distributions are especially suitable when a relatively large number of values of a variable are collected and we would like to review them in a grouped fashion. We have introduced a small number of data for illustration purposes.

DEFINITION 2.6 *A class is an interval of values of a variable.*

The concept of an interval needs clarification. An *interval* is simply a segment of the real line. Thus the first class, 0.5-3.5, forms a line segment starting at 0.5 and ending at but not including 3.5.

DEFINITION 2.7 *A class boundary (or class limit) is an end-point of a class.*

In our example, the measurements were read to one decimal and then rounded off and recorded to whole centimeters. In general, if measurements are recorded to m decimals, i.e., they are rounded off to m decimals, then to prevent values from falling on the class boundaries one uses the exact limits to set up the classes. Thus if a measurement is recorded as $a_1a_2 \cdots a_n.b_1b_2 \cdots b_m$, then class limits will appear with m + 1 decimals.

Table 2.2 Thirteen Measurements Arranged in Classes

Class boundaries	Class limits	Mid-points	Class frequency	Cumulative frequency
1-3	0.5-3.5	2	5	5
4-6	3.5-6.5	5	6	11
7-9	6.5-9.5	8	2	13

The importance of exact limits for continuous variables is quite obvious. How can we extend the notion of exact limits to the discrete case? Well, here we introduce a useful artificiality. Suppose that a discrete variable takes on integer values. If we take an integer to mean a degenerate interval (equal to any point on the real line), then most conveniently we can create a class around an integer by subtracting 0.5 from and adding 0.5 to the integer. The resulting class then has as limits the integer minus 0.5 and the integer plus 0.5. For example, in a dichotomous population a discrete variable can take on the values 0 and 1. These two values have class limits -0.5 to 0.5 and 0.5 to 1.5, respectively. Other discrete cases can be handled in a similar way. Moreover, values can be grouped. Therefore, we have unified the discrete and continuous cases with the above approach.

DEFINITION 2.8 *A class midpoint is the midpoint of an interval defining a class.*

From the above we know that in the integer-valued case the classes are the integers minus and plus 1/2 and the midpoints are the integers themselves.

DEFINITION 2.9 *The class width of a class is the difference between the higher and lower class limits.*

Thus, in our example the class width of the first class is equal to 3.5 - 0.5 = 3 cm. A similar definition holds for the width of all the classes simultaneously. This number is referred to as the *range of the classes*, which in our example is equal to 9.5 - 0.5 = 9 cm.

Therefore in the discrete case the class width is equal to 1 for every class if the values are not grouped, since each class is an integer plus and minus 0.5.

In our example each class has the same width. In practice we need not adhere to this, because if there are too few data in some adjacent classes or some classes have zero frequency, then these classes may be combined.

When setting up classes for a continuous variable with a large number of data, 10 to 20 classes seem to be about the best for visual presentation of a frequency distribution (the notion of a frequency distribution will be defined shortly). Consequently once the number of classes is decided upon and if each class is to have the same width, then the approximate width is equal to the range of the data divided by the number of classes. On the other hand, if the uniform class width is given, then one can determine the approximate number of classes. As remarked earlier, classes need not have the same width and some classes can have zero frequency.

DEFINITION 2.10 *A class frequency is the number of values of a variable falling in a given class.*

This notion can be simply stated as the number of elements or individuals whose values fall in a given class. In many cases one is not interested in the *absolute frequency* of a class but rather in the *relative frequency*, which is the class frequency divided by the total number of values under consideration. If one multiplies the relative frequency by 100%, then we speak of the *percentage relative frequency* of a class. In Table 2.2 the absolute frequency of the first class is equal to 5 while the relative frequency is equal to 5/13, so that the percentage relative frequency is $(5/13) \cdot 100\% = 38.5\%$.

DEFINITION 2.11 *The cumulative frequency of a given class is the sum of the frequencies of all classes up to and including the frequency in the given class.*

The cumulative frequency of the second class in our example is equal to 5 + 6 = 11. Note that the cumulative frequency of the last class is always equal to the total number of values under consideration. Also, notice that the relative cumulative frequency of the last class is 1.

DEFINITION 2.12 *A frequency table is a table with classes and/or their midpoints along with their frequencies.*

By this definition Table 2.2 is a frequency table.

DEFINITION 2.13 *A frequency histogram is a bar graph such that the base of each of the adjacent rectangles is determined by the class limits and the height of a rectangle is the frequency of a class.*

The data of Table 2.2, when presented in a frequency histogram, result in Figure 2.6. (A *relative frequency histogram* uses relative frequency or percentage on the left vertical scale.)

DEFINITION 2.14 *A frequency polygon is a broken line graph connecting the frequencies at the midpoints of the classes including a class at each end with zero frequency.*

The data of Table 2.2 are presented in Figure 2.7 as a frequency polygon. Note that from a frequency polygon we may always reconstruct a frequency histogram. If the number of data is large, and hence there are a large number of classes, then a frequency polygon can be made into a fluent curve. This leads us to our next definition.

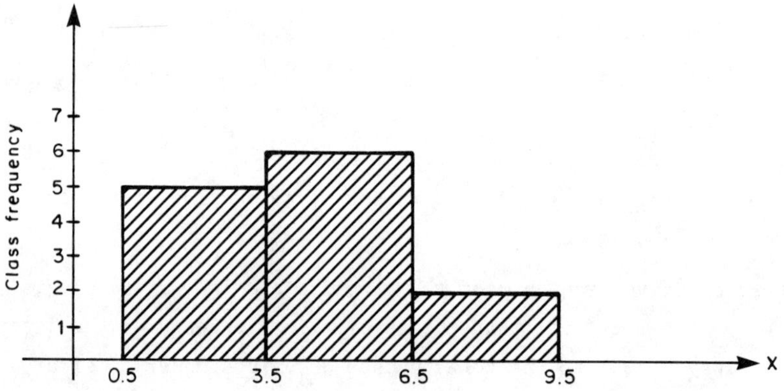

Figure 2.6 Frequency histogram of 13 values of a variable X measured in centimeters.

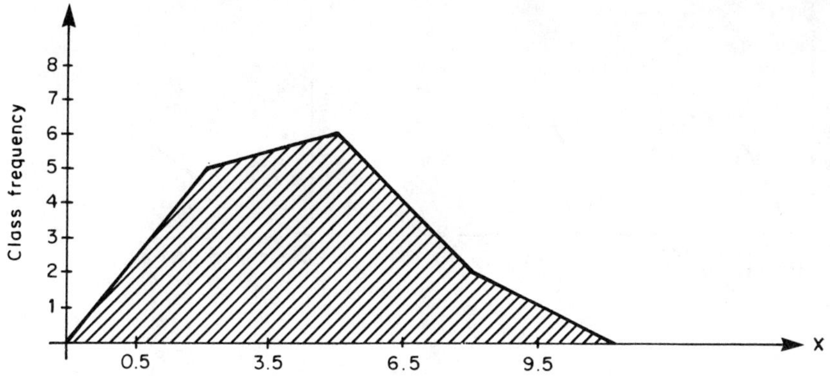

Figure 2.7 Frequency polygon of 13 values of a variable X measured in centimeters.

DEFINITION 2.15 *A frequency curve is a frequency polygon which has been drawn as a continuous curve if the number of values of a variable under consideration is large so that the number of classes is large.*

An illustration of such a curve is drawn in Figure 2.8. If the total area under a frequency curve is taken to be 100%, then

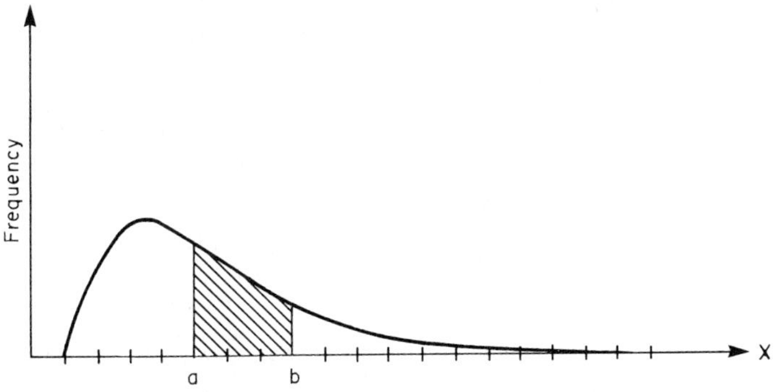

Figure 2.8 The frequency curve of a variable X for which a large number of values and a large number of classes are given.

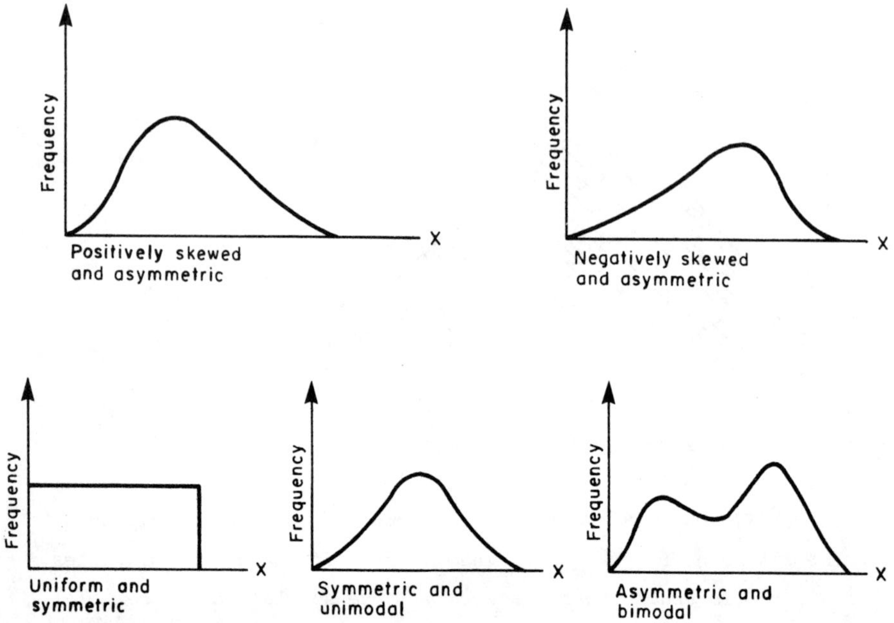

Figure 2.9 Some types of frequency curves.

the percentage of values falling in any given interval a to b is equal to the area under the curve from a to b. Such an interpretation is useful when we deal with probability distributions.

There are many shapes which a frequency curve can assume, depending on the nature of the variable being studied. Several such curves are drawn in Figure 2.9, with self-explanatory captions. Some of these curves, especially the symmetric and asymmetric unimodal ones, will find applications in the discussions of the distributions introduced in Chapters 6 and 9.

DEFINITION 2.16 *A cumulative frequency graph or ogive is a broken line graph connecting the cumulative frequencies at the upper limits of the classes.*

This concept is illustrated in Figure 2.10 for the data of Table 2.2.

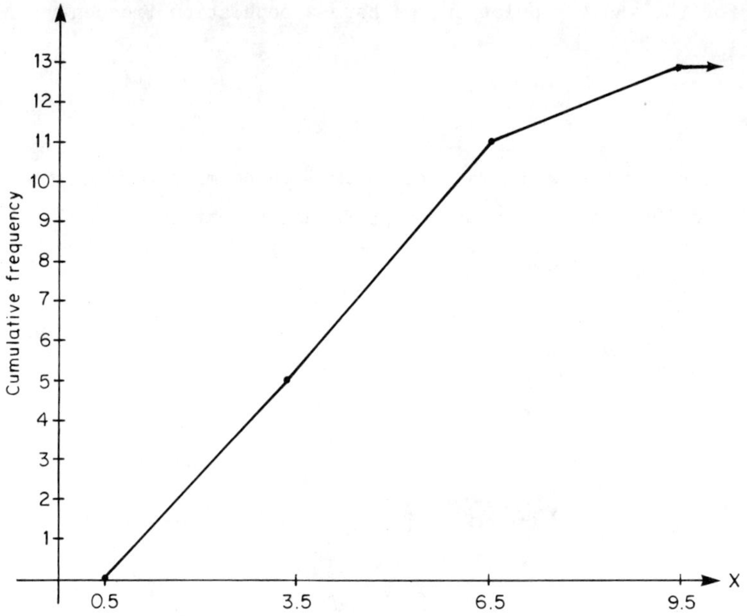

Figure 2.10 Cumulative frequency graph of 13 values of a variable X measured in centimeters.

We can define a cumulative frequency histogram in a similar fashion by letting the height of the rectangles be the cumulative frequencies of the classes. Also, instead of plotting the frequencies, we can plot the relative frequencies or percentages.

From a cumulative frequency graph or cumulative frequency histogram it is easy to make statements such as: The number of measurements less than 6.5 is equal to 11, and the number of measurements between 3.5 and 6.5 is equal to 11 - 5 = 6, which checks out with the frequency of the second class.

DEFINITION 2.17 *A frequency distribution is any table, chart, or graph consisting of classes and/or midpoints with the corresponding frequencies.*

Therefore when sample data are arranged in classes with their frequencies, we have a sample frequency distribution, and if the data are for the whole population, we have a population frequency distribution.

EXERCISES

2.1 The following two diagrams were taken with permission from the 1971 and 1974 University of Guelph Annual Reports.

1971 INCOME Year ended June 30

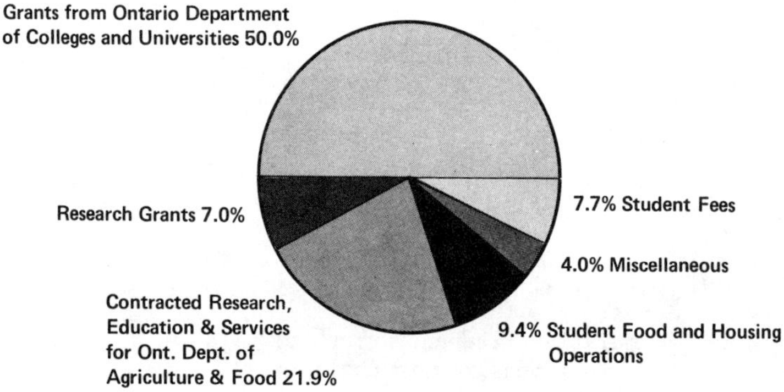

1974 INCOME Year ended April 30

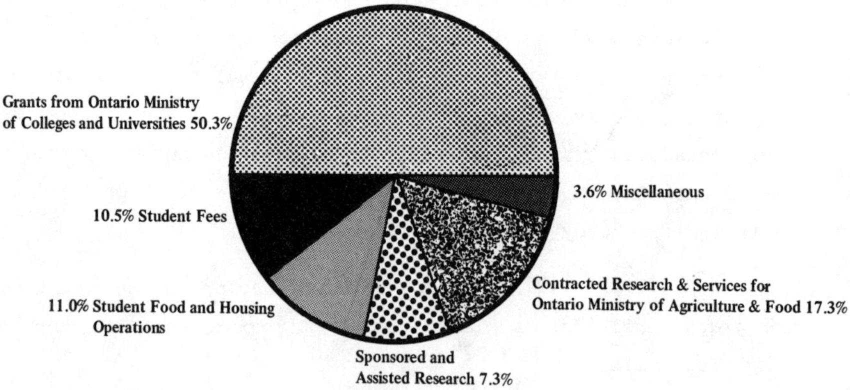

(a) Name the type of diagram used.

(b) What do these diagrams represent?

(c) What percentage of the area of the diagram refers to student fees in 1971? In 1974?

(d) What information cannot be obtained from these diagrams?

(e) The actual total income for the year ending April 30, 1974 was $56,346,000. How much income (in dollars) did the University of Guelph obtain from student food and housing operations?

2.2 Construct two bar charts in which the left vertical axis is the percentage of students for the following data.

Enrollment of Full-Time Undergraduates in the Fall Semesters of 1972 and 1973 at the University of Guelph

Degree program	1972	1973
B.A.	2647	2708
B.Sc.	1648	1963
D.V.M.	366	402
B.A.Sc.	752	916
B.Sc. (Agr.)	985	1117
B.Sc. (Eng.)	152	162
B.L.A.	95	104
B.Sc. (H.K.)	208	220
B. Comm.	163	208
Assoc. Dip.	268	275
Total	7284	8075

2.3 True or false:

(a) From a frequency table we may always derive the number of measurements less than an upper limit of a class.

(b) The cumulative frequency of the last class is always equal to the number of observations.

(c) The frequency in a class can never be zero for any kind of data.

(d) Frequency distributions are always symmetric.

(e) The relative frequency histogram is such that the sum of the relative frequencies is 1.

2.4 Make a circle chart from these data, supplied by a student who receives $2000.00 per year for university studies.

Board and lodging	$1100.00
Tuition	510.00
Entertainment	240.00
Presents for girlfriend	50.00
Miscellaneous	Remainder

2.5 A professor in statistics noted that in his 8 a.m. class in the winter semester, on the average 20% of the class is absent. Identifying 0 with "absent" and 1 with "present," draw the percentage histogram of the 8 a.m. class attendance.

2.6 A psychologist interviewed 10 undergraduate students in each of 50 universities on whether they have ever cheated or not. The results were as follows:

No. of yeses	0	1	2	3	4	5	6	7	8	9	10
No. of universities	1	11	19	9	2	2	2	1	2	1	0

(a) Plot the percentage frequency histogram for the above discrete variable.

(b) Plot the cumulative percentage frequency graph for the data above.

(c) If X denotes the number of yeses, then find how many universities had $2 \leq X \leq 7$. Also find how many universities had $X \geq 5$.

2.7 The following marks were recorded on a statistics test given to 25 students:

62, 61, 79, 81, 42, 74, 40, 55, 60, 61, 89, 43, 59,
47, 87, 66, 39, 73, 74, 75, 61, 82, 47, 77, 59

(a) Using 5 classes, plot a frequency histogram.

(b) Using 3 classes, plot a frequency histogram.

(c) Using 6 classes, plot a cumulative relative frequency histogram.

2.8 Construct a histogram with 6 classes for the following data on the variable X = length of sleep in minutes, as recorded by Webb and Hiestand (1975):

357, 453, 439, 370, 550, 494, 343, 377, 379, 404,
395, 389, 413, 458, 451, 460, 405, 227, 337, 384

2.9 Construct a diagram which you feel would best present the following data.

Distribution of Employment by Industry in 1971 for Canada

Industry	Percentage
Community, business, personal services	26.4
Manufacturing	22.6
Trade	16.2
Transportation, communication, other utilities	8.5
Construction	6.5
Agriculture	6.4
Public administration	6.3
Finance, insurance, real estate	4.4
Other primary industries	2.8

Source: Statistics Canada (1973b)

2.10 Construct a diagram which would best present the following data.

Age-Specific Fertility Rates, Canada, 1941-1971

Age group	1941	1951	1961	1971
15-19	30.7	48.1	48.2	40.1
20-24	138.4	188.7	233.6	134.6
25-29	159.8	178.8	219.2	142.3
30-34	122.3	144.5	144.9	77.4
35-39	80.0	86.5	81.1	33.6
40-44	31.6	30.9	28.5	4.4
45-49	3.7	3.1	2.4	0.6

Source: Statistics Canada (1973c)

2.11 From literature of your own choice list three methods of data presentation which have not been discussed in Chapter 2. Give an example of each.

2.12 From any publication find a set of measurement data and present them in a frequency histogram. Also construct a frequency polygon and a percentage cumulative frequency graph.

2.13 Sketch the ogive for the frequency curve for the uniform distribution in Figure 2.9.

2.14 Give an example of a variable for which you can construct a frequency curve.

2.15 For Exercise 2.7(c) complete the ogive.

2.16 Construct a broken line graph for the following data on the rates (per 100,000 inhabitants) for auto theft in Canada and the United States for the years 1964-1971 (from Michalos, 1974).

Year	U.S.A.	Canada
1964	245	209
1965	255	191
1966	285	195
1967	331	217
1968	389	244
1969	432	279
1970	454	293
1971	457	301

3

THE SUMMATION SIGN

In this chapter we introduce a tool which will be used throughout our study of statistics. This tool is the summation sign. The concept of summation is really not difficult, it is just a shorthand notation and anyone can quickly master it.

Variables are usually indicated by capital letters, e.g., U, V, X, Y, Z. Particular values of a variable, say X, are indicated by subscripts, e.g., X_1, X_2, X_3. Suppose that there are k values of a variable X under consideration, and let these values be indicated by $X_1, X_2, X_3, \ldots, X_k$. An arbitrary value of X can be indicated by X_i, which is also read as "X sub i." If we wish to talk about X_2, then clearly we are talking about X_i, with i = 2. This shows that the subscript i can go from 1 to k inclusive. The usefulness of the subscript i becomes apparent when arithmetic operations are to be carried out with the values $X_1, X_2, \ldots, X_k$. The selection of a subscript is not restricted to i, but j, k, ℓ, m, n, or any convenient letter can be used.

The simplest arithmetic operation is the *addition* of the values $X_1, X_2, X_3, \ldots, X_k$. A convenient symbol has been introduced for this purpose, namely, Σ (the Greek capital letter sigma).

DEFINITION 3.1 $\sum_{i=1}^{k} X_i = X_1 + X_2 + \cdots + X_k$, *where* $\sum$ *is called the summation sign and* $\sum_{i=1}^{k} X_i$ *is read as "summation* X_i*,* i *going from* 1 *to* k*,"* i *is the subscript or index of summation,* 1 *is the lower limit of summation, and* k *is the upper limit of summation.*

In our example from Chapter 2 (Müller-Lyer, 1889), $X_1 = 4$, $X_2 = 2$, $X_3 = 2$, $X_4 = 1$, $X_5 = 3$, $X_6 = 4$, $X_7 = 2$, $X_8 = 6$, $X_9 = 6$, $X_{10} = 6$, $X_{11} = 7$, $X_{12} = 8$, and $X_{13} = 4$. Therefore $\sum_{i=1}^{13} X_i = 4 + 2 + \cdots + 4 = 55$ cm. Note, $\sum_{i=1}^{k} X_i$ is essentially an instruction to sum all the values starting from X_1 and ending with X_k. We may start the summation at a lower limit different from 1 and end at an arbitrary limit; e.g.,

$$\sum_{i=2}^{6} X_i = X_2 + X_3 + X_4 + X_5 + X_6$$

$$= 2 + 2 + 1 + 3 + 4 = 12 \text{ cm}$$

for our example. We may delete certain values and indicate this in the summation; for example,

$$\sum_{\substack{i=1 \\ i \neq 3,4}}^{10} X_i = X_1 + X_2 + X_5 + \cdots + X_{10}$$

$$= 4 + 2 + 3 + \cdots + 6 = 33 \text{ cm}$$

for our data. We may raise the values to any power and then sum them; for example, we can square every value and then add them up. Using the summation sign, this is:

$$\sum_{i=1}^{k} X_i^2 = X_1^2 + X_2^2 + \cdots + X_k^2$$

$$= 4^2 + 2^2 + \cdots + 4^2 = 291 \text{ cm}^2$$

in our example. This quantity is known in statistics as the *raw* or *uncorrected sum of squares*.

We may add or subtract a fixed quantity from each of the values and then sum them:

$$\sum_{i=1}^{k} (X_i - a) = (X_1 - a) + (X_2 - a) + \cdots + (X_k - a)$$

Taking a = 4, we see that

$$\sum_{i=1}^{13} (X_i - 4) = (4 - 4) + (2 - 4) + \cdots + (4 - 4) = 3 \text{ cm}$$

for our example. Using a little bit of algebra on the right-hand side, we obtain:

$$\sum_{i=1}^{k} (X_i - a) = \sum_{i=1}^{k} X_i - ka$$

$$= 55 - (13)(4) = 3 \text{ cm}$$

which agrees with the earlier calculation. We may add or subtract a fixed quantity from each of the values, raise them to any power, and then add them:

$$\sum_{i=1}^{k} (X_i - a)^2 = (X_1 - a)^2 + (X_2 - a)^2 + \cdots + (X_k - a)^2$$

$$= (4 - 4)^2 + (2 - 4)^2 + \cdots + (4 - 4)^2 = 59 \text{ cm}^2$$

in our illustration. This quantity is known as the *sum of squares corrected for* a.

We may multiply or divide every value by a number a and then add them up:

$$\sum_{i=1}^{k} aX_i = aX_1 + aX_2 + \cdots + aX_k$$

Taking a = 4, we see that

$$\sum_{i=1}^{13} 4X_i = 4(4) + 4(2) + \cdots + 4(4) = 220 \text{ cm}$$

for our data. Thus the right-hand side can be written as: $a(X_1 + X_2 + \cdots + X_k)$, so that

$$\sum_{i=1}^{k} aX_i = a \sum_{i=1}^{k} X_i = 4 \sum_{i=1}^{13} X_i = 4(55) = 220 \text{ cm}$$

which verifies the above calculation. Using this, we may rewrite the sum of squares corrected for a as:

$$\sum_{i=1}^{k} (X_i - a)^2 = \sum_{i=1}^{k} X_i^2 - 2a \sum_{i=1}^{k} X_i + \sum_{i=1}^{k} a^2$$

$$= \sum_{i=1}^{k} X_i^2 - 2a \sum_{i=1}^{k} X_i + ka^2$$

$$= 291 - (2)(4)(55) + 13(4^2) = 59 \text{ cm}$$

as already calculated. Here we have used the fact that summation of a constant is k times the constant.

Note that when they are clear from the context, we delete limits of the summation and simply write ΣX_i. Also, when the subscript is clear we delete it and write ΣX.

An interesting case of the corrected sum of squares is the case where the deviations are taken from the arithmetic average or mean of the data, i.e., $\Sigma(X_i - \overline{X})^2$, where $\overline{X}$ = the arithmetic average or *mean* = $(1/k)\Sigma X_i$. In this case the *sum of squares corrected for the mean*, also known as the *sum of squares of deviations from the mean*, can be shown to be equal to:

$$\Sigma(X_i - \overline{X})^2 = \Sigma X_i^2 - \frac{1}{k}(\Sigma X_i)^2$$

To illustrate this concept, suppose that a sample of scores (out of 10) for a test written by five individuals was equal to:

6, 6, 7, 5, 8

Here $X_1 = 6$, $X_2 = 6$, $X_3 = 7$, $X_4 = 5$, and $X_5 = 8$. Using the summation sign, the sum or total of these scores is obtained and written out as:

$$\Sigma X_i = X_1 + X_2 + X_3 + X_4 + X_5 = 6 + 6 + 7 + 5 + 8 = 32$$

The raw sum of squares is equal to:

$$\Sigma X_i^2 = X_1^2 + X_2^2 + X_3^2 + X_4^2 + X_5^2 = 6^2 + 6^2 + 7^2 + 5^2 + 8^2 = 210$$

The average of the scores is:

$$\bar{X} = \frac{1}{k}\Sigma X_i = \frac{1}{5}(32) = 6.4$$

and the sum of squares corrected for the mean is:

$$\begin{aligned}\Sigma(X_i - \bar{X})^2 &= \Sigma X_i^2 - \frac{(\Sigma X_i)^2}{k} \\ &= 210 - \frac{(32)^2}{5} \\ &= 210 - 204.8 \\ &= 5.2\end{aligned}$$

Now suppose that there are two variables, say X and Y, observed for an individual, and there are k individuals under consideration;

$$\begin{aligned}\sum_{i=1}^{k} (X_i + Y_i) &= (X_1 + Y_1) + (X_2 + Y_2) + \cdots + (X_k + Y_k) \\ &= (X_1 + X_2 + \cdots + X_k) + (Y_1 + Y_2 + \cdots + Y_k) \\ &= \Sigma X_i + \Sigma Y_i\end{aligned}$$

At this stage one should easily be able to provide an illustration for these two variables. In many such situations where there are pairs of values for each individual one also computes the sum of the products of the values, i.e.,

$$\sum_{i=1}^{k} X_i Y_i = X_1Y_1 + X_2Y_2 + \cdots + X_kY_k$$

Another extension, which will be employed later on, is the use of two subscripts. Suppose there are v levels of a factor and that k values of a variable are observed at each level; then a general value of X can be indicated by X_{ij}, where X_{ij} is the jth value observed at the ith level of the factor. Here $i = 1, 2, \ldots, v$ and $j = 1, 2, \ldots, k$. If we display the data in a table, we get what is known as a one-way table, or simply a *one-way classification*.

To illustrate the use of summation in the one-way classification of data, suppose that 15 subjects (matched for other factors)

						Totals
Levels of factor	1	X_{11}	X_{12}	$\cdots$	X_{1k}	$\sum_{j=1}^{k} X_{1j}$
	2	X_{21}	X_{22}	$\cdots$	X_{2k}	$\sum_{j=1}^{k} X_{2j}$
	$\vdots$	$\vdots$	$\vdots$		$\vdots$	$\vdots$
	i	X_{i1}	X_{i2}	$\cdots$	X_{ik}	$\sum_{j=1}^{k} X_{ij}$
	$\vdots$	$\vdots$	$\vdots$		$\vdots$	$\vdots$
	v	X_{v1}	X_{v2}	$\cdots$	X_{vk}	$\sum_{j=1}^{k} X_{vj}$
						$\sum_{i=1}^{v} \sum_{j=1}^{k} X_{ij}$

were taught a skill using three different teaching methods, A, B, and C, say. The subjects were randomly assigned to the methods, and after a learning period a test was given with the following scores:

Teaching methods	A	4	6	6	7	8	31
	B	7	7	6	8	8	36
	C	8	8	7	9	8	40
							107

In this example:

$X_{11} = 4 \quad X_{12} = 6 \quad X_{13} = 6 \quad X_{14} = 7 \quad X_{15} = 8$

$X_{21} = 7 \quad X_{22} = 7 \quad X_{23} = 6 \quad X_{24} = 8 \quad X_{25} = 8$

$X_{31} = 8 \quad X_{32} = 8 \quad X_{33} = 7 \quad X_{34} = 9 \quad X_{35} = 8$

An arbitrary score is represented by X_{ij}, i = 1, 2, 3 and j = 1, 2, 3, 4, 5; i.e., X_{ij} is the score of the jth subject taught by the ith teaching method.

The sum of the values of the variable for the first level of the factor is $X_{11} + X_{12} + \cdots + X_{1k} = \sum_{j=1}^{k} X_{ij}$, which for our case is equal to $4 + 6 + 6 + 7 + 8 = \sum_{j=1}^{5} X_{ij} = 31$. The totals for the other levels of the factor are similarly obtained.

The *grand total* of all values of the variable X is equal to the sum of the *marginal totals*; i.e.,

$$\sum_{j=1}^{k} X_{1j} + \sum_{j=1}^{k} X_{2j} + \cdots + \sum_{j=1}^{k} X_{vj} = \sum_{i=1}^{v} \left(\sum_{j=1}^{k} X_{ij} \right)$$

which is usually written as $\sum_{i=1}^{v} \sum_{j=1}^{k} X_{ij}$. This grand total is also equal to the sum of all the individual values; i.e.,

$$\sum_{i=1}^{v} \sum_{j=1}^{k} X_{ij} = X_{11} + X_{12} + \cdots + X_{1k} + \cdots + X_{v1} + X_{v2} + \cdots + X_{vk}$$

In our example, the grand total is equal to

$$\sum_{j=1}^{5} X_{1j} + \sum_{j=1}^{5} X_{2j} + \sum_{j=1}^{5} X_{3j} = 31 + 36 + 40 = 107 = \sum_{i=1}^{3} \left(\sum_{j=1}^{5} X_{ij} \right)$$

As may be verified, this grand total is also equal to

$$\sum_{i=1}^{3} \sum_{j=1}^{5} X_{ij} = X_{11} + X_{12} + \cdots + X_{15} + \cdots + X_{31} + X_{32} + \cdots + X_{35}$$

$$= 4 + 6 + \cdots + 8 + \cdots + 8 + 8 + \cdots + 8 = 107$$

The expression $\sum_{i=1}^{v} \sum_{j=1}^{k} X_{ij}$ for the grand total is read as "double summation X_{ij}, i going from 1 to v and j going from 1 to k."

Note that from the totals one may obtain the averages by dividing through by k.

If there are two factors under consideration, say A and B with levels 1, 2, ..., v and 1, 2, ..., w, respectively, then X_{ij} refers to the value of the variable at the ith level of factor A and at the jth level of factor B. Such data are displayed in a two-way table or a *two-way classification*. This display with the relevant marginal totals and grand total is as follows:

		Levels of factor B						Totals
		1	2	⋯	j	⋯	w	
	1	X_{11}	X_{12}	⋯	X_{1j}	⋯	X_{1w}	$\sum_{j=1}^{w} X_{1j}$
	2	X_{21}	X_{22}	⋯	X_{2j}	⋯	X_{2w}	$\sum_{j=1}^{w} X_{2j}$
Levels of factor A	⋮	⋮	⋮		⋮		⋮	⋮
	i	X_{i1}	X_{i2}	⋯	X_{ij}	⋯	X_{iw}	$\sum_{j=1}^{w} X_{ij}$
	⋮	⋮	⋮		⋮		⋮	⋮
	v	X_{v1}	X_{v2}	⋯	X_{vj}	⋯	X_{vw}	$\sum_{j=1}^{w} X_{vj}$
		$\sum_{i=1}^{v} X_{i1}$	$\sum_{i=1}^{v} X_{i2}$	⋯	$\sum_{i=1}^{v} X_{ij}$	⋯	$\sum_{i=1}^{v} X_{iw}$	$\sum_{i=1}^{v} \sum_{j=1}^{w} X_{ij}$

One should be completely comfortable with such a display, and an example is called for before proceeding further. For this reason consider a maze performance experiment done with mice by a psychologist. Four mice in each of five litters were subjected to three concentrations of a hallucinatory drug and a placebo. These four levels of the drug were randomly allocated to the mice in each litter. The variable of interest was the time required to complete the circuit in a maze, and these values are given in seconds:

		Litter 1	2	3	4	5	Total
	Placebo	4	6	5	5	4	24
Levels	Conc. 1	7	6	8	6	7	34
of drug	Conc. 2	9	9	10	9	8	45
	Conc. 3	15	21	20	19	16	91
Total		35	42	43	39	35	194

Note that:

$X_{11} = 4$ $X_{12} = 6$ $X_{13} = 5$ $X_{14} = 5$ $X_{15} = 4$

$X_{21} = 7$ $X_{22} = 6$ $X_{23} = 8$ $X_{24} = 6$ $X_{25} = 7$

$X_{31} = 9$ $X_{32} = 9$ $X_{33} = 10$ $X_{34} = 9$ $X_{35} = 8$

$X_{41} = 15$ $X_{42} = 21$ $X_{43} = 20$ $X_{44} = 19$ $X_{45} = 16$

An arbitrary measurement is indicated by X_{ij}, $i = 1, 2, 3, 4$ and $j = 1, 2, 3, 4, 5$. The performance totals for the four levels of the drug are:

Placebo: $\sum_{j=1}^{5} X_{1j} = 24$

Conc. 1: $\sum_{j=1}^{5} X_{2j} = 34$

Conc. 2: $\sum_{j=1}^{5} X_{3j} = 45$

Conc. 3: $\sum_{j=1}^{5} X_{4j} = 91$

In terms of summation the litter totals are:

Litter 1: $\sum_{i=1}^{4} X_{i1} = 35$

Litter 2: $\sum_{i=1}^{4} X_{i2} = 42$

Litter 3: $\sum_{i=1}^{4} X_{i3} = 43$

Litter 4: $\sum_{i=1}^{4} X_{i4} = 39$

Litter 5: $\sum_{i=1}^{4} X_{i5} = 35$

The grand sum can be obtained by totaling either the drug totals or the litter totals (or by adding all the individual entries in the table). Thus:

$$\sum_{i=1}^{4} \sum_{j=1}^{5} X_{ij} = \sum_{i=1}^{4} \left(\sum_{j=1}^{5} X_{ij} \right) = 24 + 34 + 45 + 91 = 194$$

$$= \sum_{j=1}^{5} \left(\sum_{i=1}^{4} X_{ij} \right) = 35 + 42 + 43 + 39 + 35 = 194$$

$$= 4 + 6 + \cdots + 4 + 15 + 21 + \cdots + 16 = 194$$

From the totals one may calculate averages. Later on we will see how to assess the effects of the drug statistically.

EXERCISES

3.1 If $X_1 = 6$, $X_2 = 3$, $X_3 = 5$, $X_4 = 2$, and $X_5 = 2$, then find the answers to the following. ($\overline{X}$ means "average" of these 5 values of X; k = 5.)

(a) $\sum_{i=1}^{k} X_i$ (b) $\sum_{i=1}^{k} X_i^2$

(c) $\sum_{\substack{i=1 \\ i\neq 2,3}}^{k} (X_i - 5)$ (d) $\sum_{\substack{i=3 \\ i\neq 4}}^{k} (X_i + 6)$

(e) $\sum_{i=1}^{k} 3(X_i)$ (f) $\frac{1}{k}\sum_{i=1}^{k} X_i$

(g) $\frac{1}{k - 1}\sum_{i=1}^{k} (X_i - \overline{X})^2$ (h) $\sum_{i=1}^{k} (X_i - \overline{X})$

(i) $\sum_{i=1}^{k} (X_i - \overline{X})^3$ (j) $\frac{\sum_{i=1}^{k} X_i^2 - \left(\sum_{i=1}^{k} X_i\right)^2/k}{k - 1}$

(k) $\sum_{i=1}^{k} (X_i - 1)(X_i + 1)$ (l) $\sum_{j=1}^{k} X_j(X_j - 1)$

(m) $\sum_{i=1}^{k} (3X_i + 2X_i^2)$

3.2 Write in expanded form:

(a) $\sum_{i=1}^{6} (X_i - a)$ (b) $\sum_{i=1}^{6} (X_i - Y_i)$

(c) $\sum_{i=6}^{m} (X_i + a)^2$ (d) $\sum_{t=1}^{p} (X_t - a)^3$

(e) $\sum_{i=1}^{n} (b_i - X_i)$ (f) $\sum_{j=1}^{k} b_j(X_j - a)$

(g) $\sum_{a=1}^{k} X_a^a$ (h) $\sum_{i=1}^{k} (-1)^i X_iY_i$

(i) $\sum_{j=1}^{4} 2^j X_j^j Y_j^{j+1}$ (j) $\sum_{u=1}^{3} (-3)^u X_u$

(k) $\sum_{j=1}^{7} 1$ (l) $\sum_{j=0}^{3} (b_j - X_j)$

(m) $\sum_{u=1}^{N} u^2$ (n) $\sum_{n=0}^{5} (-1)^{n+1}$

(o) $\sum_{k=3}^{7} k$ (p) $\sum_{n=1}^{6} (-1)^{n+1} n$

3.3 Let $X_1, X_2, \ldots, X_k$ be a set of numbers. Express the following using the summation sign Σ:

(a) Sum of all numbers

(b) Sum of the first (k - 1) numbers

(c) Sum of squares of the numbers

(d) Sum of squares of deviations from 10

(e) Average of the numbers

(f) Sum of squares of the deviations from the average of the numbers divided by k - 1

3.4 Use the summation sign Σ to write the sums in a compact form:

(a) $Z_1 + Z_2 + \cdots + Z_6$

(b) $Y_1^2 + Y_2^2 + \cdots + Y_m^2$

(c) $X_1Y_1 + X_2Y_2 + \cdots + X_kY_k$

(d) $X_1f_1 + X_2f_2 + \cdots + X_nf_n$

(e) $X_1^2Y_1 + X_2^2Y_2 + \cdots + X_k^2Y_k$

(f) $X_1Y_1^3 + 2X_2Y_2^3 + 3X_3Y_3^3 + \cdots + kX_kY_k^3$

(g) $(X_1 - \overline{X}) + (X_2 - \overline{X}) + \cdots + (X_5 - \overline{X})$

(h) $(X_1 + X_2 + \cdots + X_k)^2/k$

(i) $X_0Y_0 + X_1Y_1 + X_2Y_2 + X_3Y_3$

(j) $(X_1 - \overline{X})^2 f_1 + (X_2 - \overline{X})^2 f_2 + \cdots + (X_k - \overline{X})^2 f_k$

(k) $a_0 + a_1 X + a_2 X^2 + \cdots + a_n X^n$

(l) $1 + 4 + 9 + 16 + 25 + 36$

3.5 Illustrate by example:

(a) $\sum_{i=1}^{k} cX_i = c \sum_{i=1}^{k} X_i$

(b) $\sum_{i=1}^{k} c = ck$

(c) $\sum_{i=1}^{k} (a + bX_i) = ak + b \sum_{i=1}^{k} X_i$

(d) $\sum_{i=1}^{k} (X_i + Y_i - Z_i) = \sum_{i=1}^{k} X_i + \sum_{i=1}^{k} Y_i - \sum_{i=1}^{k} Z_i$

(e) If $\overline{X} = \frac{1}{k} \sum_{i=1}^{k} X_i$, then $\sum_{i=1}^{k} (X_i - \overline{X}) = 0$.

(f) $\sum_{X=1}^{N} X = \frac{N(N + 1)}{2}$

(g) $\sum_{X=1}^{N} X^2 = \frac{N(N + 1)(2N + 1)}{6}$

(h) $\left(\sum_{i=1}^{N} X_i\right)^2 \neq \sum_{i=1}^{N} X_i^2$

3.6 Silverman and Battram (1975) selected a large number of Canadian students in each of four educational levels. At each level the subject was asked 10 questions about Canada and 10 questions about the United States. The entries in the table represent the mean number of correct answers out of 10 for the grade level.

Educational level	Knowledge of Canadian students About Canada	About U.S.
5	4.4	1.8
8	7.0	4.8
11	8.5	6.7
College	9.7	8.9

Using the notation developed in this chapter, find the values:

(a) Grand total (b) $\sum_{j=1}^{2} X_{2j}$ (c) $\sum_{i=1}^{4} X_{i2}$

3.7 From literature of your own choice select a one-way classification and illustrate the use of summation in obtaining marginal totals, marginal averages, and the grand total.

3.8 Do the same as in Exercise 3.7 for a two-way classification.

3.9 A sociologist was interested in investigating the phenomenon of "moonlighting" in Academia. He selected at random 25 faculty members within each rank and found the following results when the question "Have you ever moonlighted?" was asked:

Rank	Yes	No
Full professor	2	23
Associate professor	6	19
Assistant professor	11	14
Lecturer	13	12

(a) What is X_{ij}?

(b) Is this a one-way or two-way classification?

(c) Find: X_{11}, X_{12}, X_{21}, X_{22}, X_{31}, X_{32}, X_{41}, X_{42}.

(d) Write out the summation expressions for all the six marginal totals.

3.10 In Exercise 3.9 find $\Sigma\Sigma_{ij}^{2}$ and describe it verbally.

3.11 In Exercise 3.9 find $(\Sigma\Sigma X_{ij})^2$ and give a verbal description o it.

3.12 Show using elementary algebra that the following is true:

$$\sum_{i=1}^{k} (X_i - \overline{X}) = 0$$

where $\overline{X} = (1/k)\Sigma X_i$. Describe this result verbally and illustrate it using some numbers.

3.13 Prove: If $\overline{X} = (1/k) \sum_{i=1}^{c} X_i f_i$ and $k = \sum_{i=1}^{c} f_i$, then

$$\sum_{i=1}^{c} (X_i - \overline{X}) f_i = 0$$

3.14 Prove: If $k = \sum_{i=1}^{c} f_i$, then

$$\sum_{i=1}^{c} (X_i + 1)^2 f_i = \sum_{i=1}^{c} X_i^2 f_i + 2 \sum_{i=1}^{c} X_i f_i + k$$

3.15 Prove: If $k = \sum_{i=1}^{c} f_i$ and $\overline{X} = (1/k) \sum_{i=1}^{c} X_i f_i$, then

$$\sum_{i=1}^{c} (X_i - \overline{X})^2 f_i = \sum_{i=1}^{c} X_i^2 f_i - \frac{1}{k} \left(\sum_{i=1}^{c} X_i f_i \right)^2$$

3.16 Prove:

$$\sum_{i=1}^{c} (X_i - \overline{X})^2 = \sum_{i=1}^{c} X_i^2 - \frac{1}{c} \left(\sum_{i=1}^{c} X_i \right)^2$$

if $\overline{X} = (1/c) \sum_{i=1}^{c} X_i$.

3.17 The following data are from Slotnick and Bleiberg (1974) on the relationships between work motivation and personality. One result of their analysis was that those who are "low authoritarians" choose their profession for intrinsic factors. Find the values of X_{12}, X_{21}, and $\Sigma\Sigma X_{ij}$.

	Expected advantages	
Authoritarianism	Intrinsic	Extrinsic
High	14	26
Low	25	12

4

MEASURES OF CENTRAL TENDENCY AND VARIABILITY

Measures of central tendency and variability play a fundamental role in characterizing data. Suppose we have data either appearing in a raw set or organized in a frequency distribution. Looking at *all* the data simultaneously quickly becomes uninformative and cumbersome, especially when the number of data is large. Clearly then, it is desirable to have a few quantities which are "typical" and which "describe" the data at hand. This means that we need quantities which represent the data in terms of location and dispersion. Measures such as the mean and the median are frequently used as measures of location or *measures of central tendency* because they are central or middle values. The variance and standard deviation are some popular measures of dispersion, scatter, or variability. Let us discuss some of these measures systematically, because we intend to use them intensively throughout the following chapters.

4.1 THE MEAN

Let $X_1, X_2, \ldots, X_n$ be a finite set of values of the variable X. The *mean* or average of this set of values is equal to the total of the values divided by the number of values. Using the summation sign of Chapter 3, we have:

DEFINITION 4.1 *The mean or average of a finite set of* n *values* $X_1, X_2, \ldots, X_n$ *of a variable* X *is equal to* $(1/n) \sum_{i=1}^{n} X_i$ *and is denoted by* $\overline{X}$.

Suppose now that the n values are arranged in a frequency distribution with c classes such that $X_1, X_2, \ldots, X_c$ are the midpoints of the classes and $f_1, f_2, \ldots, f_c$ are the corresponding frequencies; then:

DEFINITION 4.2 *The mean* $\overline{X}$ *of the values of a variable* X *in a frequency distribution is equal to* $(1/n) \sum_{i=1}^{c} X_i f_i$.

We may rewrite the expression $(1/n) \sum_{i=1}^{c} X_i f_i$ as $\sum_{i=1}^{c} X_i f_i/n$, where f_i/n is the relative frequency of the ith class. This implies that the mean can be directly calculated from any relative frequency distribution as well.

To illustrate the calculation of the mean, consider the Müller-Lyer data of Chapter 2. The set of measurements in centimeters was: 4, 2, 2, 1, 3, 4, 2, 6, 6, 6, 7, 8, 4. The mean of these raw data is equal to

$$\frac{1}{13} \sum_{i=1}^{13} X_i = \frac{1}{13} (4 + 2 + \cdots + 4) = \frac{55}{13} = 4.23 \text{ cm}$$

The fact that the mean is a measure of location or central tendency can be seen graphically by plotting all the 13 values and the mean on the real line. From Figure 4.1 we see that the mean is located at the "center of gravity" in the sense that we have perfect balance when a bar is placed on a fulcrum at 4.23. In this interpretation the data are seen as weights placed on the bar; that is, there are one 1-lb and three 1-lb weights, etc., placed on the bar at the positions 1 and 2, etc., respectively.

To illustrate the calculations of the mean of the values in a frequency distribution, we shall use the data given in Table 2.2.

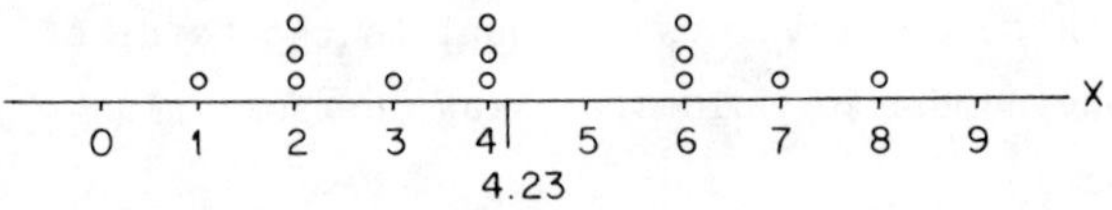

Figure 4.1 The mean as a measure of location or central tendency.

Using the formula in Definition 4.2, we have:

$$\frac{1}{13}\sum_{i=1}^{3} X_i f_i = \frac{1}{13}[(2)(5) + (5)(6) + (8)(2)] = \frac{56}{13} = 4.31$$

This calculation shows that the mean of the values in a frequency distribution need not be the same as that for the values in the original raw set. The phrase "mean of the values in a frequency distribution" is abbreviated to *mean of a frequency distribution*, because this is what we essentially find.

Finally, note that the formula in Definition 4.2 simplifies to $\Sigma X_i f_i^*$ in the case of relative frequency distributions, where $f_i^* = f_i/n$.

Although the mean is the most popular measure of central tendency, it has a drawback in that extreme values easily influence it, so that it may not represent the values under consideration at all. If, for example, the income of a millionaire is added to a set of incomes of 10 blue collar workers, the mean income will not be a good measure. In such cases a different measure is needed, such as the median.

4.2 THE MEDIAN

Suppose that n values of a variable X are ordered in increasing order of magnitude, i.e., we have $X_1, X_2, \ldots, X_n$ such that $X_1 \leq X_2 \leq \cdots \leq X_n$. We then have:

DEFINITION 4.3 *The median of a set of values in increasing order* $X_1, X_2, \ldots, X_n$, *is equal to* $X_{(n+1)/2}$ *if* n *is odd and* $(1/2)(X_{n/2} + X_{(n/2)+1})$ *if* n *is even.*

In other words, the median is that value such that 50% of the ordered values are to its left and 50% are to its right. If $X_1 = 2$, $X_2 = 3$, $X_3 = 5$, then the median is equal to 3, which is confirmed by the formula, since $X_{(n+1)/2} = X_{(3+1)/2} = X_2$ so that $X_2 = 3$ is the median. On the other hand, if $X_1 = 2$, $X_2 = 3$, $X_3 = 5$, and $X_4 = 7$,

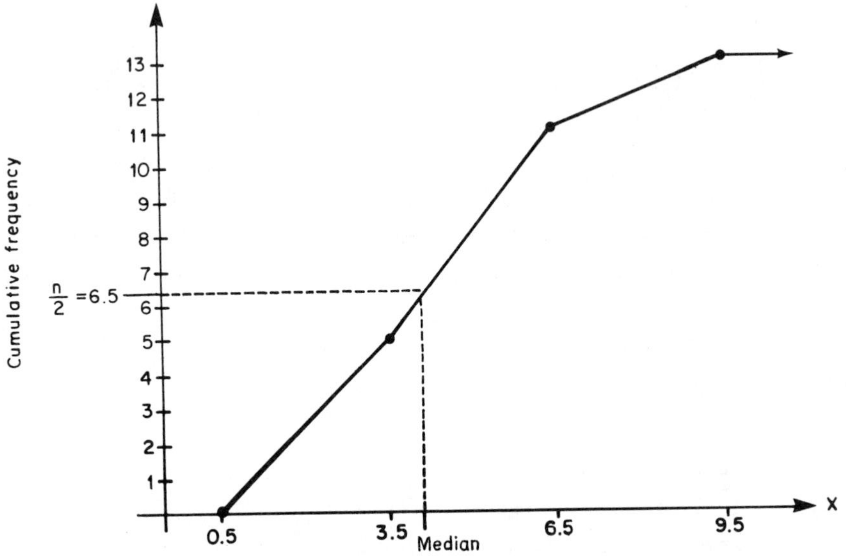

Figure 4.2 Calculation of the median using a cumulative frequency graph.

then the median value is (3 + 5)/2 = 4, which is also obtained from the formula, since $(X_{4/2} + X_{(4/2)+1})/2 = (X_2 + X_3)/2 = (3 + 5)/2 = 4$.

When the values of a variable are arranged in a frequency distribution, the best way to understand the median is to construct a cumulative frequency graph (Figure 4.2). Let us first go through the example introduced in Chapter 2, for which the cumulative frequency graph appears in Figure 2.10. If we wish to have 50% of the values lower and 50% of the values higher than the median, then in our example the median is that value such that n/2 = 13/2 = 6.5 of the cases are below it and 6.5 of the cases are above it. If we do not wish to calculate the median, then this is read off from the graph at the point where a perpendicular from the intersection point of the parallel line at 6.5 and the broken line cuts the X axis. On the other hand, a calculation of the median follows by noting that: (1) the total number of cases between 3.5 and 6.5 is equal to 11 - 5 = 6; (2) the total number of cases between 3.5 and the median is equal to 6.5 - 5 = 1.5; (3) the width of the class in which the median falls is equal to 6.5 - 3.5 = 3 cm. Hence, using interpolation, we see that the distance from 3.5 to the median is equal to (1.5/6)(3) = 0.75. Hence the median is equal to 3.5 + 0.75 = 4.25 cm.

To generalize the calculation of the median, we introduce the following:

DEFINITION 4.4 *The median class of a frequency distribution is the first class having a cumulative frequency greater than* n/2.

In our example, the median class is the class from 3.5 to 6.5 cm, because this class is the lowest class with cumulative frequency greater than n/2 = 6.5.

DEFINITION 4.5 *The median of a frequency distribution is equal to*

$$L_m + \left(\frac{n/2 - F_d}{F_m - F_d}\right)(w)$$

where L_m *is the lower limit of the median class,* F_m *and* F_d *are the cumulative frequencies of the median class and of the class next lower to the median class, respectively, and* w *is the width of the median class.*

Note that $F_m - F_d$ is the frequency of the median class, which in our example is 11 - 5 = 6.

Therefore, when the median is calculated using the relative cumulative frequency distribution, the procedure is exactly the same, except the quantity 1/2 rather than n/2 is entered on the vertical axis and F_d and F_m are replaced by F_d/n and F_m/n.

Sometimes the median is not the desired quantity in a distribution. Rather, one may ask for a quantity such that p% of the values are to its left and (100 - p)% are to its right. Such quantities are referred to as the p*th percentile of the distribution.* The median is a special case, namely, the 50th percentile. The commonly encountered percentiles are the *first (or lower) quartile* = 25th percentile, the *second quartile* = median = 50th percentile, and the *third (or upper) quartile* = 75th percentile.

The calculation of any percentile proceeds in the same manner as the median calculation, except that we enter (p • n)/100 on the vertical axis of the cumulative frequency graph and we talk about the pth percentile class rather than the median class. A formula similar to the one in Definition 4.5 will then be the result.

4.3 THE MODE

Another measure used to describe a set of values or a frequency distribution is given in the following two definitions. Before stating these we note that the mode will make sense if the data in the set do not all have the same frequency or the classes in the frequency distribution do not all have the same frequency.

DEFINITION 4.6 *The mode of a raw set of values of a variable is the most frequently occurring value in the set.*

DEFINITION 4.7 *The modal class of a frequency distribution is the class with the highest frequency and the crude mode is the midpoint of the modal class.*

The mode is an ambiguous measure of central tendency, because it may not be unique. For example, the set of values 5, 6, 6, 8, 9, 9 has two modes. (Such a set is called *bimodal*.) Secondly the mode need not be a sort of middle or central value such as the mean or the median. For example, the set of values 4, 3, 6, 8, 7, 12, 12 has as a mode the extremal value 12.

For the values introduced in Table 2.2, the modal class is the class 3.5 to 6.5 cm, because this class has the highest frequency equal to 6. The crude mode is equal to 5 cm.

The mean, median, and the mode of the data in Table 2.2 are indicated in the frequency histogram in Figure 4.3.

The relative positions of some of the measures for symmetric and asymmetric frequency curves are given in Figure 4.4. Notice the relative positions of B and C in the lower two diagrams.

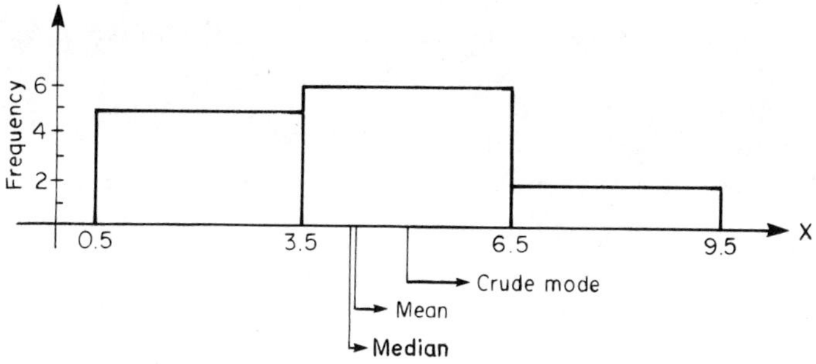

Figure 4.3 Mean, median, and the crude mode of a frequency distribution.

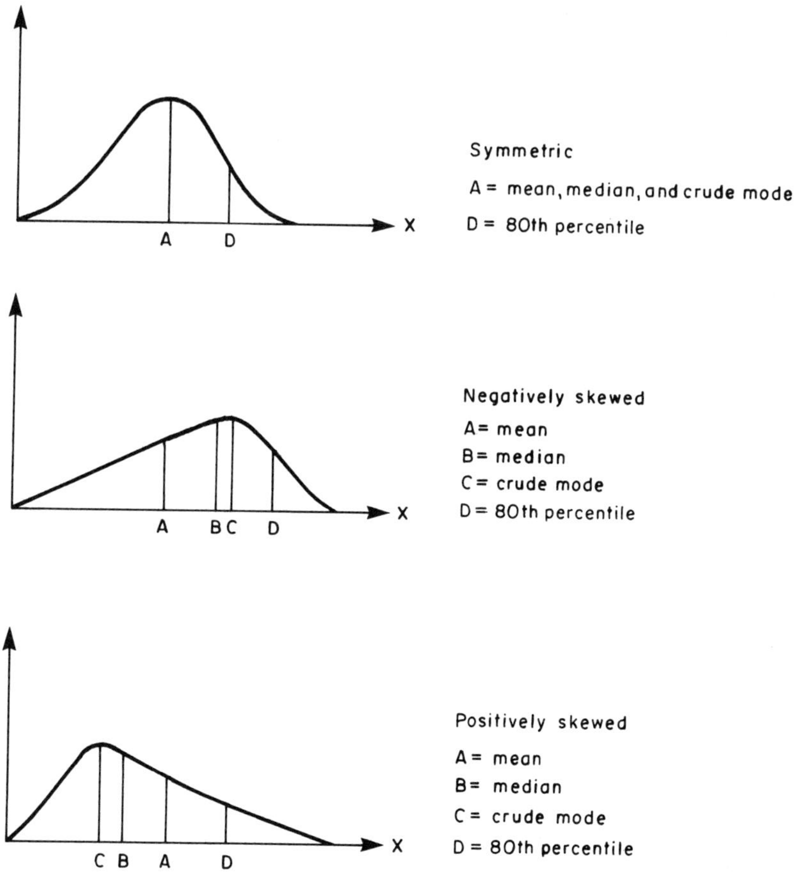

Figure 4.4 Positions of some measures of central tendency in several types of frequency curves.

Two other types of means play a role in certain specialized situations. They are the *geometric mean* and the *harmonic mean.* The former is usually applied to data following approximately a geometric progression and for averaging rates of change, while the latter is useful in average speed and average price calculations.

4.4 THE GEOMETRIC MEAN

DEFINITION 4.8 *For a set of positive numbers* $X_1, X_2, \ldots, X_n$ *the geometric mean is equal to* $\sqrt[n]{(X_1)(X_2)\cdots(X_n)}$.

DEFINITION 4.9 *for a frequency distribution the geometric mean is equal to* $\sqrt[n]{(X_1^{f_1})(X_2^{f_2})\cdots(X_c^{f_c})}$, *where* $X_1, X_2, \ldots, X_c$ *are the positive midpoints and* $f_1, f_2, \ldots, f_c$ *are the frequencies of the classes.*

To illustrate the calculation of the geometric mean, consider the term deposit of \$1000 in 1968 in an interest account for a period of 3 years at an interest rate of 4%. What is the average capital invested during this period? To solve this problem, we note first that the data follow a geometric progression; i.e., the capitals \$1000.00, \$1040.00, \$1081.60, and \$1124.86 are obtained successively by multiplying each previous amount by 1.04. Thus \$1040.00 is equal to (1.04)(\$1.000.00), \$1081.60 is equal to (1.04)(\$1040.00), and finally \$1124.86 is equal to (1.04)(\$1081.60). This means that the geometric mean is the proper mean to calculate the average amount of capital invested. This is obtained as:

$$\sqrt[4]{(1000.00)(1040.00)(1081.60)(1124.86)} = \$1060.80$$

(In general, the best way to obtain the answer to such a root is by using logarithmic tables.) Note that the arithmetic mean is

$$\frac{1}{4}(\$1000.00 + \$1040.00 + \$1081.60 + \$1124.86) = \$1061.62$$

and of course would provide an incorrect answer. By the way, this example is an illustration of the fact that the arithmetic mean is greater than or equal to the geometric mean.

Suppose now that the share of a certain company was worth \$4 in 1960, \$16 in 1961, and \$144 in 1962. The growth *rate* is 4 for the period 1960-1961 and 9 for the period 1961-1962, so that the average growth rate is $(4 + 9)/2 = 6.5$ from year to year. Using the geometric mean, we find the growth rate to be equal to $\sqrt{(4)(9)} = 6$ from year to year. If we use the arithmetic mean, we obtain $(6.5)(4) = \$26$ as a prediction for the actual value of \$16 for the year 1961, while the geometric mean gives a better prediction of $(6)(4) = \$24$ for this same year.

4.5 THE HARMONIC MEAN

DEFINITION 4.10 *For a set of* n *positive values* $X_1, X_2, \ldots, X_n$, *the harmonic mean is equal to* $n/\sum_{i=1}^{n} (1/X_i)$.

DEFINITION 4.11 *For a frequency distribution, the harmonic mean is equal to* $n/\sum_{i=1}^{c} f_i(1/X_i)$, *where* $X_1, X_2, \ldots, X_c$ *are the positive midpoints of the classes with corresponding frequencies* $f_1, f_2, \ldots, f_c$.

As an illustration of Definition 4.10 we consider a stretch of highway from A to B, which is divided into three equal parts. If a vehicle travels the first part at a speed of 50 kph, the second part at a speed of 60 kph, and the third part at a speed of 60 kph, then find the average speed for the whole distance from A to B. It can be proved that the harmonic mean is the proper mean to use in such a situation. Using the formula in Definition 4.10, the reader can verify that the average speed is equal to $3/(1/50 + 1/60 + 1/60) = 56.25$ kph.

Another use of the harmonic mean is the calculation of the average price of an item. If \$14 worth of potatoes is bought in store A for 7¢ a pound and \$14 worth is bought in store B for 10¢ a pound, then the average price is not $(7 + 10)/2 = 8.5$¢ per pound. The average price is found using the harmonic mean, namely, $2/(1/7 + 1/10) = 8.2$¢ a pound.

Before we define certain *measures of variability*, it is important to stress that when data are to be summarized, a measure of central tendency is not sufficient to describe the data; we also need a measure of variability or spread of the data. We see that when the data are all the same then there is no variability present. On the other hand, if they are not the same, then we should be able to measure how much variability there is. Many measures of variability are available in the literature. Let us look at some of them and then illustrate each one.

4.6 THE RANGE

DEFINITION 4.12 *The range of a set of data is equal to the maximum value minus the minimum value of the data.*

For example, suppose that a certain 13-week course at a university had the following number of cases of nervous breakdowns:

1, 1, 1, 2, 1, 2, 3, 2, 4, 1, 5, 7, 9

The range is equal to 9 - 1 = 8 breakdowns, and this quantity is a reasonable measure of the variability of these data.

DEFINITION 4.13 *The range of a frequency distribution is equal to the upper class limit of the highest class minus the lower class limit of the lowest class.*

The range of the frequency distribution introduced in Chapter 2 is equal to 9.5 - 0.5 = 9 cm.

Observe that the range as a measure of variability depends solely on the extremal values. It does not say how the data are dispersed around the mean.

4.7 THE MEAN ABSOLUTE DEVIATION

The ith deviation from the mean is given by $X_i - \overline{X}$. If the original values vary, then so do the deviations from the mean. It might appear we could use the average deviation from the mean, i.e.,

$(1/n) \sum_{i=1}^{n} (X_i - \overline{X})$, as a measure of variability. Unfortunately, this quantity is always equal to 0, because $\Sigma(X_i - \overline{X}) = \Sigma X_i - \Sigma\overline{X} = \Sigma X_i - n\overline{X} = \Sigma X_i - \Sigma X_i = 0$.

To escape this and still use deviations from the mean, we work with the magnitudes of the deviations and ignore the signs; i.e., we use the *absolute values* of the deviations. These absolute values are indicated by $|X_i - \overline{X}|$'s. Since every $|X_i - \overline{X}|$ is nonnegative, it follows that $\Sigma|X_i - \overline{X}|$ is nonnegative. Averaging the sum of the absolute values of the deviations leads us to:

DEFINITION 4.14 *The mean absolute deviation of a set of numbers is equal to* $(1/n) \sum_{i=1}^{n} |X_i - \overline{X}|$.

Note that when the original data vary so do the $|X_i - \overline{X}|$'s, and if all the data are the same then every $|X_i - \overline{X}|$ is equal to zero, so that the mean absolute deviation is a measure of variability. In contrast to the range, the mean absolute deviation takes into account the dispersion of the data about the mean. The calculation of the mean absolute deviation, abbreviated MAD, is, however, more involved.

To illustrate the calculations of the MAD, consider the data on nervous breakdowns introduced above. The sample mean $\overline{X} = 3$, so that

$$\begin{aligned} \text{MAD} &= \frac{1}{n} \sum_{i=1}^{n} |X_i - \overline{X}| \\ &= \frac{1}{13}(|1 - 3| + |1 - 3| + \cdots + |9 - 3|) \\ &= \frac{1}{13}(2 + 2 + \cdots + 6) \\ &= \frac{26}{13} = 2.0 \text{ nervous breakdowns} \end{aligned}$$

DEFINITION 4.15 *The mean absolute deviation of a frequency distribution is equal to* $(1/n) \sum_{i=1}^{c} |X_i - \overline{X}| f_i$, *where* X_i *is the midpoint of the ith class,* f_i *is the frequency of the ith class,* c *is the number of classes, and* $\overline{X} = (1/n) \sum_i X_i f_i$ *is the mean of the frequency distribution.*

For the data in Exercise 2.6 the mean number of yeses is:

$$\overline{X} = \frac{1}{50}[0(1) + 1(11) + 2(19) + \cdots + 10(0)]$$

$$= \frac{138}{50} = 2.76 \text{ yeses}$$

Therefore

$$\text{MAD} = \frac{1}{50}[|0 - 2.76|(1) + |1 - 2.76|(11) + \cdots + |10 - 2.76|(0)]$$

$$= \frac{1}{50}[2.76(1) + 1.76(11) + \cdots + 6.24(0)]$$

$$= \frac{73.12}{50} = 1.46 \text{ yeses}$$

One of the disadvantages of using the MAD as a measure of variability is its difficulty in algebraic manipulations. The variance, which is discussed next, is easily treated algebraically. It also leads to the most popular measure of variability, namely, the standard deviation.

4.8 THE VARIANCE

DEFINITION 4.16 *The sample variance of a set of numbers* $X_1, X_2, \ldots, X_n$ *is equal to* $s_X^2 = \frac{1}{n-1} \sum_{i=1}^{n} (X_i - \overline{X})^2$.

Notice that when all the X_i's are the same, $s_X^2 = 0$, and when the X_i's are not all equal, $s_X^2 > 0$. The value of the sample variance increases as the variability of the X_i's about $\overline{X}$ increases.

DEFINITION 4.17 *The variance of a frequency distribution is equal to* $s_X^2 = \frac{1}{n-1} \sum_{i=1}^{c} (X_i - \overline{X})^2 f_i$, *where* X_i *and* f_i *are the midpoint and frequency, respectively, of the ith class.*

The variance s_X^2 in both definitions can be more easily calculated by utilizing equivalent algebraic expressions. The computational formulas are:

$$s_X^2 = \frac{1}{n-1}\left[\sum_{i=1}^{n} X_i^2 - \frac{1}{n}\left(\sum_{i=1}^{n} X_i\right)^2\right]$$

$$s_X^2 = \frac{1}{n-1}\left[\sum_{i=1}^{c} X_i^2 f_i - \frac{1}{n}\left(\sum_{i=1}^{c} X_i f_i\right)^2\right]$$

4.9 THE STANDARD DEVIATION

The variance s_X^2 is in squared units of measurement, so that it is not compatible with the units of X. To bring this statistic back to the original units of measurement, we introduce the concept of the standard deviation.

DEFINITION 4.18 *The standard deviation of a set of numbers* $X_1, X_2, \ldots, X_n$ *is equal to* $s_X = \sqrt{s_X^2} = \sqrt{\frac{1}{n-1}\sum_{i=1}^{n}(X_i - \bar{X})^2}$.

DEFINITION 4.19 *The standard deviation of a frequency distribution is equal to* $s_X = \sqrt{s_X^2} = \sqrt{\frac{1}{n-1}\sum_{i=1}^{c}(X_i - \bar{X})^2 f_i}$, *where* X_i *and* f_i *are the midpoint and frequency, respectively, of the ith class.*

The standard deviation s_X is in the units of measurement, and as such can be given a better interpretation than the variance. Moreover, a relatively large standard deviation indicates a relatively large variability among the values in the set or in the distribution. Later on, when we discuss the normal distribution, the standard deviation will be given a better interpretation. The computational formulas for the standard deviation are:

$$s_X = \sqrt{\frac{1}{n-1}\left[\sum_{i=1}^{n} X_i^2 - \frac{1}{n}\left(\sum_{i=1}^{n} X_i\right)^2\right]}$$

and

$$s_X = \sqrt{\frac{1}{n-1}\left[\sum_{i=1}^{c} X_i^2 f_i - \frac{1}{n}\left(\sum_{i=1}^{c} X_i f_i\right)^2\right]}$$

where the second expression refers to a frequency distribution of c classes. Note that if n is very large, then $(n - 1)/n$ is close to 1, so that the standard deviation can be written using n as a divisor rather than $n - 1$. Hence, in the case of relative frequency distributions, the expression simplifies to $s_X = \sqrt{\Sigma X_i^2 f^*_i - (\Sigma X_i f^*_i)^2}$, where $f^*_i = f_i/n$.

To illustrate the calculation of the standard deviation for a set of data, consider the number of cigarettes smoked by 6 high school students during a school day, as reported to an interviewer on a certain date:

4, 6, 5, 1, 3, 2

Using the first formula, we obtain:

$$s_X = \sqrt{\frac{1}{6-1}\sum_{i=1}^{6}(X_i - \overline{X})^2}$$

$$= \sqrt{\frac{1}{5}[(4 - 3.5)^2 + (6 - 3.5)^2 + \cdots + (2 - 3.5)^2]}$$

$$= \sqrt{\frac{1}{5}[(0.5)^2 + (2.5)^2 + (1.5)^2 + (-2.5)^2 + (-0.5)^2 + (-1.5)^2]}$$

$$= \sqrt{3.5} = 1.8 \text{ cigarettes}$$

The easier computational formula gives the same result:

$$s_X = \sqrt{\frac{1}{6-1}\left[\sum_{i=1}^{6} X_i^2 - \frac{1}{6}\left(\sum_{i=1}^{6} X_i\right)^2\right]}$$

$$= \sqrt{\frac{1}{5}[(4^2 + 6^2 + 5^2 + 1^2 + 3^2 + 2^2) - \frac{(21)^2}{6}]}$$

$$= \sqrt{\frac{1}{5}(91 - 73.5)} = \sqrt{3.5} = 1.8 \text{ cigarettes}$$

The calculation of s_X for the frequency distribution introduced in Chapter 2 using the computational formula proceeds most conveniently as follows:

X_i	f_i	$X_i f_i$	$X_i^2 f_i$
2	5	10	20
5	6	30	150
8	2	16	128
Totals	13	56	298

Hence

$$s_X = \sqrt{\frac{1}{12}[298 - \frac{1}{13}(56)^2]}$$

$$= \sqrt{\frac{1}{12}(56.77)} = \sqrt{4.7308} = 2.18 \text{ cm}$$

4.10 THE COEFFICIENT OF VARIATION

When *comparing* the variability of two types of data, it is not only realistic to take into account the units of measurement, but also the magnitudes of the values. In such situations we use the coefficient of variation as a measure of variability, which is defined as follows:

DEFINITION 4.20 *The coefficient of variation (CV) for a collection of data is equal to* $(s_X/\overline{X}) \cdot 100\%$.

Note that the CV for a set of data $X_1, X_2, \ldots, X_n$ is calculated as

$$CV = \frac{\sqrt{[\Sigma X_i^2 - (1/n)(\Sigma X_i)^2]/(n - 1)}}{(\Sigma X_i)/n} \cdot 100\%$$

while for a frequency distribution the computational formula is equal to

$$CV = \frac{\sqrt{[\Sigma X_i^2 f_i - (1/n)(\Sigma X_i f_i)^2]/(n - 1)}}{(\Sigma X_i f_i)/n} \cdot 100\%$$

Let us illustrate the concept of the CV by looking at the cigarette data and the measurement data in the frequency distribution above. The CV for the cigarette data is equal to (1.8/3.5) • 100% = 50%, while that for the frequency distribution is equal to (2.18/4.3) • 100% = 51%. Hence we may conclude that for the given numbers of data points the two sets exhibit approximately the same variability.

4.11 THE INTERQUARTILE DEVIATION

When the median instead of the mean is used as a measure of central tendency, then the proper measure of variability to be used is the interquartile deviation (IQD):

DEFINITION 4.21 *The interquartile deviation (IQD) is equal to* $(q_3 - q_1)/2 = 1/2$ *(75th percentile - 25th percentile)*.

The reader is invited to calculate the interquartile deviation for the distribution introduced in Chapter 2.

Remark. Besides the measures of central tendency and measures of variability, there are measures which deal with the shape and form of a distribution. These two measures are referred to as *skewness* and *kurtosis*. We will not delve into these measures here. However, the interested reader certainly can find discussions of these measures elsewhere.

In this chapter we have not distinguished between sample and population measures of central tendency and variability. In future chapters we will do so.

EXERCISES

4.1 In 10 grocery stores of a city the prices of one 12-oz can of a certain product were: 41, 47, 43, 46, 44, 41, 43, 42, 48, 40¢.

(a) Calculate the mean, the median, the mode, and the harmonic mean.

(b) Which measure is most appropriate? Why?

(c) Calculate the range, the MAD, the standard deviation, and the CV. Decide on the merits of these measures of variability.

(d) Find what percentage of the data is between the mean ± the standard deviation and the mean ± 2 times the standard deviation.

4.2 If the distance between two stations is equal to 400 miles, and by train the first 100 miles is done at 50 mph, the second 100 miles at 60 mph, the third 100 miles at 55 mph, and the last 100 miles at 70 mph, then find the average speed.

4.3 Ten true-false questions were asked in a test given to 145 students. The results were:

No. of correct answers	No. of students
0	1
1	5
2	11
3	15
4	21
5	35
6	23
7	19
8	10
9	3
10	2

(a) Plot the relative frequency histogram.

(b) Find the mean, the median, the mode, and the 90th percentile, and indicate these quantities in the frequency histogram of (a).

(c) Calculate the standard deviation and the CV of the distribution. Compare this CV with the one in Exercise 4.1.

(d) Plot the two values $\overline{X} + 2s_X$ and $\overline{X} - 2s_X$ on the histogram in (a). Estimate the percentage of students in this interval.

4.4 Find two sets of five numbers each such that they have the same mean but the standard deviations are different.

4.5 Construct two frequency distributions such that they have the same standard deviation but different means.

4.6 True or false:

(a) If data are multiplied by a constant, then the resulting mean is the same as the mean of the original data.

(b) If a constant is subtracted from every observation, then the resulting standard deviation is less than the original standard deviation by that constant.

(c) If in a set every observation is divided by a constant, then the resulting mean is equal to the original mean divided by the constant, and the standard deviation is equal to the original standard deviation divided by the constant.

(d) The corrected sum of squares is zero when all the observations are equal to the same value.

(e) We find that two distributions, A and B say, have the same mean. The standard deviation of A is 18 and that of B is 6. We may conclude that the observations in A are grouped closer to the mean than are the observations in distribution B.

(f) If the numbers $X_1, X_2, \ldots, X_c$ occur $f_1, f_2, \ldots, f_c$ times, respectively, the arithmetic mean is

$$\overline{X} = \frac{X_1f_1 + X_2f_2 + \cdots + X_cf_c}{f_1 + f_2 + \cdots + f_c}$$

4.7 Make one short and valid statement (not the definition) concerning each of the following concepts:

(a) Standard deviation

(b) Median

(c) Variance

(d) MAD

(e) Sum of the deviations about the mean

(f) Range

4.8 Interview 25 male students and ask them their heights (in centimeters). Plot a frequency histogram with five classes. Do the same for 25 female students. Compare the two distributions with respect to the mean and the standard deviation.

4.9 The following data are from Mills and Seng (1954) on the birth weights of 307 eighth-born Chinese male infants in Singapore in 1950 and 1951.

Birth weight (oz)	No. of male births
≤71	0
72-79	4
80-87	5
88-95	19
96-103	52
104-111	55
112-119	61
120-127	48
128-135	39
136-143	19
144-151	4
152-159	0
160-167	1
≥168	0
Total	307

(a) Construct a frequency histogram and the cumulative frequency polygon.

(b) Calculate the mean, median, and "crude" mode.

(c) Graphically estimate the median and 10th percentile.

4.10 An interesting empirical relationship between the three measures of location which appears to hold for unimodal distributions of moderate asymmetry is: mean - mode = 3(mean - median). Verify this relationship for the following data, where X is the number of typographical errors per page and f is the observed number of pages in the sample of 100 pages:

X	0	1	2	3	4	5	6
f	12	34	29	16	8	1	0

4.11 It is a useful mnemonic to observe that mean, median, and mode occur in the same order (or reverse order) as in the dictionary, and that the median is nearer to the mean than to the mode, just as the corresponding words are nearer together in the dictionary. For the data in Exercise 4.10, construct the frequency polygon and indicate the relative positions of these values on the X axis.

4.12 For the data of Exercise 4.8, compute the range, variance, standard deviation, coefficient of variation, and the interquartile deviation.

4.13 A child is born to Mary every year for 7 consecutive years. Compute the standard deviation of the children's ages (a) when the youngest is 1 year old, and (b) when the youngest is 8 years old. In a few sentences discuss the implications of (a) and (b).

4.14 For the data in Exercise 4.1:

(a) Show that if $a = 1/12$, then the standard deviation of aX equals a times the standard deviation of X and that the mean of aX equals a times the mean of X.

(b) Show that if $a = -40$, then the variance of X + a equals the variance of X. Also what is the mean of aX?

4.15 (a) If $X_1 = X_2 = \cdots = X_n$, show by example that the arithmetic mean (AM) equals the geometric mean (GM) equals the harmonic mean (HM).

(b) For the integers 1, 3, 5, 7, 9 verify the relationship that AM ≥ GM ≥ HM.

4.16 Compute the mean number of contributions per author to the first 15 volumes of the journal, *Canadian Mathematical Bulletin*. Here g(X) = numbers of authors with exactly X contributions (from Hubert, 1975).

X	1	2	3	4	5	6	7	8	9	11	12	14	16	18	19
g(X)	508	132	44	28	10	8	4	6	1	2	1	1	1	1	1

4.17 Compute the ungrouped and grouped mean and variance for the data in Exercise 2.8 and comment on your answer.

4.18 Eight subjects were presented with a verbal stimulus (a number) during sleep, and 15 seconds after each presentation the subject was awakened and asked if he had heard anything. If not, he was presented a test including the stimulus and another three numbers. If X is the number of correct answers out of 64 such trials, use the following data (obtained by Lasaga and Lasaga, 1973) to compute the mean, median, mode, range, and standard deviation.

$$X = 21, 18, 20, 18, 19, 20, 15, 24$$

5

BASIC PROBABILITY THEORY

5.1 PRELIMINARY CONCEPTS

In Chapter 1 we have indicated that statistics as a discipline relies heavily on the theory of probability, because data are obtained from samples. Since repeated samples from a given population produce different results, it is desirable to attach a measure of reliability to quantities calculated from a single sample. It is the idea of repeated sampling which brings in the theory of probability as a necessary tool, and strangely, the concepts of probability itself are best understood from the viewpoint of sampling or experimentation.

Suppose that a population consists of 49 elements such that 10 of them have property A. Consider an experiment of drawing randomly one element from this population and noting whether it does or does not have property A. We may repeat this experiment by replacing the drawn unit and making a new drawing.

When we repeat this experiment an *indefinite* or *infinite* number of times, then the proportion or *relative frequency* of outcomes having property A will be equal to 10/49. With this approach in mind, we refer to 10/49 as the *probability of the event that a randomly selected element will have property* A.

Note that for a particular number of repetitions the observed relative frequency will be approximately equal to 10/49. Phrased differently, we see that the probability of an event is the observed relative frequency of that event for an infinite number of repetitions of an experiment.

Similarly, if an experiment consists of tossing a fair coin once, and we consider an infinite number of repetitions of this experiment, then the relative frequency of getting a "head" will be 1/2. However, for a particular number of repetitions, 60 say, the observed relative frequency of "heads" might be 32/60 or some other number rather than 1/2.

When using the repetitive experimentation approach, one speaks of *statistical probability*, in contrast to *personal* or *subjective* probability. Typical of this latter approach are statements such as "The probability that my daughter will get married is 0.8" or "The probability that I will become rich is 0.2." Such probabilistic statements vary from person to person, in contrast to obtaining a unique value in the repetitive experimentation approach. In the remainder the subjective probability approach will not be considered.

To facilitate the calculations of probabilities we now introduce certain basic definitions which arise in the calculus of probability.

DEFINITION 5.1 *The set of outcomes of an experiment is called the sample space.*

In the first example, introduced at the beginning of this chapter, the set of outcomes of drawing one individual randomly from among 49 individuals can be written down by letting $A_1, A_2, \ldots, A_{10}$ indicate the elements having property A and by letting $B_1, B_2, \ldots, B_{39}$ indicate the rest of the elements, which do not have property A. It is convenient to indicate the sample space by the letter S, so that in compact notation the sample space in our case is written as:

$$S = \{A_1, A_2, \ldots, A_{10}, B_1, B_2, \ldots, B_{39}\}$$

In the coin tossing example the sample space of the experiment of tossing a coin is described conveniently by:

$$S = \{H, T\}$$

where H stands for the outcome "head" and T for the outcome "tail."

Consider the experiment of drawing randomly two individuals to form a committee. If the individuals are indicated by A, B, C, D, E, then the sample space is:

$$S = \{(A, B), (A, C), (A, D), (A, E), (B, C), (B, D), (B, E), (C, D), (C, E), (D, E)\}$$

where, for example, the two-tuple (A, B) indicates the outcome consisting of the individuals A and B.

In each of these examples, we see that the sample space is the totality of all simple or elementary outcomes (those which cannot be subdivided into a finer set of results) of the experiment.

A sample space need not be finite. Consider the experiment of drawing randomly (if at all feasible) a positive integer from among all the positive integers. Then the sample space is infinite and can be written as:

$$S = \{1, 2, 3, \ldots\} \qquad \text{or} \qquad S = \{x: x = 1, 2, \ldots\}$$

The first way is called the *list method*; the second way is called the *roster method*.

Another example of an infinite sample space is the set of positions which can be taken by a spinner when it is spun on a wheel of fortune. Any point on the circumference is a possible outcome. We will come back to this example when we deal with *continuous* sample spaces.

DEFINITION 5.2 *An event is a subset or part of a sample space.*

The fundamental outcomes or building blocks of a sample space are referred to in the literature as *elementary events*. A collection of elementary events is called a *compound event*. Events are traditionally depicted using capital letters, such as A, B or A_1,

A_2. *Elementary events are always such that they are disjoint and their totality exhausts the sample space.* (A precise definition of disjoint events is given in Definition 5.7.)

Suppose that two fair dice are tossed. The sample space for this experiment is given in Table 5.1. Note that each of the 36 outcomes is an elementary event, because they are the basic building blocks of the sample space. A compound event A of getting a double is described by the subset:

$$A = \{(1, 1), (2, 2), (3, 3), (4, 4), (5, 5), (6, 6)\}$$

Also, the compound event B of getting an outcome such that the sum of the faces is equal to 8 is:

$$B = \{(2, 6), (3, 5), (4, 4), (5, 3), (6, 2)\}$$

DEFINITION 5.3 *An event having no outcomes is called the impossible event, and the whole sample space is called the sure event.*

DEFINITION 5.4 *The union of two events* A *and* B, *written as* $A \cup B$, *is the event having outcomes belonging to* A *or* B *or possibly both.*

In the dice tossing example above we note that:

$$A \cup B = \{(1, 1), (2, 2), (2. 6), (3, 3), (3, 5), (4, 4), (5, 3), (5, 5), (6, 2), (6, 6)\}$$

Table 5.1 The Sample Space of Tossing Two Dice Once

		Outcomes of die 2					
		1	2	3	4	5	6
	1	(1, 1)	(1, 2)	(1, 3)	(1, 4)	(1, 5)	(1, 6)
	2	(2, 1)	(2, 2)	(2, 3)	(2, 4)	(2, 5)	(2, 6)
Outcomes	3	(3, 1)	(3, 2)	(3, 3)	(3, 4)	(3, 5)	(3, 6)
of die 1	4	(4, 1)	(4, 2)	(4, 3)	(4, 4)	(4, 5)	(4, 6)
	5	(5, 1)	(5, 2)	(5, 3)	(5, 4)	(5, 5)	(5, 6)
	6	(6, 1)	(6, 2)	(6, 3)	(6, 4)	(6, 5)	(6, 6)

since the outcomes in $A \cup B$ satisfy Definition 5.4. Also, the event of getting a sum greater than 12 is an impossible event, because there are no outcomes such that the sum of the faces is greater than 12.

DEFINITION 5.5 *The intersection of two events* A *and* B, *written as* $A \cap B$, *is the event whose outcomes belong to both* A *and* B.

Let A and B be the events as described in the dice-tossing example above, and let C be the event that the product of the faces on the outcomes is even; then

$$\begin{aligned} C = \{&(1, 2), (1, 4), (1, 6), (2, 1), (2, 2), (2, 3), (2, 4), \\ &(2, 5), (2, 6), (3, 2), (3, 4), (3, 6), (4, 1), (4, 2), \\ &(4, 3), (4, 4), (4, 5), (4, 6), (5, 2), (5, 4), (5, 6), \\ &(6, 1), (6, 2), (6, 3), (6, 4), (6, 5), (6, 6)\} \end{aligned}$$

Therefore,

$$A \cap C = \{(2, 2), (4, 4), (6, 6)\}$$

and

$$B \cap C = \{(2, 6), (4, 4), (6, 2)\}$$

The reader should describe the events $A \cap C$ and $B \cap C$ verbally.

DEFINITION 5.6 *The complement of an event* A *in* S, *denoted by* $\bar{A}$, *is the event whose outcomes do not belong to* A.

As an illustration, the complement of C in the dice-throwing example is:

$$\begin{aligned} \bar{C} = \{&(1, 1), (1, 3), (1, 5), (3, 1), (3, 3), \\ &(3, 5), (5, 1), (5, 3), (5, 5)\} \end{aligned}$$

that is, $\bar{C}$ is the event having outcomes such that the product of the two faces is odd.

DEFINITION 5.7 *Two events* A *and* B *are said to be disjoint or mutually exclusive if the intersection* $A \cap B$ *is the impossible event.*

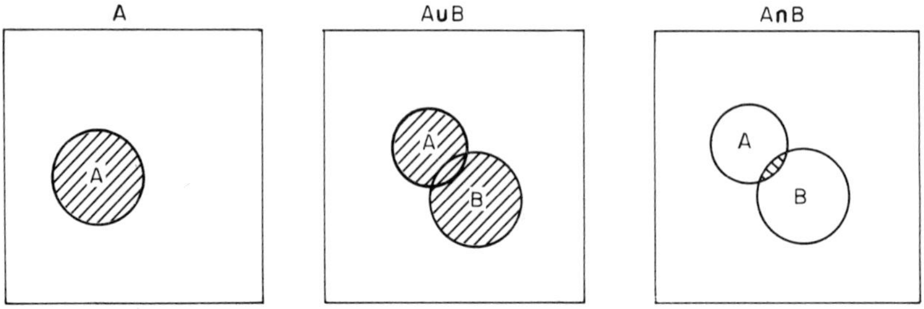

Figure 5.1 Some Venn diagrams of events.

In the dice-throwing example, the events C and $\overline{C}$ are clearly disjoint. Note that elementary events of a sample space are always disjoint events. Events can be depicted using *Venn diagrams* (see Figure 5.1), which were named after the English logician, John Venn (1843-1923).

5.2 DEFINITION OF PROBABILITY

After these preliminary notions we may define probability in a formal way. The interpretation, however, remains as outlined at the beginning of this chapter; i.e., we look at probability from the viewpoint of repeated experimentation.

DEFINITION 5.8 *To any event* A *in the sample space* S *we assign the number* P[A], *called the probability of the event* A, *such that the following conditions are satisfied: (1)* $P[A] \geq 0$; *(2) for a series of mutually exclusive events* A_1, A_2, ... *we have* $P[A_1 \cup A_2 \cup \cdots] = \Sigma P[A_i]$; *and (3)* $P[S] = 1$.

In the dice-tossing example, each of the elementary events is equally likely. Therefore, if we assign to each of these events the probability 1/36, then the two other conditions in Definition 5.8 are automatically satisfied. Hence in this example:

$$P[A] = P[\text{getting a double}] = \frac{6}{36} = \frac{1}{6}$$

$$P[B] = P[\text{sum of the faces is 8}] = \frac{5}{36}$$

$$P[A \cup B] = \frac{10}{36} = \frac{5}{18}$$

$$P[C] = P[\text{product of the faces is even}] = \frac{27}{36} = \frac{3}{4}$$

$$P[A \cap C] = \frac{3}{36} = \frac{1}{12}$$

and

$$P[B \cap C] = \frac{3}{36} = \frac{1}{12}$$

This example suggests the general rule

$$P[A \cup B] = P[A] + P[B] - P[A \cap B]$$

because

$$P[A \cup B] = \frac{6}{36} + \frac{5}{36} - \frac{1}{36} = \frac{10}{36} = \frac{5}{18}$$

As another illustration consider the case of tossing two fair or unbiased coins once. The elementary events are {(H, H)}, {(H, T)}, {(T, H)}, and {(T, T)}. A reasonable approach is to assign each of the four events the same probability, 1/4. An unreasonable approach would be to distinguish only three elementary events, namely, {2 heads}, {a head and a tail}, and {2 tails}, and assign each of these events the probability 1/3. Why is this last approach unreasonable? If you toss two fair coins 100,000 times we would expect each of the four elementary elements in the first set-up to occur 25,000 times, and this can be borne out by actual repeated experimentation, while in the second case, repeated experimentation will refute the probability assignments.

5.3 CONDITIONAL PROBABILITY

DEFINITION 5.9 *If* A *is an event such that* $P[A] > 0$, *then the conditional probability of the event* B *given that the event* A *has occurred, written as* $P[B|A]$, *is equal to* $P[A \cap B]/P[A]$.

Here we are essentially restricting ourselves to a sample space, which consists of all those outcomes belonging to A, and we are asking for the probability of the event B. Also, we assume A not to be the impossible event. Let us illustrate this concept by an example.

Suppose that an individual is selected at random from a set of 25 individuals. Assume further that these individuals were classified as follows:

	Liberal (L)	Conservative (C)	Independent (I)	Total
Over 40 = W	3	4	3	10
Under 40 = U	6	7	2	15
Total	9	11	5	25

The sample space here consists of six elementary events: S = {(W, L), (W, C), (W, I), (U, L), (U, C), (U, I)}. Using {W}, {U}, {L}, {C}, and {I} to depict the events that the randomly selected individual is over 40, under 40, Liberal, Conservative, and Independent, respectively, we obtain the following probabilities:

$$P[\{W\}] = \frac{10}{25} \qquad P[\{U\}] = \frac{15}{25} \qquad P[\{L\}] = \frac{9}{25}$$

$$P[\{C\}] = \frac{11}{25} \qquad P[\{I\}] = \frac{5}{25} \qquad P[\{W|L\}] = \frac{3}{9}$$

$$P[\{U|L\}] = \frac{6}{9} \qquad P[\{W|C\}] = \frac{4}{11} \qquad P[\{U|C\}] = \frac{7}{11}$$

$$P[\{W|I\}] = \frac{3}{5} \qquad P[\{U|I\}] = \frac{2}{5} \qquad P[\{L|W\}] = \frac{3}{10}$$

$$P[\{C|W\}] = \frac{4}{10} \qquad P[\{I|W\}] = \frac{3}{10} \qquad P[\{L|U\}] = \frac{6}{15}$$

$$P[\{C|U\}] = \frac{7}{15} \qquad P[\{I|U\}] = \frac{2}{15}$$

This last probability is stated verbally as P[that a randomly selected individual is an Independent given that he is under 40].

DEFINITION 5.10 *Two events* A *and* B *are said to be independent if the conditional probability of one event given the other is equal to the unconditional probability; i.e.,* $P[A|B] = P[A]$ *or* $P[B|A] = P[B]$.

In the case $P[A] = 0$ or $P[B] = 0$, the conditional probabilities will be undefined, and hence independence of events will be meaningless.

To illustrate the notion of independence, consider the example where the 25 individuals were classified according to age and political outlook. Since $P[\{W|L\}] = 3/9$ does not equal $P[\{W\}] = 10/25$, then the events $\{W\}$ and $\{L\}$ are not independent; i.e., they are dependent.

Suppose a coin and a die are tossed once and let A be the event "head" on the coin and B the event of a "6" on the die. Clearly $P[B] = 1/6 = P[B|A] = (1/12)/(1/2) = 1/6$, and hence A and B are independent. This conclusion could have been reached without any calculation whatsoever, because the probability of the occurrence of one event does in no way depend on the occurrences (or indeed the nonoccurrence) of the other event.

From Definition 5.10 it can be shown that the two events will be independent if $P[A \cap B] = P[A] \cdot P[B]$. This convenient condition does not require the calculation of conditional probabilities.

So far in our examples of probability calculations we have dealt with sample spaces having a finite number of fundamental outcomes or elementary events. Suppose now that the experiment consists in drawing randomly a number from all positive real numbers. Assuming equal probabilities for the positive real numbers (not just integers), one sees quickly that P[of obtaining a particular positive number] = 0, because there are infinitely many positive numbers. (There are a countably infinite number of positive integers, and there are an uncountably infinite number of real numbers.)

If the arrival times of a bus are within 10 minutes of every hour, then the same reasoning as above will show that P[that a bus will arrive exactly at time x] = 0.

A similar situation arises when we deal with a "continuous" sample space. To explain this notion in a simple manner, consider

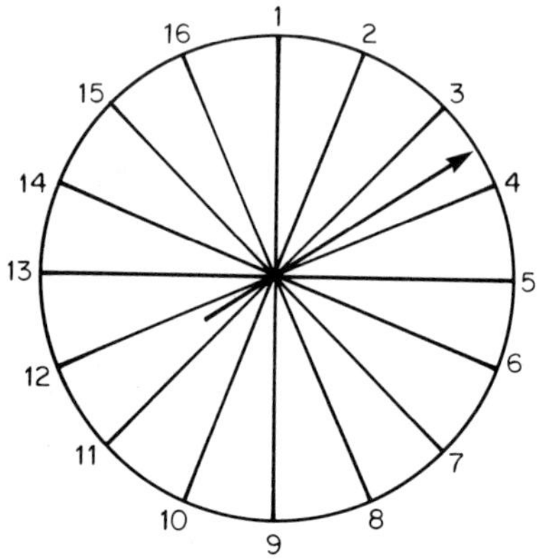

Figure 5.2 Wheel with spinner.

a wheel marked off in 16 equal divisions (see Figure 5.2). A pointer can be spun around freely and it can stop anywhere along the circumference of the wheel. The sample space for a spin of the wheel is a *continuous* sample space, because a whole interval of positions, i.e., the whole circumference of the wheel, can be assumed by the pointer. The probability that the pointer will stop at a particular position is zero, because there are infinitely many (and uncountable) number of positions. However, the probability that the pointer will be in any segment can be set equal to 1/16, because there are 16 segments and the relative frequency approach implies that in repeated spins one would get a relative frequency close to 1/16 for the spinner to land in any segment.

5.4 PERMUTATIONS AND COMBINATIONS

In order to facilitate counting in calculations of probability, we now introduce the concepts of permutations and combinations. (For the remainder of this chapter we shall assume that the objects are distinguishable.)

DEFINITION 5.11 *A permutation of* n *objects* $A_1, A_2, \ldots, A_n$ *is an ordered arrangement of the objects.*

As an example, consider the arrangement of three objects A_1, A_2, and A_3. If we arrange these objects side by side we get the following permutations:

$A_1A_2A_3$ $A_3A_2A_1$

$A_2A_1A_3$ $A_3A_1A_2$

$A_1A_3A_2$ $A_2A_3A_1$

DEFINITION 5.12 n *factorial, written as* n!, *is equal to* $(n)(n - 1)(n - 2) \cdots (2)(1)$ *or equivalently* $1 \cdot 2 \cdot 3 \cdot \cdots \cdot (n - 1) \cdot n$.

To make the factorial notation consistent, we make the convention $0! = 1$. Other examples of factorial are:

$1! = 1$ $4! = 1 \cdot 2 \cdot 3 \cdot 4 = 24$

$2! = 1 \cdot 2 = 2$ $5! = 1 \cdot 2 \cdot 3 \cdot 4 \cdot 5 = 120$

$3! = 1 \cdot 2 \cdot 3 = 6$ $6! = 1 \cdot 2 \cdot 3 \cdot 4 \cdot 5 \cdot 6 = 720$

Using Definition 5.12 we now show that there are n! permutations of n distinct objects. The first object in an ordered arrangement can be chosen in n ways, the second object can be chosen in (n - 1) ways, the third object can be chosen in (n - 2) ways, and finally the nth object can be chosen in one way. Hence we can form precisely

$$(n)(n - 1)(n - 2) \cdots (2)(1) = n!$$

ordered arrangements from n objects.

We now have a method to find the number of permutations for any n. We have already exhaustively shown that there are 6 permutations of 3 objects. This agrees with the general result that $n! = 3! = 6$.

DEFINITION 5.13 *A permutation of* r *objects from among* n *distinct objects is an ordered arrangement of* r *objects.*

For example, suppose there are 5 objects, i.e., $n = 5$, and we consider permutations of 2 objects at a time. An exhaustive enumeration produces the following ordered arrangements:

A_1A_2	A_2A_1	A_2A_4	A_4A_2
A_1A_3	A_3A_1	A_2A_5	A_5A_2
A_1A_4	A_4A_1	A_3A_4	A_4A_3
A_1A_5	A_5A_1	A_3A_5	A_5A_3
A_2A_3	A_3A_2	A_4A_5	A_5A_4

We conclude that there are exactly 20 permutations possible of 2 objects from among 5 objects. The task of writing down these permutations can quickly become horrendous. For example, try writing out the permutations of 6 objects out of 20.

Let us now develop the counting formula for the number of permutations when r and n are arbitrary. In a permutation of r objects from among n, the first object can be selected in n ways, the second in $(n - 1)$ ways, the third in $(n - 2)$ ways, etc., until the rth selection is from among the remaining $n - (r - 1) = n - r + 1$ objects. Hence the total number of permutations of r objects from among n is equal to $(n)(n - 1)(n - 2)(n - 3) \cdots (n - r + 1)$. This number, indicated by the symbol P_r^n, is equal to:

$$\frac{(n)(n - 1)(n - 2) \cdots (n - r + 1)(n - r)(n - r - 1) \cdots (2)(1)}{(n - r)(n - r - 1)(n - r - 2) \cdots (2)(1)}$$

$$= \frac{n!}{(n - r)!}$$

Hence the total number of permutations of r objects from among n is given by the formula:

$$P_r^n = \frac{n!}{(n - r)!}$$

Verifying the example above, we see that the number of permutations of 2 objects from among 5 is equal to

$$P_2^5 = \frac{5!}{(5 - 2)!} = \frac{5!}{3!} = (5)(4) = 20$$

DEFINITION 5.14 *A combination of* r *objects from among* n *objects is an unordered arrangement of* r *objects.*

This definition says that when one writes out the arrangement, the order in which the objects appear is not important. Taking the "2 out of 5" case given above, we notice that there are 20 permutations, but disregarding order, we find the following combinations:

A_1A_2	A_2A_4
A_1A_3	A_2A_5
A_1A_4	A_3A_4
A_1A_5	A_3A_5
A_2A_3	A_4A_5

We see that each combination gives rise to 2! permutations, or, vice versa, to each 2! permutations there corresponds one combination. Hence there are $P_2^5/2! = 20/2 = 10$ combinations.

In general r! permutations give rise to one combination, so that the number of combinations, indicated by the symbol C_r^n, is equal to:

$$C_r^n = \frac{P_r^n}{r!} = \frac{n!}{r!(n - r)!}$$

[Other equivalent notations are C(n, r) or $\binom{n}{r}$.]

Suppose that 4 people are chosen randomly from among 20 to form a committee. In how many ways can such a committee be formed? Clearly, the order in which the 4 people are drawn is not of importance. Hence the solution is found by calculating the number of combinations of 4 people out of 20 people, and this is found to be:

$$C_4^{20} = \frac{20!}{4!16!} = \frac{(20)(19)(18)(17)}{(4)(3)(2)(1)} = 4845$$

5.5 TWO SAMPLING PROCEDURES

Permutations and combinations can be used to discuss two common types of sampling procedures. Suppose we draw elements one at a time from a population having n elements, numbered 1 to n, until r elements are drawn. If after each draw the element is returned to the population, then the drawing is said to be done with replacement, and the process of obtaining the sample is called *sampling with replacement*. On the other hand, the process is called *sampling without replacement* if the element is not returned to the population after each draw.

From the above it follows that when ordered samples in the without replacement case are considered, then the number of possible samples is equal to

$$P_r^n = \frac{n!}{(n - r)!}$$

while the number of unordered samples is equal to

$$C_r^n = \frac{n!}{r!(n - r)!}$$

In this last instance the probability that a particular random sample will be the selected one is equal to

$$\frac{1}{C_r^n} = \frac{r!(n - r)!}{n!}$$

Thus when a sample of size 3 is drawn from a population consisting of 10 elements, then the chance that a particular sample will be the selected one, when sampling is done without replacement, is equal to

$$\frac{1}{C_3^{10}} = \frac{3!7!}{10!} = \frac{1}{120}$$

When sampling is done with replacement, then the first element can be selected in n ways, the second element can be selected in n ways, etc., and the rth element can be selected in n ways. Hence the number of ways in which a sample with replacement can be selected is equal to

$$(n)(n) \cdots (n) = n^r$$

where there are r factors on the left.

For example, if 3 balls are drawn with replacement from an urn containing 4 different colored balls, then the number of possible samples is equal to $4^3 = 64$.

If there are k different sets of objects with $n_1, n_2, \ldots, n_k$ objects, respectively, then there are $(n_1)(n_2) \cdots (n_k)$ ways in which one object may be selected from each set. This rule is known as the *multiplication rule* (which was used in Definition 5.12).

For example, suppose a restaurant menu lists 4 soups, 8 meat dishes, 4 desserts, and 2 beverages. In how many ways can a meal (consisting of soup, meat dish, dessert, and beverage) be ordered? The answer is obtained from the multiplication rule, and equals $(4)(8)(4)(2) = 256$ choices, since a soup can be chosen in 4 ways, a meat dish in 8 ways, a dessert in 4 ways, and finally a beverage in 2 ways.

Suppose that an urn contains 3 white balls numbered W_1, W_2, W_3 and 2 red balls numbered R_1, R_2. If 2 balls are drawn without replacement, then what is the probability that the sample has exactly 1 white ball? The answer to this question can be found either dealing with combinations or permutations. Using combinations we see that the sample space has exactly $C_2^5 = 10$ possible outcomes. The event that the sample has exactly 1 white ball has $C_1^3\, C_1^2$ outcomes by the multiplication rule, because 1 white ball can be selected in C_1^3 ways from among the 3 white ones and the remaining ball, which must be red, can be selected in exactly C_1^2 ways. Hence the required probability is:

$$\frac{N[A]}{N[S]} = \frac{C_1^3\, C_1^2}{C_2^5} = \frac{(3)(2)}{10} = \frac{6}{10} = \frac{3}{5}$$

(Here, the notation N[D] denotes the size of any event D, i.e., the number of outcomes in D.) The reader is invited to answer the same question for the case of sampling with replacement.

In his paper "Probability Preferences in Gambling," W. Edwards (1953) has the following story: A farmer brought a carved whalebone with which he claimed that he could locate hidden sources of water. To test him out, he was taken into a room which contained 10 covered cans, 5 containing water and 5 empty. The farmer was challenged to divide the 10 cans into two equal groups, one group containing the cans with water and the other containing the empty cans. What is the probability that the farmer put 3 or more cans in the water group just by chance? The answer to this question is obtained by noting that $N[S] = C_5^{10}$. The event A that 3 or more cans are put in the water group has size $N[A] = C_3^5 C_2^5 + C_4^5 C_1^5 + C_5^5 C_0^5$. Therefore:

$$P[A] = \frac{C_3^5 C_2^5 + C_4^5 C_1^5 + C_5^5 C_0^5}{C_5^{10}}$$

$$= \frac{(10)(10) + (5)(5) + (1)(1)}{252} = 0.5$$

which is a sizeable probability!

If an urn contains M identified elements of two types, say "white" and "red," such that A of them are white and B = M - A are red, and when a random sample of n elements is extracted without replacement, then the probability of the event A_k that the sample has exactly k white balls is given by the formula:

$$P[A_k] = \frac{C_k^A C_{n-k}^B}{C_n^M} \qquad k = 0, 1, 2, \ldots, t$$

where t is the minimum of A and n.

In the case of sampling with replacement,

$$P[A_k] = C_k^n \frac{A^k B^{n-k}}{M^n} = C_k^n p^k (1 - p)^{n-k}$$

where $p = A/M$, and $k = 0, 1, 2, \ldots, n$.

In general, one refers to the event A_k as the event of *scoring exactly* k *successes in a sample of size* n.

EXERCISES

5.1 A committee of 2 people is chosen from among 6 individuals. Give a sample space explicitly.

5.2 Two fair coins are flipped once. Find S.

5.3 Five drivers were arrested for speeding on a certain day. List the outcomes in a sample space if we are interested only in whether they are previous offenders or not.

5.4 Three balls are extracted with replacement from an urn containing two balls numbered 1 and 2. Find a sample space.

5.5 Two cards are selected without replacement from a deck of 52 cards. How many points (outcomes) are in your sample space?

5.6 If three coins are tossed once and A is the event of obtaining exactly 2 heads, then find the outcomes in A. Also find the outcomes in the event B of obtaining exactly 1 tail. Are A and B identical events?

5.7 Estimate the probability of twins occurring, from the data below on single and multiple births. (Data are for Germany, 1901-1935; from Dufrenoy, 1938.)

Birth type	No. of cases
1	41,570,685
2	531,541
3	5,364
4	67
≥5	0
Total	42,107,657

5.8 Two fair dice are tossed. Let A denote the event such that the sum of the faces is greater than 9, and B the event such that the faces are the same. Find the outcomes in the following events:

(a) $\overline{A}$	(b) $\overline{B}$	(c) $A \cup B$
(d) $A \cap B$	(e) $\overline{A} \cup \overline{B}$	(f) $\overline{A} \cap \overline{B}$
(g) $\overline{A \cap B}$	(h) $\overline{A \cup B}$	(i) $A \cap S$

Draw Venn diagrams of these nine events.

5.9 Forty-seven high school graduates were classified in the following way:

Type of secondary school	Type of position: Executive	Nonexecutive	Total
Private	15	10	25
Public	9	13	22
Total	24	23	47

If an individual is selected at random and A is the event that he attended private school, B is the event that he attended public school, C is the event that he is an executive, and D is the event that he is not an executive, then calculate: $P[A]$, $P[B]$, $P[C]$, $P[D]$, $P[A|B]$, $P[A|C]$, $P[A|D]$, $P[B|A]$, $P[B|C]$, $P[B|D]$, $P[C|A]$, $P[C|B]$, $P[C|D]$, $P[D|A]$, $P[D|B]$, $P[D|C]$. Are any two of the four events A, B, C, and D independent?

5.10 A man has 3 pairs of shoes, 5 ties, 6 shirts, 4 jackets, and 5 pairs of pants. If an outfit consists of one item from among each of the five, then how many outfits can he create?

5.11 Given that 4 airlines provide service between Toronto and Calgary, in how many distinct ways can a person select airlines for a trip from Toronto to Calgary and back if:

(a) She must travel both ways by the same airline?

(b) She can, but need not travel both ways by the same airline?

(c) She cannot travel both ways by the same airline?

5.12 A building contractor offers 2-, 3-, and 4-bedroom houses, which may be had in 6 different exterior finishes and with or without a garage. How many different choices can be made?

5.13 Six scientists meet at a party. How many handshakes are exchanged if each scientist shakes hands with every other scientist?

5.14 In how many ways can a prospective driver answer a 10-question true-false test if she marks half of the questions true and the other half false?

5.15 If license plates are to have 3 letters followed by 3 digits, and the first letter cannot be O, how many different plates can be made?

5.16 In how many ways can 8 persons be seated in a row?

5.17 In how many ways can 8 persons be seated in a row of 12 seats?

5.18 A student is required to answer 12 out of 15 questions on an exam.

(a) How many choices has he?

(b) How many, if he must answer the first three questions?

(c) How many, if he must answer at least 6 of the first 8 questions?

5.19 If 4 cards are drawn at random successively without replacement from a deck of 52 playing cards, in how many ways can one obtain 1 card from each suit?

5.20 How many different luncheons consisting of soup, an entrée, 2 vegetables, dessert, and beverage can be ordered from a menu which lists 3 soups, 5 entrées, 4 vegetables, 7 desserts, and 3 beverages?

5.21 How many different "words" can be made by using the letters of the word "statistics"?

5.22 How many 4-letter "words" can be constructed from the 26 letters of the alphabet?

5.23 (a) If $P_r^5 = C_r^5$, then find r.

(b) If $4P_3^n = 5P_3^{n-1}$, then find n.

(c) If $C_{12}^n = C_8^n$, then find n.

5.24 Give a relative frequency interpretation to the following statement: P[that it will rain tomorrow] = 0.80.

5.25 Calculate the probabilities of the events in Exercise 5.8.

5.26 A class of 10 students in a certain course has 2 students who have smoked marijuana at least once and 8 who have never smoked. If 3 students are selected for a committee, then find the probability that at least 1 has smoked marijuana.

5.27 A sample of 3 balls from an urn containing 40% red balls and 60% purple balls is extracted with replacement. Find the probability that the sample has 1 or more purple balls.

5.28 Five fair coins are tossed. Let X denote the number of heads among the five coins.

(a) Find $P[X = i]$, for $i = 0, 1, 2, 3, 4, 5$.

(b) Plot these probabilities as a relative frequency histogram.

(c) Calculate $P[X \geq 1]$, $P[X \leq 3]$, and $P[2 \leq X \leq 5]$.

5.29 Let A_1 and A_2 be independent events such that $P[A_1] = P[A_2] = 0.4$. Find $P[A_1 \cup A_2]$.

5.30 Using the fact that $P[A \cup B] = P[A] + P[B] - P[A \cap B]$, show that

$$P[A \cup B \cup C] = P[A] + P[B] + P[C] - P[A \cap B] - P[A \cap C] - P[B \cap C] + P[A \cap B \cap C]$$

5.31 A random sample of two is selected with replacement from a population which has 40% science students and 60% arts students. Find the probability that both are arts students.

5.32 A bridge hand is drawn from a deck of 52 cards. What is the probability that the hand has at least two aces?

5.33 The faculty of a certain college is composed of 60% "over-forty" and 40% "under-forty" members. If 40% of "over-forty" are "drinkers" and 60% of the "under-forty" are "nondrinkers," then find the probability that a randomly selected faculty member who is a "drinker" is "over-forty."

5.34 The probability that a patient visiting his dentist will have a tooth extracted is 0.07, the probability that he will have a cavity filled is 0.19, and the probability that he will have a tooth extracted as well as a cavity filled is 0.02. What is the probability that a patient visits his dentist but has neither a tooth extracted nor a cavity filled?

5.35 Among the 200 students enrolled in a program, 85 are enrolled in a course in philosophy, 69 are enrolled in a course in English Literature, and 73 are not taking either course. How many of these 200 students must be enrolled in both courses?

5.36 Explain the error in the following statements:

(a) The probabilities that there will be 0, 1, 2, 3 or more accidents at a certain intersection in Guelph are, respectively: 0.48, 0.25, 0.12, and 0.05.

(b) The probability that a laboratory technician will make exactly one mistake during a chemical analysis is 0.05, and the probability that he will make at least one mistake is 0.03.

5.37 If A_1, A_2, A_3, A_4 are four mutually exclusive events such that $P[A_i] = 1/4$ for $i = 1, 2, 3, 4$, then define two events B and C in terms of the A_i's such that B and C are independent events.

5.38 From a group of 50 sociologists, 3 are to be chosen to attend a conference. Find the probability that none of the men in a particular group of 10 will be among the 3 chosen.

5.39 Find the probability that a bridge hand will consist of:

(a) 5 spades, 4 clubs, 3 diamonds, and 1 heart

(b) 2 aces, a king, a queen, 2 tens, and no other honor cards

5.40 Five cards are drawn from a deck. Find the probability that:

(a) Four are aces.

(b) Four are aces and one is a king.

(c) Three are aces and two are jacks.

(d) A 9, 10, jack, queen, king are obtained in any order.

(e) Three are of any one suit and two are of another.

(f) At least one ace is obtained.

5.41 Consider a family with two children. Find the conditional probability that both children are boys given that (a) the older child is a boy and (b) at least one of the children is a boy. (Assume the probability of a male birth is 1/2.)

5.42 Draw 3 cards from a deck of 52 cards. If each card is replaced before the next drawing, what is the probability that at least one of the cards drawn is a spade?

5.43 A bag contains 4 white, 5 red, and 6 black pens. Three are drawn at random without replacement. Find the probability that:

(a) No pen chosen is black.

(b) Exactly two pens chosen are white.

5.44 A green urn contains 3 white and 2 black balls; a red urn contains 2 white and 1 black ball.

(a) If we choose an urn at random and then draw a ball from it, what is the probability of getting a white ball?

(b) All the balls are poured into the red urn and a ball is drawn; what is the probability of getting a white ball?

5.45 If the letters a, b, c, c, d are rearranged at random, what is the probability that the c's stand together?

5.46 Peter has three cards; the first is painted red on both sides, the second is painted red on one side and white on the other side, and the last card is painted white on both sides. Peter then shuffles the cards well and randomly draws one of the cards and shows Paul only one side of this card. The color is white. What is the probability that the other side is also white?

5.47 The probability of a man aged 60 dying within 1 year is 0.025 and the probability of a woman aged 55 dying within 1 year is 0.011. If a man and his wife are 60 and 55, respectively, find the probability:

(a) Of their both living a year

(b) Of at least one of them living a year

5.48 If a box contains 40 good and 10 defective light bulbs, and if 10 bulbs are selected, what is the probability that they will all be good?

5.49 Prove the following results:

(a) $r!\, C^n_r = P^n_r$ (b) $C^n_r = C^n_{n-r}$

(c) $C^n_r = 1$ if $r = 0$ (d) $C^n_1 = n$

(e) $C^n_{r-1} + C^n_r = C^{n+1}_r$

5.50 Prove that if two events are independent, the complements of these two are also independent.

5.51 A fair die has faces marked 1, 2, 2, 3, 3, 3. A second fair die has faces marked 2, 2, 2, 3, 3, 4. Let X = sum of the numbers turning up on the two dice.

(a) What are the possible numerical values which X can take?

(b) Find the probability with which X assumes each of these possible values.

5.52 A fair die is rolled 3 times. What is the probability that the three results include exactly one 5 and one 6?

5.53 What is the probability of obtaining more than three 4's on five tosses of an ordinary fair die?

5.54 If you toss a dime and nickel into the air and then you are told that at least one of the coins turned up "heads," then what is the probability that you tossed two "heads"?

5.55 A bag contains 3 white and 2 black balls; another bag contains 2 white and 1 black ball; and still another bag contains 1 white and 1 black ball. If you choose a bag at random and then draw a ball from it, then what is the probability that the color of the ball is white?

5.56 Suppose that you draw a ball from a box which contains 4 red balls and 6 black balls. What is the probability that the color of the ball is red? Suppose you do not look at this first ball and put it aside. Suppose further that you repeat this two more times. What is the probability that the color of the fourth ball drawn is red?

5.57 If A and B are two mutually exclusive events, then verify that

$$P[\bar{B}|\bar{A}] = 1 - \frac{P[B]}{1 - P[A]}$$

5.58 Deutscher (1964) defined "postparental life" as the time between the marriage of the youngest child and the death of one or both partners. He estimated that on the average parents have about 14 years of postparental life. Also, he asked a representative sample of postparental couples to rate the postparental against the preparental and parental phases; the results for two socioeconomic groups were:

Evaluation	Lower-middle class (%)	Upper-middle class (%)
Favorable	68	86
Neutral	21	5
Unfavorable	11	9

(a) If we pick a parent from a population of N postparents, estimate the probability, according to these data, that:

(i) She will rate these postparental years favorably.

(ii) She will rate these postparental years unfavorably.

(iii) If she rates them unfavorably, she is of the lower-middle class.

(b) Answer these three questions if you had chosen a postparent from a community having 2/3 lower-middle-class and 1/3 upper-middle-class parents.

6

RANDOM VARIABLES AND THREE POPULAR PROBABILITY DISTRIBUTIONS

6.1 RANDOM VARIABLES

At the end of Chapter 5 we introduced the event A_k, which meant "scoring exactly k successes in a sample." If we let X denote the number of successes in a sample, then it follows that the probability of the event A_k can be written as

$$P[A_k] = P[X = k]$$

This means that a probability statement dealing with events has been transformed to a probability statement which deals with a quantitative variable. Such variables are called *random variables* in probability theory. There are two types of random variables, depending on the structure of the sample space underlying an experiment. If the events in the sample space admit probability statements in terms of a variable which is discrete, then we speak of a *discrete sample space*, and the variable is called a *discrete random variable*. On the other hand, if the probability statements about events in the sample space are made in terms of a continuous variable, then the sample space is called a *continuous sample space*, and the variable is called a *continuous random variable*.

Before defining these two types of random variables more precisely, let us give some illustrations.

If a fair coin is tossed, then we know that $P[\{H\}] = P[\{T\}] = 1/2$. Let X denote the number of heads when tossing a fair coin once. Then X can take only two values, namely, 0 heads and 1 head. Therefore P[{H}] = P[X = 1] = 1/2 and P[{T}] = P[X = 0] = 1/2. Hence, X is a discrete random variable.

Suppose that five fair coins are tossed once. Let X denote the number of heads; then the possible values of X are 0, 1, 2, 3, 4, and 5. Let A_k be the event of scoring k heads; then

$$P[A_0] = P[\{(T,T,T,T,T)\}] = \frac{1}{32} = P[X = 0]$$

$$P[A_1] = P[\{(H,T,T,T,T,), (T,H,T,T,T), (T,T,H,T,T), (T,T,T,H,T), (T,T,T,T,H)\}] = P[X = 1] = \frac{5}{32}$$

Similarly,

$$P[A_2] = P[X = 2] = \frac{10}{32}$$

$$P[A_3] = P[X = 3] = \frac{10}{32}$$

$$P[A_4] = P[X = 4] = \frac{5}{32}$$

and

$$P[A_5] = P[X = 5] = \frac{1}{32}$$

Hence X is a discrete random variable. A convenient way to summarize the values of a discrete random variable and the associated probabilities is in a table:

k	P[X = k]
0	1/32
1	5/32
2	10/32
3	10/32
4	5/32
5	1/32
Total	32/32

The number of aces in a hand of 13 cards drawn without replacement from a bridge deck is a discrete random variable, because the event A_k, that the sample of 13 cards has exactly k aces, corresponds to the fact that X takes on the value k. More specifically:

$$P[A_0] = P[X = 0] = \frac{C_0^4 C_{13}^{48}}{C_{13}^{52}}$$

$$P[A_1] = P[X = 1] = \frac{C_1^4 C_{12}^{48}}{C_{13}^{52}}$$

$$P[A_2] = P[X = 2] = \frac{C_2^4 C_{11}^{48}}{C_{13}^{52}}$$

$$P[A_3] = P[X = 3] = \frac{C_3^4 C_{10}^{48}}{C_{13}^{52}}$$

$$P[A_4] = P[X = 4] = \frac{C_4^4 C_{9}^{48}}{C_{13}^{52}}$$

Consider the fortune wheel introduced in Chapter 5. The circumference was divided into 16 equal parts, and we had observed the fact that the probability of the event that the spinner would land in any part was equal to 1/16. Let X denote the position of the spinner on the circumference. Then it follows that P[{that the spinner lands on a particular point on the circumference}] = P[X = particular real number] = 0, and P[{that the spinner lands in any part}] = P[X is between two given numbers] = 1/16. Therefore X is a continuous random variable. More examples of continuous random variables will be given later.

The notions of discrete and continuous variables (see Chapter 1) and sample spaces (see Chapter 5) can be combined with the notion of a random variable as introduced above, leading us formally to the following definitions:

DEFINITION 6.1 *A random variable* X *is said to be a discrete random variable if it is a discrete variable defined on the sample space such that with each possible real value there is a probability which is greater than or equal to* 0 *and the sum of all the probabilities is equal to* 1.

DEFINITION 6.2 *A random variable* X *is called a continuous random variable if it is a continuous variable defined on the sample space such that for any given real value there is a probability equal to* 0, *but for any interval of real values there is a probability which is greater than or equal to* 0 *and the probability for the whole interval of real values is equal to* 1.

6.2 PROBABILITY DISTRIBUTIONS

The relative frequency distribution of Chapter 2 refers to data collected in nature on a variable X. Probability distributions are relative frequency distributions of a random variable when the basic underlying experiment is repeated an infinite number of times. Therefore, probability distributions are theoretical distributions and serve only as a model to analyze data in nature. We now have:

DEFINITION 6.3 *A probability distribution is said to be discrete (continuous) if the underlying random variable is discrete (continuous).*

Note that in agreement with the developments of Chapter 2, probability histograms make sense only in the discrete case, while in the continuous case the probability curve will provide the probabilities. In both situations the probabilities are given by areas, and the total area is equal to 1 or 100%. Since probabilities are given interpretation in terms of infinite number of repetitions of experiments, it follows that to each probability distribution there corresponds an *infinite population*.

We will now discuss certain celebrated probability distributions which are often used as infinite population models to analyze phenomena in nature.

6.3 THE BERNOULLI PROBABILITY DISTRIBUTION

If a fair coin is tossed once, then the corresponding discrete variable X takes on the values 0 and 1, with $P[X = 0] = P[X = 1] = 1/2$. The probability histogram for this simple example is shown in Figure 6.1. Note that $P[X = 0] = 1/2$ can be read directly from the vertical axis and is equal to the area of the rectangle with base equal to 1 and height equal to 1/2.

Now suppose that it is unknown whether the coin is fair or not, i.e., $P[X = 1] = p$ and $P[X = 0] = 1 - p = q$, say. Note that p is a number between 0 and 1 and that when $p > 1/2$, the coin will be biased toward heads and when $p < 1/2$, the coin will be biased toward tails. The probability distribution of a "general" coin is a "two-point" or "dichotomous" distribution and is known in the literature as the *Bernoulli probability distribution*. [Named after Jacob Bernoulli (1654-1705), whose main work *Ars Conjectandi* was published posthumously in 1713.] Its histogram for $p > 1/2$ is given in Figure 6.2.

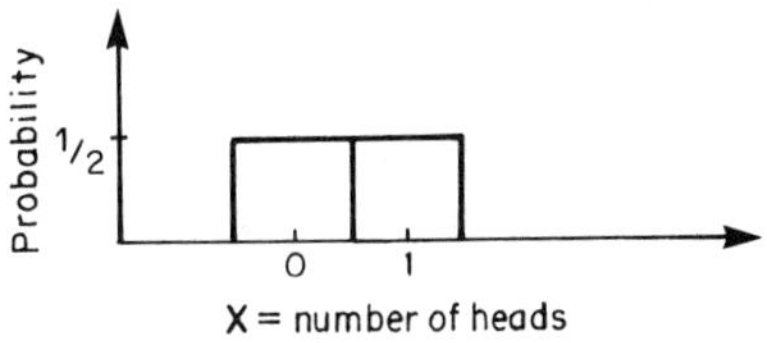

Figure 6.1 Probability distribution of the number of heads when tossing a fair coin once.

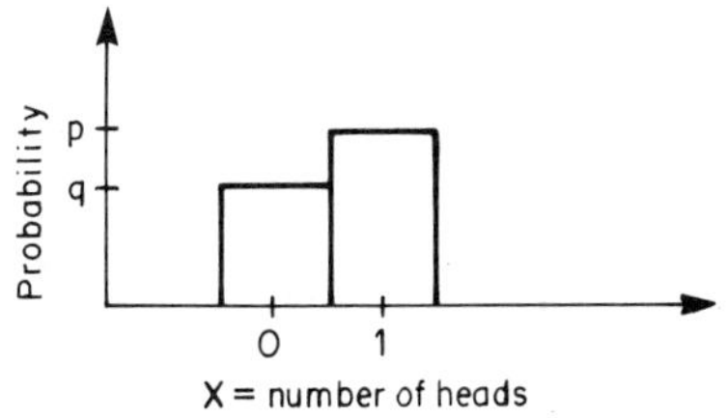

Figure 6.2 The Bernoulli probability distribution for $p > 1/2$.

In nature many "success-failure," "yes-no," "defective-nondefective," etc., types of variables are assumed to have a Bernoulli distribution. For example, if a voter is interviewed with respect to a specific question which can be answered by a "yes" or a "no," then we assume the response of the voter to be a Bernoulli variable such that P[{yes}] = P[X = 1] = p and P[{no}] = P[X = 0] = q, p + q = 1.

DEFINITION 6.4 *The mean,* μ_X*, and variance,* σ_X^2*, of a discrete random variable* X *are defined by* $\mu_X = \Sigma X_i \cdot P[X = X_i]$ *and* $\sigma_X^2 = \Sigma(X_i - \mu_X)^2 \cdot P[X = X_i]$*, respectively.*

It can be shown that $\sigma_X^2 = \Sigma x_i^2 \cdot P[X = x_i] - \mu_X^2$. Since μ_X and σ_X^2 are quantities calculated from the whole infinite population, they are parameters and we will be concerned with estimating them via a sample in Chapter 10.

For the Bernoulli distribution, we have:

$$\mu_X = 0 \cdot q + 1 \cdot p = p$$

and

$$\sigma_X^2 = (0^2)q + (1^2)p - p^2 = p - p^2 = pq$$

6.4 THE BINOMIAL PROBABILITY DISTRIBUTION

While the Bernoulli probability distribution dealt with tossing a coin once, a problem of a more complex nature requires finding the probability distribution of a number of heads when n coins are tossed once, given that each coin has probability p of coming up heads. Such a problem has been resolved at the beginning of this chapter for the case n = 5 fair coins. Using combinations, we see that scoring 2 heads when tossing 5 coins once can be obtained in $C_2^5 = 10$ ways, because exactly two of the positions in (•••••) must be occupied by 2 H's. Hence

$$P[X = 2] = C_2^5\left(\frac{1}{2}\right)\left(\frac{1}{2}\right)\left(\frac{1}{2}\right)\left(\frac{1}{2}\right)\left(\frac{1}{2}\right) = C_2^5\left(\frac{1}{2}\right)^2 \cdot \left(\frac{1}{2}\right)^3$$

In the same way, the complete list of probabilities is obtained as:

$$P[X = 0] = C_0^5\left(\frac{1}{2}\right)^0\left(\frac{1}{2}\right)^5 = \frac{1}{32}$$

$$P[X = 1] = C_1^5\left(\frac{1}{2}\right)^1\left(\frac{1}{2}\right)^4 = \frac{5}{32}$$

$$P[X = 2] = C_2^5\left(\frac{1}{2}\right)^2\left(\frac{1}{2}\right)^3 = \frac{10}{32}$$

$$P[X = 3] = C_3^5\left(\frac{1}{2}\right)^3\left(\frac{1}{2}\right)^2 = \frac{10}{32}$$

$$P[X = 4] = C_4^5\left(\frac{1}{2}\right)^4\left(\frac{1}{2}\right)^1 = \frac{5}{32}$$

$$P[X = 5] = C_5^5\left(\frac{1}{2}\right)^5\left(\frac{1}{2}\right)^0 = \frac{1}{32}$$

Now, consider n coins each having probability p for heads, and suppose that our objective is to find the probability of scoring k heads, i.e., $P[X = k]$, where $k = 0, 1, 2, \ldots, n$. Arguing as above, we observe that among the n positions exactly k of them must be occupied by H's and $(n - k)$ of them by T's. This can be done in C_k^n ways. Hence:

$$P[X = k] = C_k^n \underbrace{(p)(p)\cdots(p)}_{k}\ \underbrace{(q)(q)\cdots(q)}_{(n-k)}$$

$$= C_k^n p^k q^{n-k} \qquad k = 0, 1, 2, \ldots, n$$

This formula is known as the *binomial probability law*, which in turn completely specifies the *binomial probability distribution* for tossing n coins, each having probability p of coming up heads. Let us plot the binomial distribution for $n = 5$ and $p = 1/2$.

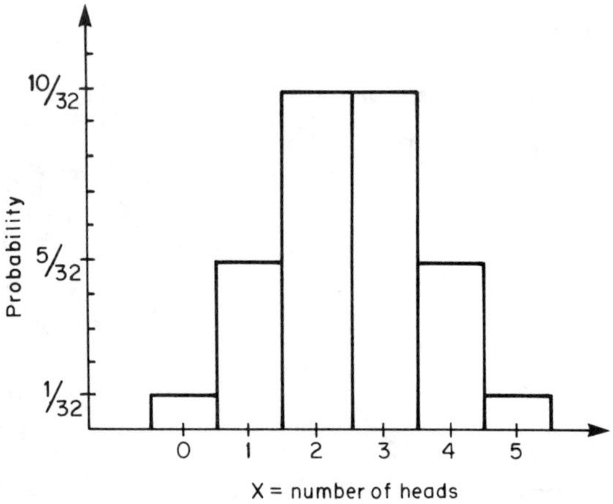

Figure 6.3 The binomial probability distribution for n = 5 and p = 1/2.

The binomial distribution is symmetric if $p = 1/2$, is positivetively skewed if $p < 1/2$, and is negatively skewed when $p > 1/2$. Also, the name binomial distribution comes from the fact that one may obtain all the probabilities by using the binomial expansion,

$$(q + p)^n = C_0^n p^0 q^n + C_1^n p^1 q^{n-1} + C_2^n p^2 q^{n-2} + \cdots + C_n^n p^n q^0$$

For example, when $n = 5$, $p = 1/4$, so that $q = 3/4$, then we have

$$\left(\frac{3}{4} + \frac{1}{4}\right)^5 = C_0^5\left(\frac{1}{4}\right)^0\left(\frac{3}{4}\right)^5 + C_1^5\left(\frac{1}{4}\right)^1\left(\frac{3}{4}\right)^4 + C_2^5\left(\frac{1}{4}\right)^2\left(\frac{3}{4}\right)^3$$

$$+ C_3^5\left(\frac{1}{4}\right)^3\left(\frac{3}{4}\right)^2 + C_4^5\left(\frac{1}{4}\right)^4\left(\frac{3}{4}\right)^1 + C_5^5\left(\frac{1}{4}\right)^5\left(\frac{3}{4}\right)^0$$

$$= \frac{243}{1024} + \frac{405}{1024} + \frac{270}{1024} + \frac{90}{1024} + \frac{15}{1024} + \frac{1}{1024}$$

Hence $P[X = 0] = 243/1024$, $P[X = 1] = 405/1024$, $P[X = 2] = 270/1024$, etc.

To facilitate the calculation of binomial probabilities we have included Table A.1 (see the Appendix). The table is self-explanatory.

The binomial distribution models many variables which arise in nature. Two common instances are: when sets of voters are interviewed with respect to a "yes-no" type of question; and when batches of a product are inspected for "defective-nondefective" classification purposes.

The mean and standard deviation of the binomial distribution can be shown to be equal to:

$$\mu_X = np$$

$$\sigma_X = \sqrt{npq}$$

For n = 5 and p = 1/2, we see that the mean $\mu_X = (5)(1/2) = 2.5$, and the standard deviation is $\sqrt{(5)(1/2)(1/2)} = 1.12$.

The parameters μ_X and σ_X in most practical settings are unknown and the problem usually is to estimate them. This is done in Chapter 10.

6.5 THE NORMAL PROBABILITY DISTRIBUTION

One of the most celebrated and popular probability distributions is given by the normal curve. [First presented in a 1733 paper by Abraham DeMoivre (1667-1754).] Many continuous variables in nature tend to have this type of distribution. (See Exercises 6.16 and 6.17 for two classical examples.) Also, many discrete distributions can be successfully approximated by the normal distribution.

The formula to draw the normal probability distribution or curve is given by the equation:

$$Y = \frac{1}{\sqrt{2\pi}\sigma_X} e^{-(1/2)(X-\mu_X)^2/\sigma_X^2}$$

where π is approximately equal to 3.14, and e is the base of the Napierian logarithm, roughly equal to 2.718. (The formula is provided for mathematically inclined readers and for completeness' sake. Of course one should not try to learn this and later formulas by heart.) The mean of the distribution is equal to μ_X and the standard deviation is σ_X. The sketch of a general normal distribution is shown in Figure 6.4. The normal curve has the following properties:

1. The normal curve is a bell-shaped distribution.
2. The normal curve is symmetric and unimodal.
3. The normal curve is asymptotic to the X axis.
4. Total area under the normal curve is equal to 1.
5. For each pair of values for μ_X and σ_X we have a particular distribution; i.e., we can draw the normal curve any time the mean and standard deviation are given. For example, when $\mu_X = 10$ and $\sigma_X = 2$, then

$$Y = \frac{1}{2\sqrt{2\pi}} e^{-(1/2)(X-10)^2/4}$$

6. The mean is equal to the median and also is equal to the crude mode.
7. The values of X at which the curvature of the curve changes are a distance σ_X away from the mean.

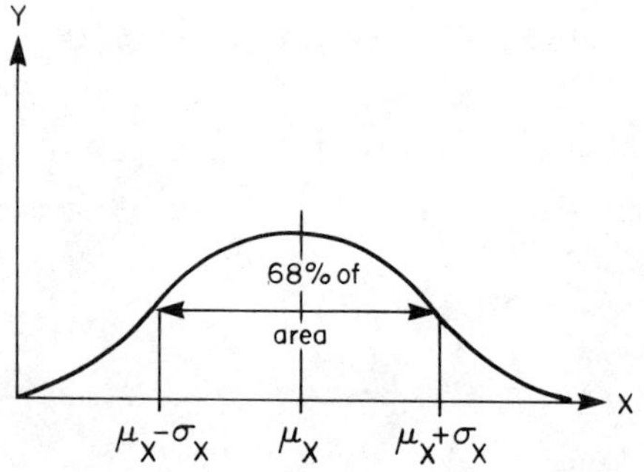

Figure 6.4 Sketch of the normal distribution.

8. The shape of the normal distribution depends on the magnitude of σ_X, i.e., if two normal curves differ only with respect to the standard deviation, then the curve with the larger standard deviation will be flatter than the other (see Figure 6.5).

Various areas under the normal curve are popular in usage. For example, the area under the curve between the mean minus the standard deviation and the mean plus the standard deviation (this is often referred to as the area under the curve within one standard deviation of the mean) is equal to 0.6827; i.e.,

$$P[\mu_X - \sigma_X < X < \mu_X + \sigma_X] = 0.6827 \text{ or } 68.27\%$$

Similarly:

$$P[\mu_X - 2\sigma_X < X < \mu_X + 2\sigma_X] = 95.45\%$$

and

$$P[\mu_X - 3\sigma_X < X < \mu_X + 3\sigma_X] = 99.73\%$$

Other values are given in Table 6.1.

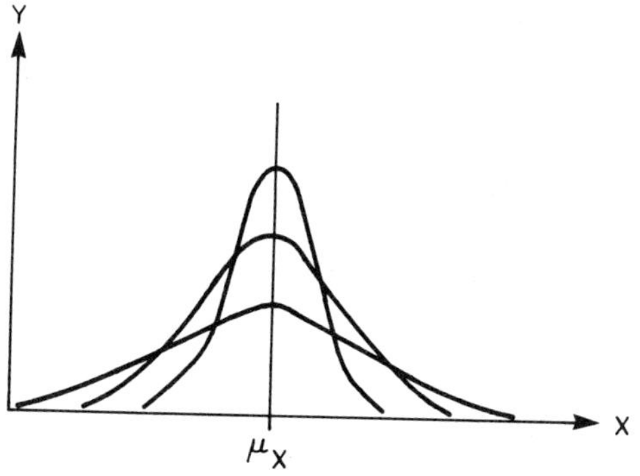

Figure 6.5 Three normal curves with the same mean but different standard deviations.

Table 6.1 Areas under the Normal Curve within k Standard Deviations of the Mean

k	Area	k	Area
0.0	0.0000	1.6	0.8904
0.1	0.0797	1.7	0.9109
0.2	0.1585	1.8	0.9281
0.3	0.2358	1.9	0.9426
0.4	0.3108	2.0	0.9545
0.5	0.3829	2.1	0.9643
0.6	0.4515	2.2	0.9722
0.7	0.5161	2.3	0.9786
0.8	0.5763	2.4	0.9836
0.9	0.6319	2.5	0.9876
1.0	0.6827	2.6	0.9907
1.1	0.7287	2.7	0.9931
1.2	0.7699	2.8	0.9949
1.3	0.8064	2.9	0.9963
1.4	0.8385	3.0	0.9973
1.5	0.8664	3.1	0.9999

Many times our interest lies in finding the number of standard deviation units on each side of the mean such that the area under the curve is a specified percentage. Table 6.2 provides these for many commonly used percentages.

Table 6.2 Specified Areas under the Curve and the Corresponding Interval

Area under the curve (%)	Interval
50	$\mu_X \pm 0.6745\sigma_X$
80	$\mu_X \pm 1.282\sigma_X$
90	$\mu_X \pm 1.645\sigma_X$
95	$\mu_X \pm 1.960\sigma_X$
98	$\mu_X \pm 2.326\sigma_X$
99	$\mu_X \pm 2.576\sigma_X$

Note that from Tables 6.1 and 6.2 many other probability statements can be made, in addition to those previously stated. For example, $P[X > \mu_X + 1.645\sigma_X] = 5\%$, $P[X < \mu_X - 1.5\sigma_X] = (100 - 86.64)/2 = 13.36/2 = 6.68\%$.

6.6 THE STANDARD NORMAL DISTRIBUTION

A special case of the normal curve is the standard normal curve or Z curve. Consider a variable X, which has a normal distribution. Suppose that we do not want to work with the original variable X, but instead with the new variable:

$$Z = \frac{X - \mu_X}{\sigma_X}$$

This new variable measures the deviation of X from the mean μ_X in terms of standard deviation units. The variable Z is known as the *standard normal variable*. In the social sciences it is commonly referred to as the *Z score*. It has a normal distribution with mean and standard deviation equal to:

$$\mu_Z = 0 \qquad \sigma_Z = 1$$

The formula to draw the standard normal curve (for those who think mathematically) is given by the equation:

$$Y = \frac{1}{\sqrt{2\pi}} e^{-(1/2)Z^2}$$

A sketch of this curve is shown in Figure 6.6.

Note that there is only one standard normal curve. Also, any probability statement dealing with the original variable X can be transformed to a probability statement in terms of Z as long as we are given what μ_X and σ_X are.

As an example, suppose that the normal model is assumed for the grades of a test given in a large class of students. If the mean $\mu_X = 70$ and the standard deviation $\sigma_X = 10$ are typical for the test,

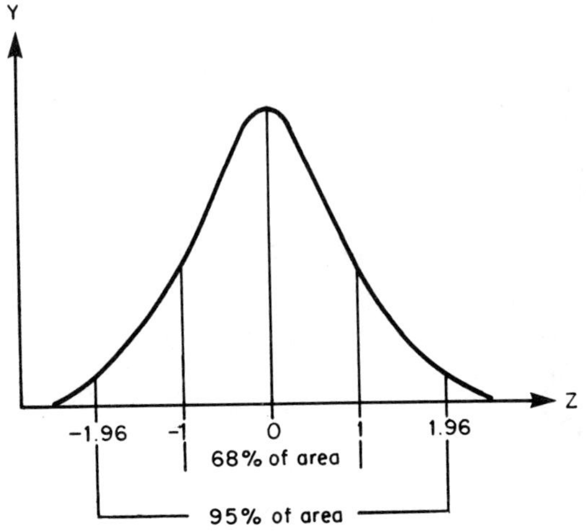

Figure 6.6 The standard normal curve.

then what are our chances of scoring a grade higher than 86 in a forthcoming test? Using the Z standardization we obtain:

$$P[X > 86] = P\left[\frac{X - \mu_X}{\sigma_X} > \frac{86 - \mu_X}{\sigma_X}\right]$$

$$= P\left[Z > \frac{86 - \mu_X}{\sigma_X}\right]$$

$$= P\left[Z > \frac{86 - 70}{10}\right]$$

$$= P[Z > 1.6]$$

$$= \frac{1 - 0.8904}{2} = 0.0548$$

from Table 6.1. Therefore, knowing the probabilities for the Z distribution will allow the calculation of any probability concerning the normal random variable X as long as the mean μ_X and the standard deviation σ_X are given.

From Table 6.1 it also follows, for example, that:

$P[-1 < Z < 1] = 0.6827$

$P[-2 < Z < 2] = 0.9545$

$P[-2.5 < Z < 2.5] = 0.9876$

etc., and from Table 6.2:

$P[-1.96 < Z < 1.96] = 0.95$

$P[-2.58 < Z < 2.58] = 0.99$

etc. Table A.2 (see the Appendix) is an extended table of the Z distribution. From this table one may obtain any probability dealing with the Z variable, where the Z values are up to two decimals. For example, $P[Z > 2.51] = 0.5 - 0.4940 = 0.0100$. The reader should be completely comfortable with this table.

As an example of the use of the normal distribution, suppose that scores in a course are to be "graded on the curve" (or "belled" as is commonly known among students) in the following way:

Z score	Grade	
$Z > 1.96$	A	Pass
$1 \le Z < 1.96$	B	Pass
$0 \le Z < 1$	C	Pass
$-1 \le Z < 0$	D	Pass
$Z < -1$	Fail	

Suppose that the mean of a very large class turns out to be 64, with the standard deviation equal to 6; then what will John get if his final mark is 71?

John's standard score, or Z score, is equal to:

$$Z = \frac{X - \mu_X}{\sigma_X} = \frac{71 - 64}{6} = 1.1$$

so that he will get a B because 1.1 falls in the second category.

EXERCISES

6.1 It is hypothesized that the dichotomous variable X has a probability distribution with $\mu_X = 0.90$.

(a) Plot the hypothesized probability distribution.

(b) Find the mean and standard deviation of (a).

6.2 Seven fair coins are tossed once. Plot the probability distribution of the number of heads. Also, find the mean and standard deviation of the probability distribution. Finally, calculate the probability of obtaining:

(a) At least 3 heads

(b) At least 2 heads

(c) At least 2 tails

6.3 An urn contains white and red balls in the proportions p and q (= 1 - p), respectively. A sample of 7 balls is drawn with replacement from the urn. Find the probability distribution of the number of white balls in the sample; i.e., obtain $P[X = k]$. Plot the distribution for $p = 1/2$. Compare this result with that of Exercise 6.2. Conclusion?

6.4 If an urn contains 10 white and 5 black balls, and if 3 balls are drawn without replacement, what is the mean number of black balls that will be obtained?

6.5 Plot the cumulative binomial probability distribution for $p = 1/4$ and $n = 5$. From this obtain $P[1 \le X \le 4]$.

6.6 If the production of parts by a machine contains on the average 20% defectives, is it more likely to have (a) no defectives among 10 parts or (b) at most one defective among 20 parts?

6.7 In a 10-question true-false examination, suppose that a student uses the following strategy. If the coin falls heads, he answers "true," if it falls tails he answers "false." Find the probability that he answers at least 6 questions correctly.

6.8 Five fair dice are tossed once. Calculate the probability of getting five 6's. What is the probability of not obtaining five 6's?

6.9 Eight dice are tossed. Calling a 5 or 6 a success, find the probability of getting:

(a) Three successes

(b) At most three successes

6.10 Imagine we have dice made in the form of a regular tetrahedron (four faces) and marked with one to four dots on the faces. The number of dots on the bottom face is the number we get when we toss the dice. Now if we toss two such dice, what is the probability of getting exactly one ace (1)? At least one ace? Exactly two aces? What are the probabilities of getting the various possible total number of dots?

6.11 Two fair dice are rolled. What is the probability of obtaining an even total? An odd total? What total number has the greatest probability? What is the probability of obtaining a total not exceeding 9?

6.12 Suppose that weather records show that on the average 3 out of the 30 days in September are rainy days.

(a) Assuming a binomial distribution with each day of September as an independent trial, find the probability that next September will have at most two rainy days.

(b) Give reasons why you may not be justified in using the binomial distribution in solving (a).

6.13 A pair of dice is tossed 6 times. Find the probability of getting a total of 9:

(a) Twice

(b) At least twice

6.14 If the probability of a defective bolt is 0.1, find:

(a) The mean

(b) The standard deviation for the distribution of defective bolts in a total of 400

6.15 The expected frequency of an event is defined to be the number of trials times the probability of the event. Suppose that an optimistic candidate thinks that 80% of the voters will be on "his side" in an upcoming election. On the basis of his hypothesis being true, calculate the expected frequency for obtaining at least 8 on "his side" when 10 interviews are done.

6.16 The following data represent the "frequency distribution of statures for adult males born in the United Kingdom, including the whole of Ireland," and were found in the final report of the Anthropometric Committee to the British Association, 1883, p. 256. (Actual measurements were taken to the nearest 1/8 in.) Plot a frequency histogram. What characteristics does this distribution possess?

Height without shoes (in.) [class midpoint]	Number of men
57	2
58	4
59	14
60	41
61	83
62	169
63	394
64	669
65	990
66	1223
67	1329
68	1230
69	1063
70	646
71	392
72	202
73	79
74	32
75	16
76	5
77	2
Total	8585

6.17 The following data represent the "frequency distribution of the heights of maize plants" (Emerson and East, 1913). Construct the frequency histogram. What percentages of the area lie within 1 and within 2 standard deviations of the mean? Why would the normal curve be a good model for these data?

Height of plant (decimeters) [class midpoint]	Observed frequency
7	1
8	3
9	4
10	12
11	25
12	49
13	68
14	95
15	96
16	78
17	53
18	26
19	16
20	3
21	1
Total	530

6.18 Assume that the height (in centimeters) of a person aged 20 is a random variable with a normal distribution having parameters $\mu_X = 170$ cm and $\sigma_X = 5$ cm. Calculate the probability that the height of a person aged 20 will be greater than 180 cm. Sketch the normal probability distribution and the standard normal probability distribution, showing the areas corresponding to the probability.

6.19 A machine produces bolts for which the length, X, in millimeters, is such that it follows a normal probability model with parameters $\mu_X = 5$ and $\sigma_X = 0.10$. The specifications for the bolt call for items with a length (in millimeters) equal to 5.05 ± 0.12. A bolt not meeting these specifications is called defective. What is the probability that a bolt produced by this machine will be defective?

6.20 If the Z score is equal to -2.4, then find the X score if $\mu_X = 65$ and $\sigma_X = 7$. Find and sketch $P[Z < -2.4 \text{ or } Z > 2.4]$.

6.21 If the Minnesota Paper-Form Board Test yields scores which are normally distributed with a mean of 43 and standard deviation of 8, then find the expected number (see Exercise 6.15) of students who fall between the scores of 36 and 53, if the test is administered to 200 students.

6.22 If IQ scores of grade 8 students have a normal distribution with mean equal to 100 and standard deviation equal to 16, then find the 50th, 80th, and the 99th percentiles.

6.23 True or false:

(a) A random variable with a normal curve is a continuous random variable.

(b) If an X value in a normal distribution is $2\sigma_X$ to the right of the mean, then the area under the curve to the left of $\mu_X - 2\sigma_X$ is equal to half the area between μ_X and $\mu_X + 2\sigma_X$.

(c) The binomial probability model applies essentially to a variable which is the sum of n independent Bernoulli random variables, each having a mean equal to p.

(d) The total area for the binomial probability distribution is equal to 100%.

(e) The value of z such that $P[Z < z] = P[Z > z]$ (i.e., the probability that the Z score is greater than z is equal to the probability that the Z score is less than z) is 0.

(f) A Z score may be defined as X/σ_X only when $\mu_X = 0$.

(g) The standard deviation may be regarded as a measure of variability in the population.

(h) The total area under the normal curve is 0.999.

(i) The standard deviation of the standard normal distribution is a known quantity.

(j) $X = \mu_X + Z\sigma_X$.

6.24 Using a Z table, find the area under the standard normal curve which lies:

(a) Between 0 and 1.50
(b) Between -0.61 and 0
(c) To the left of 2.36
(d) To the right of 1.05
(e) To the left of -0.91
(f) To the right of 1.50
(g) Between 1.25 and 1.73
(h) Between -0.75 and 1.75
(i) Between -1.30 and 1.42
(j) Between -1.96 and 1.96

6.25 From the statistical literature, list one other discrete probability distribution and one other continuous probability distribution. Give their formulas and sketch the distributions. Also, give examples from your own area which obey the distributions.

6.26 Assume that the response of 10 housewives regarding preference or no preference of "baby bonus" obeys the binomial probability law with p = 0.9. Find the probability that at least 9 housewives prefer "baby bonus."

6.27 If X is normally distributed with $\mu_X = 1$ and $\sigma_X = 1/2$, find:

(a) $P[X > 2]$.

(b) $P[0 < X < 1]$

6.28 If X is normally distributed with $\mu_X = 1$ and $\sigma_X = 2$, find a number X_0 such that:

(a) $P[X > X_0] = 0.10$

(b) $P[X > -X_0] = 0.20$

6.29 The mean weight of 500 male students at a certain college is 151 lb and the standard deviation is 15 lb. Assuming that weights are normally distributed, find the expected frequency of students (see Exercise 6.15) weighing:

(a) Between 120 and 155 lb

(b) More than 185 lb

(c) Less than 128 lb

(d) Less than or equal to 128 lb

6.30 In Exercise 6.29 find the weight such that 90% of the students weigh less than it.

6.31 For the binomial distribution verify that

$$P[X = k + 1] = \frac{(n - k)p}{(k + 1)q} P[X = k]$$

for $k = 0, 1, \ldots, n - 1$.

7

NORMAL APPROXIMATION TO THE BINOMIAL AND FITTING THE NORMAL DISTRIBUTION

As stated in the previous chapter, the normal distribution is the most widely used distribution, because the relative frequency distributions of many variables in nature exhibit such a structure. It is also referred to as the Gauss-Laplace or Gaussian distribution, after K. F. Gauss (1777-1855). A continuous random variable which has a normal probability distribution is called a *normal variable*. Since a normal probability distribution deals with an infinite population (because that is the interpretation we have adopted for probability) one refers to the normal distribution as a *normal population*.

The normal distribution can be used to approximate other distributions, even discrete ones. Indeed, the binomial distribution is a discrete distribution and for a wide range of values of n and p it can be successfully approximated by the normal distribution.

In this chapter we show how to approximate the binomial distribution with the normal and how to fit the normal distribution to a frequency histogram.

7.1 NORMAL APPROXIMATION TO THE BINOMIAL

You may recall that the binomial probability distribution is given by the formula

$$C_X^n p^X q^{n-X} \qquad X = 0, 1, 2, \ldots, n, \; 0 < p < 1, \; q = 1 - p$$

We also showed that it models the results of the following experiment: Toss n coins where each coin has a probability p of coming up heads. For n = 10 and p = 1/2, the formula is equal to:

$$\frac{10!}{X!(10 - X)!}\left(\frac{1}{2}\right)^{X}\left(\frac{1}{2}\right)^{10-X} \qquad X = 0, 1, 2, \ldots, 10$$

and a plot of these 11 probabilities is given in Figure 7.1. You will also notice that we superimposed a normal curve on this binomial distribution by drawing the line through the top and middle of each rectangle. The equation of this curve is obtained by using the fact that the parameters of the binomial distribution are given by

$$\mu_X = np = (10)\,\frac{1}{2} = 5.0$$

$$\sigma_X = \sqrt{npq} = \sqrt{(10)\left(\frac{1}{2}\right)\left(\frac{1}{2}\right)} = 1.6$$

and then inserting these into the formula for the normal distribution. So, the equation for the superimposed curve becomes

$$Y = \frac{1}{\sqrt{2\pi}(1.6)}\, e^{-(1/2)[(X-5)/1.6]^2}$$

Let us now calculate some probabilities using this normal distribution. First of all, notice that the exact binomial probability for scoring, say 7 or more heads, is given by:

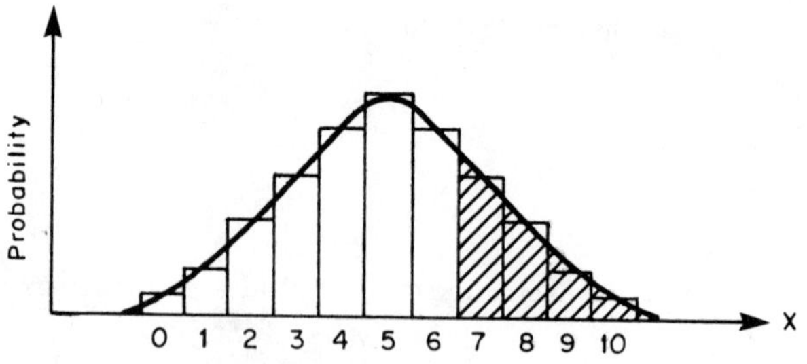

Figure 7.1 Normal approximation for the binomial with n = 10, p = 1/2.

$$P[X \geq 7] = \text{area of the four rectangles on 7 to 10}$$

$$= \sum_{X=7}^{10} C_X^{10}\left(\frac{1}{2}\right)^X\left(\frac{1}{2}\right)^{10-X}$$

$$= C_7^{10}\left(\frac{1}{2}\right)^{10} + C_8^{10}\left(\frac{1}{2}\right)^{10} + C_9^{10}\left(\frac{1}{2}\right)^{10} + C_{10}^{10}\left(\frac{1}{2}\right)^{10}$$

$$= 0.172$$

This probability is approximated by the shaded area under the normal curve, which we can, in fact, calculate:

$$P[X \geq 7] = \text{area under normal curve to the right of 6.5}$$

$$= P[X > 6.5] = P\left[Z > \frac{6.5 - 5.0}{1.6}\right]$$

$$= P[Z > 0.94]$$

$$= 0.174$$

A good approximation! In the calculation we have used the Z transformation, i.e., the standard normal variable, given by:

$$Z = \frac{X - \mu_X}{\sigma_X} = \frac{X - np}{\sqrt{npq}}$$

Thus to change the probability statement regarding X into an equivalent Z statement we have to subtract the mean and divide the result by the standard deviation. This is the reason why 5.0 is subtracted from 6.5 and the result is divided by 1.6.

It is now clear how to go about using the normal approximation. One may now calculate various approximate probabilities, such as

$$P[X < k] = P\left[Z < \frac{k - (1/2) - np}{\sqrt{npq}}\right]$$

$$P[X = k] = P\left[\frac{k - (1/2) - np}{\sqrt{npq}} < Z < \frac{k + (1/2) - np}{\sqrt{npq}}\right]$$

and

$$P[a \leq X \leq b] = P\left[\frac{a - (1/2) - np}{\sqrt{npq}} < Z < \frac{b + (1/2) - np}{\sqrt{npq}}\right]$$

In these expressions k, a, and b are arbitrary nonnegative integers between 0 and n. For fairly large n and p close to 1/2 one often deletes the so called *continuity correction* of 1/2.

The normal approximation to the binomial will be good if p is close to 1/2 for n moderate to large. If p or q are small compared to each other, then experience has shown that np and nq should be at least equal to 5 for application of the normal approximation.

How did we sketch the normal curve over the binomial distribution? The easiest way to go about this is by utilizing the standard normal distribution. In Figure 7.2 we have drawn the values of the Z variable below those of the variable X. Since the standard normal variable Z has the equation

$$Y = \frac{1}{\sqrt{2\pi}} e^{-(1/2)Z^2}$$

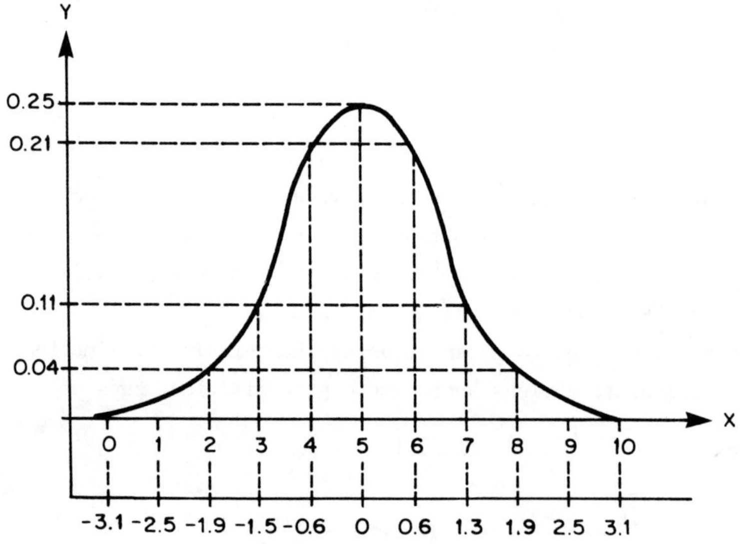

$$Z = \frac{X - \mu_X}{\sigma_X} = \frac{X - 5}{1.6}$$

Figure 7.2 Sketch of the normal distribution with $\mu_X = 5$ and $\sigma_X = 1.6$.

it follows that the Y values for given X values are obtained by dividing the Y values corresponding to the Z values by $\sigma_X = 1.6$. Tables (see Table A.3) are available which provide ordinates (i.e., Y values) for the standard normal distribution, and we have used some of these values for sketching our normal curve for this example:

Ordinate Values Corresponding to Z and X

Z	Y	X	Y
0.0	0.3989	5	$0.3989/1.6 \doteq 0.25$
0.6	0.3332	6	$0.3332/1.6 \doteq 0.21$
1.3	0.1714	7	$0.1714/1.6 \doteq 0.11$
1.9	0.0656	8	$0.0656/1.6 \doteq 0.04$
2.5	0.0175	9	$0.0175/1.6 \doteq 0.01$
3.1	0.0033	10	$0.0033/1.6 \doteq 0.00$

Let us go through one more illustration of applying the normal approximation to the binomial distribution. To find out the attitudes of housewives with respect to the abortion issue, 1000 were randomly selected. On the *assumption* that 40% of them favor abortion, we calculate:

$$\mu_X = np = (1000)(0.40) = 400$$

$$\sigma_X = \sqrt{npq} = \sqrt{(1000)(0.40)(0.60)} = 15.5$$

Ignoring the continuity correction of 1/2, we readily obtain from the normal approximation that the probability that X will be between $\mu_X - 1.96\sigma_X$ and $\mu_X + 1.96\sigma_X$ is 95%; i.e., in repeated surveys of this kind (i.e., n = 1000 and p = 0.40), 95% of the time the number of housewives favoring abortion will be between 400 - (1.96)(15.5) = 369.6 and 400 + (1.96)(15.5) = 430.4. The sketch of this probability is shown in Figure 7.3.

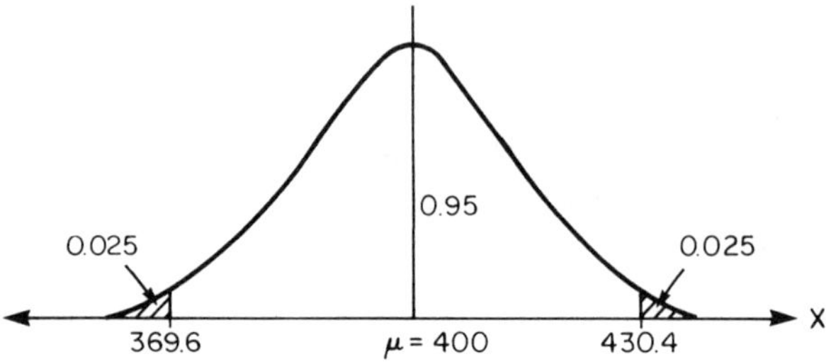

Figure 7.3 The normal distribution for the abortion survey.

7.3 FITTING THE NORMAL DISTRIBUTION TO AN OBSERVED FREQUENCY DISTRIBUTION

When fitting the normal distribution to a binomial distribution the crux of the matter was the specification of p and hence of $\mu_X = np$ and $\sigma_X = \sqrt{npq}$. In practice the need often arises to superimpose the normal distribution over a frequency histogram, where μ_X and σ_X are estimated from the data rather than specified. Let us now show how to fit a normal distribution to a histogram.

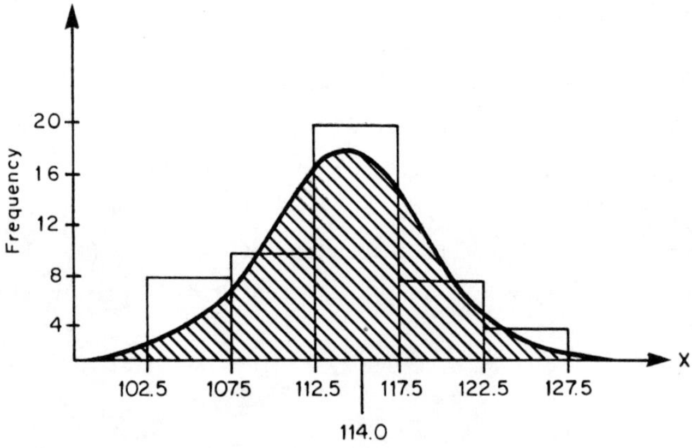

Figure 7.4 Observed versus expected frequencies using the normal distribution.

Table 7.1 Observed and Expected Frequencies

Midpoint	Observed frequency	Expected frequency
105	8	5.245
110	10	13.515
115	20	16.585
120	8	10.140
125	4	2.950

The data in the frequency histogram in Figure 7.4 relate to IQ scores obtained from 50 sixth graders of the so-called "faster" or "A" group. Although the number of data points is not very large, let us see how the normal fit will be. From the histogram the observed frequencies are obtained, and these are given in column 2 of Table 7.1. Since μ_X and σ_X are not specified for the normal distribution of IQ scores of such children, we use the natural estimates from the sample data, namely,

$$\overline{X} = \frac{1}{50}\sum_{i=1}^{5} X_i f_i = 114.00$$

$$s_X = \sqrt{\frac{1}{49}\left[\sum_{i=1}^{5} X_i^2 f_i - \frac{1}{50}\left(\sum_{i=1}^{5} X_i f_i\right)^2\right]} = 5.71$$

where the X_i's are the five midpoints and the f_i's are the five observed frequencies.

To fit the normal curve to our frequency histogram, we divide the ordinate values of the standard normal distribution (see Table A.3) by 5.71 and multiply the resulting values by the total area under the histogram, which is equal to (5)(50) = 250 square units. For *sketching* the curve, we use the following convenient values:

Ordinate Values Corresponding to Z and X

Z	Y	X	Frequency
0	0.3989	114	$(0.3989)\left(\frac{250}{5.71}\right) = 17.46$
$\frac{119 - 114}{5.71} = 0.88$	0.2709	119	$(0.2709)\left(\frac{250}{5.71}\right) = 11.86$
$\frac{124 - 114}{5.71} = 1.75$	0.0863	124	$(0.0863)\left(\frac{250}{5.71}\right) = 3.78$
$\frac{129 - 114}{5.71} = 2.63$	0.0126	129	$(0.0126)\left(\frac{250}{5.71}\right) = 0.55$

Remark. Note that in the case of the binomial it was not necessary to multiply the Y values corresponding to the X values by the total area, since the total area was equal to 1. Also, if the experiment with the 10 coins was conducted N times and the expected frequencies (i.e., N times the probabilities) were graphed, then for superimposing the normal curve the ordinates should be multiplied by N, because the total area is then equal to N.

After obtaining the Y values, the normal curve is sketched in Figure 7.4. It is obvious that a better picture will be obtained if more X values are selected. A major purpose of fitting the normal distribution to a frequency histogram is to compare the observed frequencies against *expected frequencies*. These are obtained by calculating the probabilities for the classes and then multiplying each by 50, the value of N.

The required probabilities are obtained by utilizing the normal distribution, and the reader may verify from the normal probability tables that:

$$P[102.5 < X < 107.5] = 0.1049$$
$$P[107.5 < X < 112.5] = 0.2703$$
$$P[112.5 < X < 117.5] = 0.3317$$
$$P[117.5 < X < 122.5] = 0.2028$$
$$P[122.5 < X < 127.5] = 0.0590$$

The expected frequencies are indicated in the third column of Table 7.1. From Figure 7.4 and Table 7.1 one may visually judge the normal fit. An objective method of deciding whether the discrepancies between the observed and expected frequencies are "large" or not (i.e., whether the normal fit should be rejected or accepted) will be discussed in the chapter on chi-square tests.

EXERCISES

7.1 A random sample of 7 voters in each of 128 cities was asked whether they would vote for Candidate A or not. Assuming that the probability of a "yes" vote is 1/2 for each voter, approximate the resulting binomial distribution by the normal distribution. In your answer you must draw the binomial distribution in terms of absolute frequencies and then superimpose on this the normal distribution. For each value of X compare the exact and expected frequencies by finding the percentage relative errors. (Often a "comparison" is made by stating the *percentage relative error* (PRE), which is defined by

$$\text{PRE} = \frac{|A - E|}{E}$$

where A = approximate value and E = exact value. By the way, a 10% or smaller PRE is considered quite good.)

7.2 One hundred 5-choice questions are asked in a test. On the basis that the correct answer is guessed, find the following probabilities by normal approximation:

(a) Exactly 50 are answered correctly.

(b) Sixty or more are answered correctly.

7.3 Make 150 measurements on units of your choice. Draw a frequency histogram and superimpose on this the normal distribution. (You must show all details of your computations.) Compare the observed against the expected frequencies.

7.4 Use the normal approximation to find the probability of obtaining exactly 6 sixes in 18 tosses of a fair die. Compare it with the exact probability obtained from the binomial.

7.5 If p is much smaller than 1/2 in the binomial distribution, then for which values of X do you think that the normal approximation will show marked discrepancies from the exact probabilities?

7.6 For the data of Table 7.1, measure the degree of agreement (or goodness-of-fit) between the observed frequencies (o_i) and the expected frequencies (e_i), where i = 1, 2, ..., c (where c = 5 = number of comparisons), by computing the quantity

$$H = \sum_{i=1}^{c} \frac{(o_i - e_i)^2}{e_i}$$

In general:

(a) What is the possible range of values of this measure H?

(b) When would H = 0?

(c) Why would $\Sigma(o_i - e_i)$ not be a good measure of the goodness-of-fit?

7.7 Find the probability that a student can guess correctly the answers to (a) 12 or more out of 20, (b) 24 or more out of 40, (c) 48 or more out of 80 questions on a true-false examination. (d) If passing requires a mark of at least 60%, which of the above examinations would you like to write? Why?

7.8 Queen Elizabeth wishes to fight a battle with the Spanish Armada off the coast of Africa. She knows that on the average, 10% of the ships become shipwrecked in traveling one way between Britain and Africa. She also knows that 100 ships or more are needed to defeat the Armada. If 120 ships are sent from Britain:

(a) What is the expected number of British ships that will reach the Armada to do battle?

(b) If it happened that no British ships were sunk in the battle, what would be the expected number of ships to return to Britain?

(c) What is the probability that the British fleet will triumph?

(d) In order to answer the above, what assumptions were necessary?

7.9 From the data in Exercise 6.16 construct a frequency table with 7 classes with midpoints 58, 61, 64, ..., 76, instead of 21, as given there. Then construct the frequency histogram. Fit a normal distribution to these data. Compute the H measure defined in Exercise 7.6. Compare the 7 percentage relative errors (see Exercise 7.4). Comment on the goodness-of-fit. (This is a crucial problem in your understanding of the normal distribution.)

7.10 For the data of Exercise 6.17 on the height of corn plants, construct a frequency table with only five classes. (*Hint:* Use class midpoints 8, 11, 14, 17, 20.) Fit and measure the goodness-of-fit of a normal distribution. (*Hint*: See Exercise 7.6; $\overline{X} = 14.5$, $s_X = 2.4$.)

8

RANDOM SAMPLING FROM NORMAL POPULATIONS
The Sampling Distribution of the Sample Mean and the Sampling Distribution of the Difference between Two Sample Means

As noted before, the normal distribution represents an infinite population. It is a model used in many practical applications, because many variables in nature have a histogram which is bell shaped. When the normal distribution is superimposed on a histogram and the discrepancies between the histogram and the curve are "small," then we make the assumption that the underlying variable possesses a normal distribution. Once such an assumption is made, the frequency histogram of the variable is described *mathematically* (this is again for those who are mathematically inclined) by the equation of the normal distribution, i.e., by:

$$Y = \frac{A}{\sqrt{2\pi}\sigma_X} e^{-(1/2)[(X-\mu_X)/\sigma_X]^2}$$

where A is the total area in the histogram. If the total area is made to be equal to 1 then we have the earlier discussed equation of the normal probability distribution:

$$Y = \frac{1}{\sqrt{2\pi}\sigma_X} e^{-(1/2)[(X-\mu_X)/\sigma_X]^2}$$

When a population is described by either of the equations, it is called a *normal population*.

No matter which of the two forms is used, the tremendous advantage is that we can derive mathematically the behavior of certain statistics when repeated samples are drawn from a normal population. It is our intention in this chapter to discuss the sampling behavior of the mean and the difference between two sample means, when sampling from a normal population. Several examples are given to illustrate the results.

8.1 SAMPLING DISTRIBUTIONS

Sampling distributions are based on the concept of randomness. We now require the following definition:

DEFINITION 8.1 *A simple random sampling procedure or technique is a method of extracting a sample of* n *units from a population such that each possible sample has an equal probability of selection.*

Consequently a *random sample* is a sample generated by a simple random sampling procedure. Let us for the moment not worry about how to extract a random sample. This will be done extensively in the chapter on sample surveys. Instead, assume that there is a mechanism or device which produces for us on demand a random sample. Note that in the case of an infinite population the number of possible random samples is infinite, no matter whether we draw them with or without replacement. In the finite population case, we know from the chapter on probability that there are precisely C_n^N and N^n possible samples in the without and with replacement cases, respectively. Thus the chance for any sample being selected is equal to $1/C_n^N$ and $1/N^n$, respectively.

Assume now that we are dealing with a normal population and imagine extracting all possible samples of size n by a simple random sampling procedure. Schematically this process is pictured in Figure 8.1.

You may have already asked yourself why we should go through such a theoretical set up. Indeed, the normal population is infi-

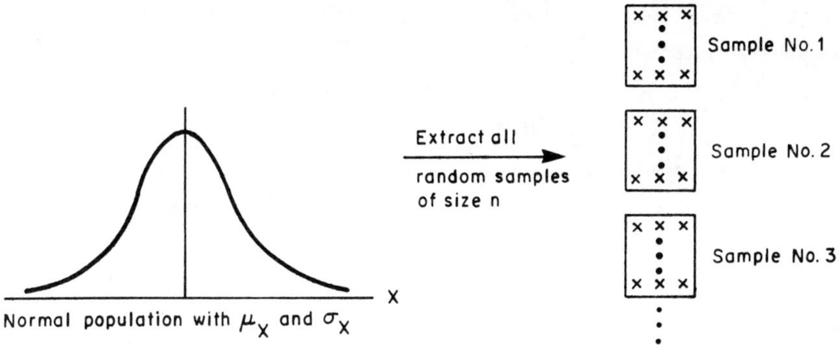

Figure 8.1 A simple random sampling procedure.

nite, and hence there will be an infinite number of samples, and no practical value to all this seems in sight. The truth of the matter is that we can predict with *certainty* how the distributions of certain statistics will look without even bothering to draw all possible random samples! The basis for this statement is the fact that the normal population is described by a mathematical equation, and we can mathematically deduce the equations of the distributions of certain important statistics. These deduced distributions, or *sampling distributions*, can serve as models for answering practical questions, just as the normal distribution did.

8.2 THE SAMPLING DISTRIBUTION OF THE MEAN

Let us look at the most important statistic, namely, the sample mean $\bar{X}$. We calculate for each sample the mean and list these as:

Sample no.	Sample mean $\bar{X}$
1	$\bar{X}_1$
2	$\bar{X}_2$
3	$\bar{X}_3$
4	$\bar{X}_4$

Now, note that all the possible sample means form a population by themselves. The process of drawing all possible samples and calculating the resulting means has created this new population, and the question to be answered is: "What is the distribution of the random variable $\overline{X}$?" That is, what does the distribution look like when we make a histogram with very small class widths (we can do this, because the population of $\overline{X}$'s is infinite) for all the possible sample means? Also what are the mean and standard deviation of this distribution of sample means?

Fortunately the sampling distribution of $\overline{X}$ for a normal population has been settled a long time ago and we state this fundamental result as:

THEOREM 8.1 *If the random variable* X *has a normal distribution, then the random variable* $\overline{X} = (1/n) \sum_{i=1}^{n} X_i$, *obtained from a random sample* $X_1, X_2, \ldots, X_n$, *also has a normal distribution with the same mean* μ_X *but with standard deviation* $\sigma_X\sqrt{n}$.

In shorthand notation Theorem 8.1 is written as: If X is $N(\mu_X, \sigma_X)$ then $\overline{X}$ is $N(\mu_{\overline{X}} = \mu_X, \sigma_{\overline{X}} = \sigma_X/\sqrt{n})$.

Let us reflect a little bit on Theorem 8.1. It says that if you plot the frequency distribution of all possible sample means and you make the class widths as small as possible, you will see that the distribution is normal (i.e., bell shaped) and it is centered at μ_X with the inflection points occurring at $\mu_X \pm \sigma_X/\sqrt{n}$. Stated differently, Theorem 8.1 tells us that the population of sample means is normal and its mean and standard deviation are given by:

$$\mu_{\overline{X}} = \mu_X$$

$$\sigma_{\overline{X}} = \frac{\sigma_X}{\sqrt{n}}$$

Notice that the standard deviation $\sigma_{\overline{X}}$ is inversely proportional to $\sqrt{n}$. This means that when $n > 1$ the standard deviation of $\overline{X}$ is smaller than that of X, which in turn implies a higher peaked and more concentrated distribution for $\overline{X}$. As n becomes larger and larger, this becomes more and more pronounced.

The equation (for those readers thinking mathematically) describing the normal probability distribution of $\overline{X}$ is given by:

$$Y = \frac{1}{\sqrt{2\pi}\sigma_{\overline{X}}} e^{-(1/2)[(\overline{X}-\mu_{\overline{X}})/\sigma_{\overline{X}}]^2}$$

where $\mu_{\overline{X}}$ and $\sigma_{\overline{X}}$ have already been given above. In Figure 8.2 we sketch this equation for various sample sizes.

We have stated that once we know the distribution of $\overline{X}$, we can use this as a model and make, for example, probabilistic statements concerning $\overline{X}$. Since $\overline{X}$ possesses a normal distribution, we can standardize it to a Z variable or standard normal variable, and then use the tabled values. For example,

$$P[a < \overline{X} < b] = P\left[\frac{a - \mu_{\overline{X}}}{\sigma_{\overline{X}}} < Z < \frac{b - \mu_{\overline{X}}}{\sigma_{\overline{X}}}\right] = P\left[\frac{a - \mu_X}{\sigma_X/\sqrt{n}} < Z < \frac{b - \mu_X}{\sigma_X/\sqrt{n}}\right]$$

Pictorially we have the graphs shown in Figure 8.3.

The probability statement regarding $\overline{X}$ has an interpretation in the sense that if we draw all possible samples from a normal population, then the proportion of sample means lying between a and b can be evaluated by using the Z table. Another way of saying this

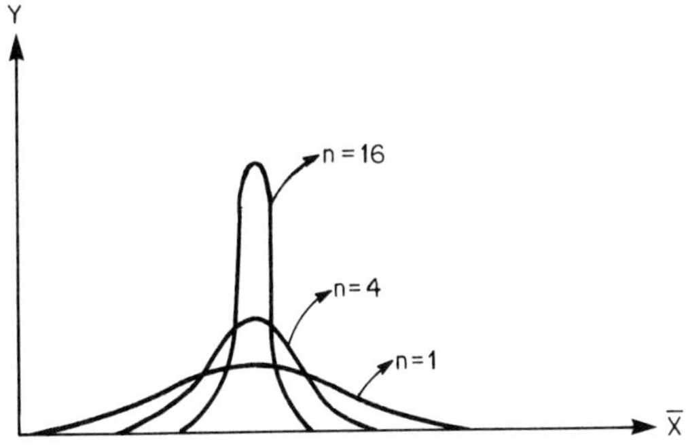

Figure 8.2 Sampling distribution of $\overline{X}$ for various sample sizes.

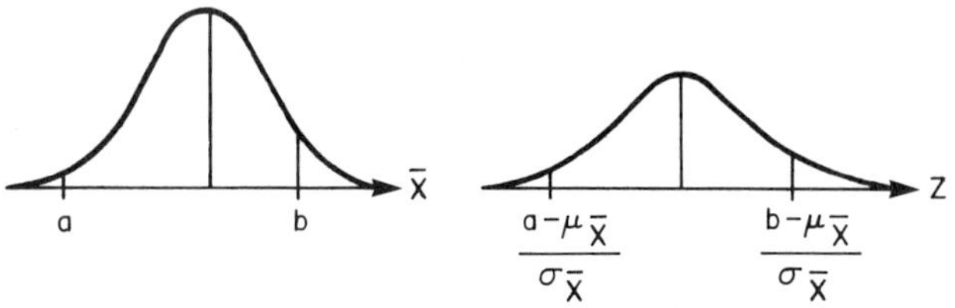

Figure 8.3 Standardization of the sampling distribution of $\overline{X}$.

is, that if we draw a random sample from a normal population, then the probability that the corresponding mean will lie between a and b can be calculated from the Z table.

Let us now proceed with some illustrative examples from several areas.

Example 8.1 Assume that the IQ scores of a population of sixth graders are normally distributed with mean $\mu_X = 103$ and $\sigma_X = 15$. If a random sample of 25 children is selected, then find the probability that the sample mean $\overline{X}$ will exceed 109. (Such data are quite common for WISC scores; see, for example, Black, 1974.)

Since the IQ scores are N(103, 15) we observe that by Theorem 8.1 the sample mean $\overline{X}$ is N(103, 3). Therefore

$$P[\overline{X} > 109] = P\left[Z > \frac{109 - \mu_{\overline{X}}}{\sigma_{\overline{X}}}\right]$$

$$= P\left[Z > \frac{109 - 103}{3}\right]$$

$$= P[Z > 2]$$

$$= 0.0227$$

Example 8.2 Suppose that the number of years from marriage to the birth of the first child for first marriages is normally distributed with the mean equal to 30 months and a standard deviation of 10 months. What is the probability that in repeated sampling of size 49 the mean $\overline{X}$ will be between 28 and 32 months?

Again, using Theorem 8.1, the required probability is:

$$P[28 < \bar{X} < 32] = P\left[\frac{28 - \mu_{\bar{X}}}{\sigma_{\bar{X}}} < Z < \frac{32 - \mu_{\bar{X}}}{\sigma_{\bar{X}}}\right]$$

$$= \left[\frac{28 - 30}{10/\sqrt{49}} < Z < \frac{32 - 30}{10/\sqrt{49}}\right]$$

$$= P[-1.39 < Z < 1.39]$$

$$= 2P[0 < Z < 1.39]$$

$$= 0.8354$$

Example 8.3 If it is true that the heights of men are normally distributed with a standard deviation of 8 cm, then how large a sample should be taken in order to be 95% sure that the sample mean does not differ from the population mean by more than 2 cm in absolute value?

Assuming that the reader is familiar with the notion of absolute value, we see that the probability that the sample mean will differ from the population mean by more than 2 cm in absolute value is equal to 1 - 0.95 = 0.05. Therefore

$$P[|\bar{X} - \mu_X| > 2] = 0.05$$

i.e.,

$$P[\bar{X} - \mu_X < -2 \text{ or } \bar{X} - \mu_X > 2] = 0.05$$

$$2P[\bar{X} - \mu_X > 2] = 0.05$$

$$2P\left[Z > \frac{2}{8/\sqrt{n}}\right] = 0.05$$

$$P\left[Z > \frac{2}{8/\sqrt{n}}\right] = 0.025$$

But from the Z table we know that $P[Z > 1.96] = 0.025$. So by identification,

$$\frac{2}{8/\sqrt{n}} = 1.96 \qquad \text{or } n = 61.46$$

Therefore, we should take a sample of 62 men.

8.3 THE SAMPLING DISTRIBUTION OF THE DIFFERENCE OF TWO MEANS

The process of drawing repeated samples from one normal population can be generalized to two independent normal populations. Imagine the extraction of all possible samples of size n_1 from normal population 1 and all possible samples of size n_2 from normal population 2. At *each draw* of a sample from population 1 and a sample from population 2 form the difference between the two sample means. This will lead us to an infinite population of differences between the sample means. Schematically the situation looks like Figure 8.4.

As before, our interest lies in finding the distribution for the population of differences, i.e., the probability distribution of the random variable $\overline{X} - \overline{Y}$. The answer is provided in Theorem 8.2:

THEOREM 8.2 *If* X *and* Y *are two independent normal random variables with means* μ_X *and* μ_Y *and standard deviations* σ_X *and* σ_Y, *respectively, then the difference* $\overline{X} - \overline{Y}$, *obtained from a pair of random samples* $X_1, X_2, \ldots, X_{n_1}$ *and* $Y_1, Y_2, \ldots, Y_{n_2}$, *is normally distributed with mean*

$$\mu_{\overline{X}-\overline{Y}} = \mu_X - \mu_Y$$

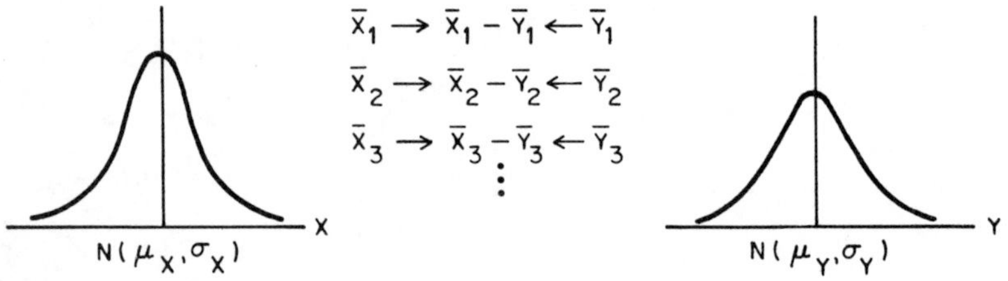

Figure 8.4 Drawing samples to form the difference of two means.

and standard deviation

$$\sigma_{\overline{X}-\overline{Y}} = \sqrt{\sigma^2_{\overline{X}} + \sigma^2_{\overline{Y}}} = \sqrt{\frac{\sigma^2_X}{n_1} + \frac{\sigma^2_Y}{n_2}}$$

We can now make probability statements regarding the differences between two sample means; e.g., 68% of the differences between the two sample means will be between $\mu_{\overline{X}-\overline{Y}} - \sigma_{\overline{X}-\overline{Y}}$ and $\mu_{\overline{X}-\overline{Y}} + \sigma_{\overline{X}-\overline{Y}}$, and 95% of the differences will be between $\mu_{\overline{X}-\overline{Y}} - 1.96\sigma_{\overline{X}-\overline{Y}}$ and $\mu_{\overline{X}-\overline{Y}} + 1.96\sigma_{\overline{X}-\overline{Y}}$.

We can transform a probability statement concerning $\overline{X} - \overline{Y}$ to a Z statement by the usual standardization:

$$P[a < \overline{X} - \overline{Y} < b] = P\left[\frac{a - \mu_{\overline{X}-\overline{Y}}}{\sigma_{\overline{X}-\overline{Y}}} < Z < \frac{b - \mu_{\overline{X}-\overline{Y}}}{\sigma_{\overline{X}-\overline{Y}}}\right]$$

The reader is encouraged to make a sketch of this probability statement by indicating the proper area under the normal curve of $\overline{X} - \overline{Y}$ and under the Z distribution.

Let us now illustrate the use of Theorem 8.2 through some examples.

Example 8.4 Student smokers in public high schools in a city smoke on the average 16 cigarettes per day with a standard deviation of 4, while those in private high schools smoke 12 with a standard deviation of 6. If random samples of sizes 25 and 9 are selected from these two populations and the numbers of cigarettes smoked in both cases are assumed to be normally distributed, then find the probability that in repeated sampling the difference between the two sample means will be more than 5 cigarettes.

Observe that $\mu_X = 16$, $\sigma_X = 4$, $\mu_Y = 12$, and $\sigma_Y = 6$. Hence the mean and standard deviation of $\overline{X} - \overline{Y}$ are:

$$\mu_{\overline{X}-\overline{Y}} = \mu_X - \mu_Y = 16 - 12 = 4$$

$$\sigma_{\overline{X}-\overline{Y}} = \sqrt{\frac{\sigma^2_X}{n_1} + \frac{\sigma^2_Y}{n_2}} = \sqrt{\frac{16}{25} + \frac{36}{9}} = 2.15$$

The desired probability is obtained as follows:

$$P[|\overline{X} - \overline{Y}| > 5] = P\left[|Z| > \frac{5 - \mu_{\overline{X}-\overline{Y}}}{\sigma_{\overline{X}-\overline{Y}}}\right]$$

$$= P\left[|Z| > \frac{5 - 4}{2.15}\right]$$

$$= P[|Z| > 0.47] \doteq 0.64$$

Example 8.5 Male and female workers in factories have mean monthly salaries of \$852 and \$749 with standard deviations of \$90 and \$85, respectively. If random samples of 100 and 75 are selected from the two populations, then under the assumption of normality, find the probability that the absolute difference between $\overline{X} - \overline{Y}$ and $\mu_{\overline{X}-\overline{Y}}$ will be more than \$25.

We do the standard calculations first and then find the required probability:

$$\mu_X = \$852 \qquad \mu_Y = \$749 \qquad \sigma_X = \$90 \qquad \sigma_Y = \$85$$

$$\mu_{\overline{X}-\overline{Y}} = \mu_X - \mu_Y = \$852 - \$749 = \$103$$

$$\sigma_{\overline{X}-\overline{Y}} = \sqrt{\frac{\sigma_X^2}{n_1} + \frac{\sigma_Y^2}{n_2}} = \sqrt{\frac{8100}{100} + \frac{7225}{75}} = 13.32$$

Hence:

$$P[|(\overline{X} - \overline{Y}) - \mu_{\overline{X}-\overline{Y}}| > 25]$$

$$= P\left[\left|\frac{(\overline{X} - \overline{Y}) - \mu_{\overline{X}-\overline{Y}}}{\sigma_{\overline{X}-\overline{Y}}}\right| > \frac{25}{13.32}\right]$$

$$= P[|Z| > 1.88] = P[Z < -1.88 \text{ or } Z > 1.88]$$

$$= 2P[Z > 1.88] \doteq 6\%$$

Theorems 8.1 and 8.2 will be applied in further chapters, where topics on interval estimation and hypothesis testing are developed.

EXERCISES

8.1 Assume that the faculty income at universities follows a normal distribution with mean income equal to \$14,725 and standard deviation of \$6219. Find the probability that in a random sampling of 25 faculty members the sample mean will exceed \$25,000. Sketch this probability under the distributions of both $\overline{X}$ and Z.

8.2 If IQ scores have a normal distribution with mean $\mu_X = 110$ and standard deviation $\sigma_X = 10$, then find the sample size n such that the sample mean $\overline{X}$ differs from the population mean either way by at least 2, 5% of the time.

8.3 Ages of first-year male and female undergraduate students are normal and have means $\mu_X = 21$ and $\mu_Y = 20$ years and standard deviations 1.2 and 1.3 years, respectively. Find the probability that in random samples of size 25 each the sample mean $\overline{X}$ will exceed $\overline{Y}$ by 1 year, and make a sketch of this probability in terms of the distributions of $\overline{X} - \overline{Y}$ and Z.

8.4 Provide a repeated sampling interpretation for $P[\overline{X} > k] = \alpha$, where $\overline{X}$ is $N(\mu_X, \sigma_{\overline{X}})$.

8.5 Provide a repeated sampling interpretation for the probability statement $P[|(\overline{X} - \overline{Y}) - (\mu_{\overline{X}} - \mu_{\overline{Y}})| < a] = \gamma$, where $(\overline{X} - \overline{Y})$ is $N(\mu_{\overline{X}-\overline{Y}}, \sigma_{\overline{X}-\overline{Y}})$.

8.6 If the number of courses failed by graduating B.A. students is a normal random variable with mean equal to 2.35 and standard deviation equal to 1.12, then find the probability that the sample mean from a random sample of size 16 will be less than 2. Make a sketch of this probability and give a repeated sampling interpretation.

8.7 True or false:

(a) If X is N(4, 1), then $\sigma_{\overline{X}} = 1/\sqrt{n}$.

(b) $P[\mu_{\overline{X}} - \sigma_{\overline{X}} < \overline{X} < \mu_{\overline{X}} + \sigma_{\overline{X}}] = 2P[\mu_X < \overline{X} < \mu_X + \sigma_X]$ if X is normal.

(c) $P[\overline{X} > a] \neq P[Z > (a - \mu_{\overline{X}})/\sigma_{\overline{X}}]$ if X is normal.

(d) The Z standardization of $\overline{X}$ for $\mu_{\overline{X}} = 5$, $\sigma_{\overline{X}} = 1.2$ is equal to $(\overline{X} - 1.2)/5$.

(e) The sampling distribution of $\overline{X}$, when X is normal, has a standard deviation which is less than that of X if $n > 1$.

(f) In general, if random samples of a fixed n are drawn from a normal population, as n increases the sampling distribution of the sample means remains normal.

(g) By treating each of the sample means as raw scores, it is possible to calculate the variance of the mean.

(h) As n becomes larger and larger, the peak of the normal distribution of $\overline{X}$ is less and less pronounced.

(i) The mean of any single sample is more likely to be closer to the mean of the population as sample size increases.

(j) Given a population of size N, and we take samples of size n, our sampling distribution would contain $N \cdot n$ sample means.

(k) $\overline{X}$ is a number when it takes on a particular value, and $\overline{X}$ is a random variable when it possesses a distribution.

8.8 At a certain community there are four Psychology Experimental Stations with one or more types of animals for research, as follows:

Station	No. of animal types
1	1
2	2
3	3
4	4

(a) If we consider these four stations as a population, then what is the population mean number of animal types? What is the population variance? (*Note:* Population is finite.)

(b) List all possible random samples of size $n = 2$ from the population of animal types. (*Note:* Sample without replacement.)

(c) Find the mean and variance of each sample.

(d) Find the mean of the sampling distribution of the mean, $\mu_{\overline{X}}$, and compare with μ_X. Conclusion?

(e) Calculate the variance of the sampling distribution of the mean and compare with σ_X^2/n. Conclusion?

8.9 Pay scales for all metropolitan areas should be on the average (in terms of percentage) a standard rate of 100. A standard of pay is computed on the basis of the average weekly salary (or hourly rate) for the occupations in all standard metropolitan areas combined; then the rates for particular metropolitan areas are expressed as percentages of the standard rate. For clerical workers, it is argued that large-sized (X) metropolitan areas are most like the standard than moderate-sized (Y) metropolitan areas. Random samples of 49 and 24 were taken, respectively. Assuming that pay scales are normally distributed with $\mu_X = 100 = \mu_Y$ and variance of X is 6 and variance of Y is 6.63:

(a) Find the probability that the difference between the two sample means will be more than 5.

(b) Make a sketch of the probability in terms of the distribution of $\overline{X} - \overline{Y}$ and Z.

8.10 Suppose the distribution of scores for children from a standard "Expressive Faces Test" has a mean of 10 with variance 6. (This is not uncommon; see, for example, Bain, 1973.) Find the probability that a random sample of 20 such children yields a mean of 8 or less. What assumption is necessary in order to evaluate such a probability?

8.11 (a) What is the relationship between the sample size and the magnitude of the standard deviation of the mean?

(b) If X is $N(\mu_X = 5, \sigma_X = 2)$, find the value of n such that $P[\overline{X} > 7] = 0.02$.

(c) Plot the graph of $\sigma_{\overline{X}} = \sigma_X/\sqrt{n}$, for $\sigma_X = 5$, by taking n as the abscissa and $\sigma_{\overline{X}}$ as the ordinate. What happens with the distribution of $\overline{X}$ as n becomes infinitely large in the normal case?

(d) Consider three independent normal populations, i.e., $N(\mu_X, \sigma_X)$, $N(\mu_Y, \sigma_Y)$, and $N(\mu_Z, \sigma_Z)$. Set up the repeated sampling process and find the distribution of $\overline{X} - \overline{Y} - \overline{Z}$. Generalize your result to k independent normal populations.

8.12 Suppose it is known that the distribution of scores based on the Piers-Harris Children's Self Concept Test for normal elementary school children is normal with mean 60 and standard deviation 23. Black (1974) considered a random sample of 25 retarded readers from elementary school and found a mean of 44.2. Using this information, find the probability $P[\overline{X} < 44.2]$. If this probability is small, say less than 5%, then you have verified Black's conclusion that retarded readers do not come from the same population.

9

SOME IMPORTANT SAMPLING DISTRIBUTIONS DERIVED FROM THE NORMAL DISTRIBUTION
The Chi-Square, t, and F Distributions

In Theorem 8.1 we have given the distribution of $\overline{X}$, which was concerned with means of samples obtained through the repeated sampling process. Let us now study the behavior of three statistics other than the mean, again drawing samples repeatedly from a normal population with mean μ_X and standard deviation σ_X. The three statistics, namely, chi-square, t, and F, have become so popular in applications, that without them one would feel deprived of some important tools. The correspondingly named distributions of these statistics were developed, respectively, by three eminent statisticians; Karl Pearson (1857-1936), a biostatistician at University College, London; Student [pseudonym for W. S. Gosset (1876-1937)], a mathematical consultant for the Guinness Brewery in Dublin; and G. W. Snedecor (1881-1974), a statistician at Iowa State University.

9.1 THE CHI-SQUARE DISTRIBUTION

We begin the development with the chi-square distribution. Consider repeated samples from a normal population with mean μ_X and standard deviation σ_X. Now, from each sample we can calculate the sample variance s_X^2, and if σ_X^2 is *known*, we can calculate for each sample the ratio $(n - 1)s_X^2/\sigma_X^2$. Schematically this process is shown in Figure 9.1.

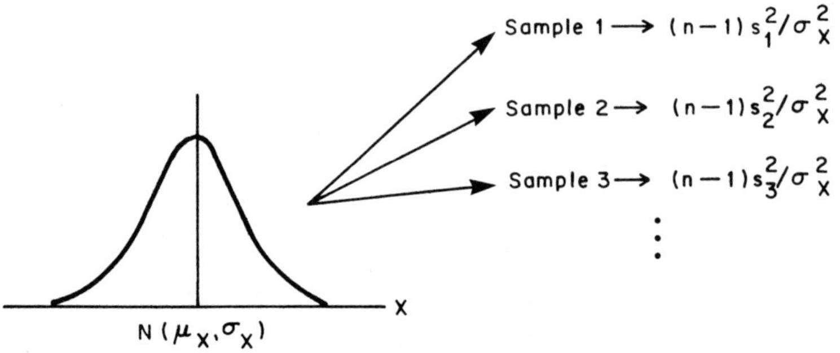

Figure 9.1 Sampling from a normal population with mean μ_X and standard deviation σ_X.

Note that we have created a new population consisting of an infinite number of values of the variable $(n - 1)s_X^2/\sigma_X^2$, and we are, of course, interested in the probability distribution of this random variable. This problem was resolved mathematically(i.e., without actually using the argument of drawing all possible samples) a long time ago by Karl Pearson. The following theorem is due to him:

THEOREM 9.1 *If the random variable* X *has a normal distribution with mean* μ_X *and standard deviation* σ_X, *then the statistic* $(n - 1)s_X^2/\sigma_X^2$, *where* s_X^2 *is obtained from a random sample* $X_1, X_2, \ldots, X_n$, *has the chi-square distribution with* $n - 1$ *degrees of freedom.*

It is customary to indicate the statistic $(n - 1)s_X^2/\sigma_X^2$ by χ^2, which is pronounced "kai-square," and for this reason it is called the chi-square statistic. Note that $(n - 1)s_X^2$ is equal to the sum of squares $\Sigma(X_i - \overline{X})^2$. The mathematical equation for drawing the χ^2 distribution is given by the rather complicated looking equation (which we should not try to learn by heart):

$$Y = \frac{1}{[(\nu/2) - 1]!2^{\nu/2}} (\chi^2)^{(\nu/2)-1} e^{-(1/2)\chi^2}$$

Note that χ^2 is a nonnegative random variable and that the shape of the distribution depends on $\nu = n - 1$, which is called *degrees of freedom* (d.f.). This term derives its name from the fact that among n independent quantities one can form only n - 1 independent differences. For example, among three independent values, X_1, X_2, and X_3 one can form only two independent differences, i.e., $D_1 = X_1 - X_2$ and $D_2 = X_1 - X_3$. It is clear that the difference $X_2 - X_3$ depends on D_1 and D_2 and in fact is equal to $D_2 - D_1$.

In Figure 9.2 we have sketched the χ^2 distribution for various values of ν. As can be observed, the larger the ν value, the more the chi-square distribution looks like the normal distribution. This fact is related to the most fundamental theorem in statistics, namely, the central limit theorem, which is discussed in Chapter 10.

Theorem 9.1 makes it possible to make probability statements regarding the sample variance or sample standard deviation when repeated samples are drawn from a normal population. Questions such as the following can now be readily answered: What are the chances that in a repeated sampling of size 7 from a normal population with variance $\sigma_X^2 = 4$, the sample variance s_X^2 will be between the two specified limits, 3 and 5? The answer is:

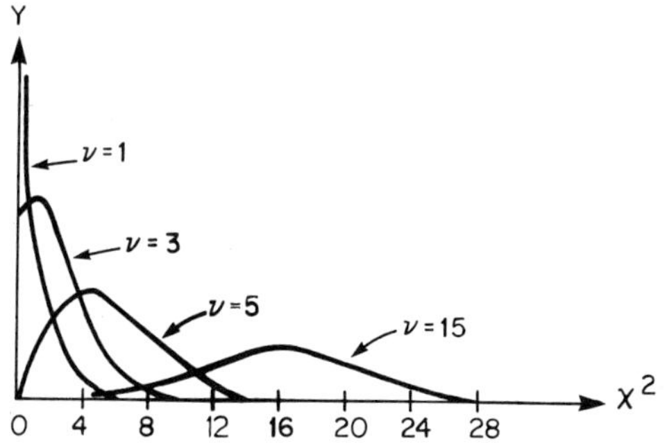

Figure 9.2 The chi-square distribution with various degrees of freedom.

$$P[3 < s_X^2 < 5] = P[(n - 1)3 < (n - 1)s_X^2 < (n - 1)5]$$

$$= P\left[\frac{18}{4} < \frac{(n - 1)s_X^2}{\sigma_X^2} < \frac{30}{4}\right]$$

$$= P[4.5 < \chi^2 < 7.5]$$

= area under the chi-square curve with 6 degrees of freedom from 4.5 to 7.5

This probability is given in Figure 9.3. It is customary to indicate the degrees of freedom as a subscript of χ^2. Thus χ_ν^2 means a chi-square variable with ν degrees of freedom. This notation has been utilized in the sketch for the chi-square variable with 6 degrees of freedom.

The chi-square distribution has been tabulated. Table A.4 (see the Appendix) can be used to find probabilities in actual calculations. The first column indicates the degrees of freedom ν, and the rest of the columns relate to chi-square values, giving certain probabilities in the tail of the distribution. For example, for $\nu = 6$ one finds the following values:

ν	0.99	0.98	0.95	0.90	0.80	0.70	0.50	0.30	0.20	0.10	0.05	0.02	0.01	0.001
6	0.87	1.13	1.64	2.20	3.07	3.83	5.35	7.23	8.56	10.64	12.59	15.03	16.81	22.46

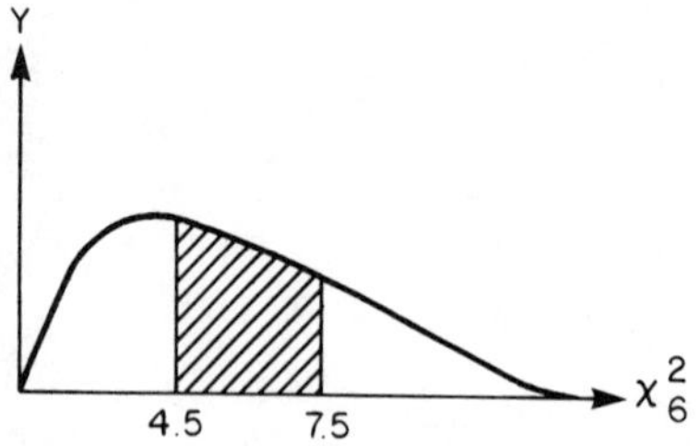

Figure 9.3 Chi-square probability with 6 degrees of freedom from 4.5 to 7.5.

The number 0.87 is the value of χ_6^2 such that $P[\chi_6^2 > 0.87] = 0.99$; i.e., the area under the chi-square distribution with 6 degrees of freedom to the right of 0.87 is equal to 0.99. Similarly, the entry 1.13 is such that the area under the same distribution to the right of 1.13 is equal to 0.98. Thus, if you know ν (the degrees of freedom) and if you specify a tail probability, then you can find the value of the χ^2 variable.

A sketch of the chi-square distribution with 6 degrees of freedom such that 5% of the area is in the right-hand tail is given in Figure 9.4. The reader is encouraged to draw sketches for the other entries.

Returning to our problem of finding $P[4.5 < \chi_6^2 < 7.5]$, we observe from Figure 9.4 that this probability is equal to $P[\chi_6^2 > 4.5] - P[\chi_6^2 > 7.5]$. In the chi-square table we see that 4.5 falls between 0.50 and 0.70. To obtain an approximate value we do linear interpolation, i.e., by carrying out a proportional split in the following way:

$$0.50 + \frac{(5.35 - 4.5)}{5.35 - 3.83}(0.70 - 0.50) = 0.61$$

Hence $P[\chi_6^2 > 4.5] = 0.61$. Similarly an approximate value of $P[\chi_6^2 > 7.5]$ is obtained by linear interpolation; i.e.,

$$0.20 + \frac{(8.56 - 7.50)}{8.56 - 7.23}(0.30 - 0.20) = 0.28$$

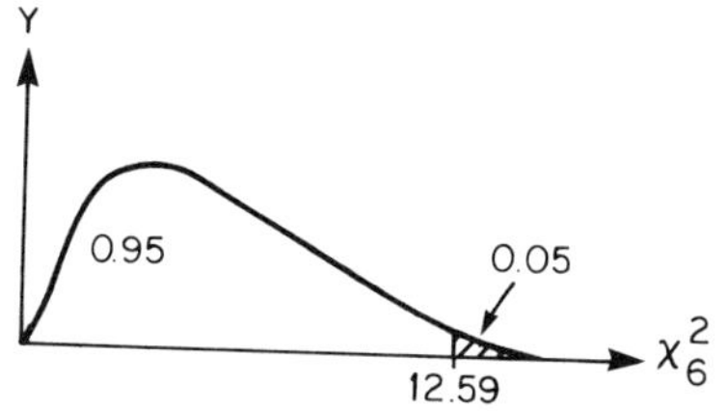

Figure 9.4 Chi-square distribution with 6 degrees of freedom.

Therefore the required probability, $P[4.5 < \chi_6^2 < 7.5]$, is approximately equal to 0.61 - 0.28 = 0.33.

Let us now apply the chi-square distribution in making a probability statement regarding the sample standard deviation s_X.

Example 9.1 In psychological studies subjects are typically given a stimulus and their reaction time is measured. (See, for example, Liebert and Baumeister, 1973.) Assume that the reaction time is a normal random variable with standard deviation $\sigma_X = 0.06$ sec. If a random sample of 15 subjects is to be experimented on, and from this the sample standard deviation is to be calculated, then find the probability that in repeated sampling it will exceed σ_X by more than 0.01 sec.

The required probability is equal to:

$$
\begin{aligned}
P[s_X - \sigma_X > 0.01] &= P[s_X - 0.06 > 0.01] \\
&= P[s_X > 0.07] \\
&= P[s_X^2 > 0.0049] \\
&= P[(n - 1)s_X^2 > (14)(0.0049)] \\
&= P\left[\frac{(n - 1)s_X^2}{\sigma_X^2} > \frac{(14)(0.0049)}{0.0036}\right] \\
&= P[\chi_{14}^2 > 19.06] \\
&\doteq 0.17
\end{aligned}
$$

Besides the use of the chi-square distribution in probability calculations concerning the sample standard deviation and sample variance, it is also extremely popular in testing goodness-of-fit of distributions and in contingency table analysis. Chapter 13 illustrates these techniques.

9.2 THE t DISTRIBUTION

Let us now proceed with the development of the distribution of a very important statistic, namely, the t statistic, which is defined as:

$$t = \frac{\bar{X} - \mu_{\bar{X}}}{s_{\bar{X}}} = \frac{\bar{X} - \mu_X}{s_{\bar{X}}}$$

where $\bar{X}$ and $s_{\bar{X}} = s_X/\sqrt{n}$ are the sample mean and standard deviation of the sample mean, obtained from a random sample of size n from a normal population with given mean μ_X and unknown, finite standard deviation σ_X. Before going through the repeated sampling argument and discussing the distribution of the t statistic, let us establish the following interesting result:

$$t = \frac{\bar{X} - \mu_{\bar{X}}}{s_{\bar{X}}}$$

$$= \frac{\bar{X} - \mu_X}{s_{\bar{X}}}$$

$$= \frac{(\bar{X} - \mu_X)/\sigma_{\bar{X}}}{s_{\bar{X}}/\sigma_{\bar{X}}}$$

$$= \frac{(\bar{X} - \mu_X)/\sigma_{\bar{X}}}{\sqrt{s_{\bar{X}}^2/\sigma_{\bar{X}}^2}}$$

$$= \frac{Z}{\sqrt{\chi_{n-1}^2 \cdot [1/(n - 1)]}}$$

The fact that $s_{\bar{X}}^2/\sigma_{\bar{X}}^2$ is a chi-square random variable times $1/(n - 1)$ is seen by noting that

$$\frac{s_{\bar{X}}^2}{\sigma_{\bar{X}}^2} = \frac{s_X^2/n}{\sigma_X^2/n} = \frac{(n - 1)s_X^2}{\sigma_X^2} \cdot \frac{1}{n - 1}$$

and this random variable is, by Theorem 9.1, a chi-square variable with n - 1 degrees of freedom multiplied by 1/(n - 1). But the numerator is just the standard normal variable, which we are quite familiar with by now! Hence we see that the t variable is, up to a constant, the ratio of a standard normal variable and the square root of a chi-square variable.

Another interesting fact is that the t variable differs from the Z variable only in the denominator; i.e., we have replaced $\sigma_{\overline{X}}$ in $Z = (\overline{X} - \mu_X)/\sigma_{\overline{X}}$ by its sample estimator $s_{\overline{X}}$. This is needed in many sampling situations where the standard deviation σ_X is not known. This is estimated by s_X, and hence $\sigma_{\overline{X}}$ is estimated by $s_{\overline{X}} = s_X/\sqrt{n}$. Let us remember this difference between the Z and t random variables!

As has usually been the case, we are naturally interested in the distribution of t. To visualize what we are after, consider again the process of drawing random samples repeatedly from a normal population with given mean μ_X and unknown standard deviation σ_X. For each sample drawn we calculate the quantity $(\overline{X} - \mu_X)/s_{\overline{X}}$, where the parameter μ_X is given. Schematically we have the situation shown in Figure 9.5. We have by the process of repeated sampling created a population consisting of the $(\overline{X} - \mu_X)/s_{\overline{X}}$ values; i.e., we have a random variable $t = (\overline{X} - \mu_X)/s_{\overline{X}}$.

The probability distribution of the statistic t was settled mathematically by Student in 1908, and is given in the following theorem:

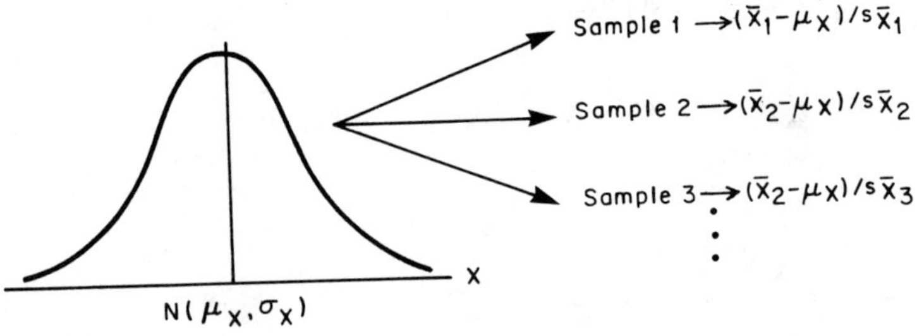

Figure 9.5 Drawing random samples.

THEOREM 9.2 *If X is a normal random variable with given mean* μ_X *and unknown finite standard deviation* σ_X*, then the random variable* $t = (\overline{X} - \mu_{\overline{X}})/s_{\overline{X}}$*, where* $\overline{X}$ *and* $s_{\overline{X}} = s_X/\sqrt{n}$ *are obtained from a random sample* $X_1, X_2, \ldots, X_n$*, has Student's t distribution with* $\nu = n - 1$ *degrees of freedom.*

For the more sophisticated and mathematically inclined reader, the explicit formula for sketching Student's t distribution is given by the equation:

$$Y = \frac{[(\nu - 1)/2]!}{\sqrt{\nu\pi}[(\nu - 2)/2]!} \cdot \left(1 + \frac{t^2}{\nu}\right)^{-(\nu+1)/2}$$

The range for the t variable is the same as for the Z variable, namely, any real number between $-\infty$ and ∞. Notice that the equation depends on n, or better, on the degrees of freedom $\nu = n - 1$. In Figure 9.6, we provide a sketch of Student's t distribution for various degrees of freedom.

This diagram suggests that the distribution is symmetric, and as the sample size becomes larger the closer it approaches the standard normal distribution. This is quite understandable, because $s_{\overline{X}}$ becomes a better estimate of $\sigma_{\overline{X}}$ with increasing n. As n approaches ∞ the two curves coincide.

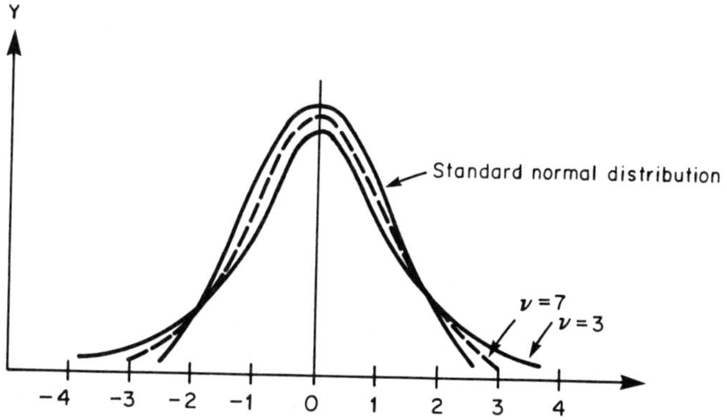

Figure 9.6 Student's t distribution for various degrees of freedom.

Theorem 9.2 provides the basis for making probabilistic statements involving the sample mean $\overline{X}$ in case σ_X, and hence $\sigma_{\overline{X}}$ is unknown. Table A.5 provides certain probabilities in the right-hand tail of the distribution. The first column is the familiar degrees of freedom column, and the other columns are headed by probabilities. The entries in a particular column headed by probability α are such that the area in the right-hand tail of the t distribution is equal to α. To illustrate how to use the t table, let $\nu = 8$, then we have:

ν	$\alpha = 0.10$	$\alpha = 0.05$	$\alpha = 0.025$	$\alpha = 0.01$	$\alpha = 0.005$
8	1.397	1.860	2.306	2.896	3.355

The number 1.397 is such that $P[t_8 > 1.397] = 0.10$; i.e., in a repeated sampling of size 9 from a normal distribution with given mean and unknown standard deviation, 10% of the calculated t values will be greater than 1.397, and hence 90% of them will be less than 1.397. Put in a different way, if a random sample of size 9 is taken from a normal population with given mean and unknown standard deviation, then the probability that the corresponding t value will exceed 1.397 is equal to 0.10. A pictorial representation is shown in Figure 9.7.

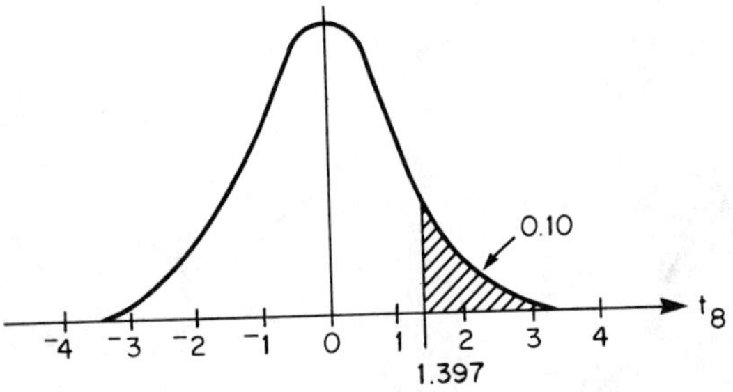

Figure 9.7 Probability that the t variable with 8 degrees of freedom exceeds 1.397.

Because the t distribution is symmetric (just like the Z distribution) we have the added bonus that we can also make the following statements:

$$P[t_8 < -1.397] = 0.10$$

$$P[t_8 > -1.397] = 0.90$$

$$P[t_8 > 1.397 \text{ or } t_8 < -1.397] = 0.20$$

$$P[-1.397 < t_8 < 1.397] = 0.80$$

The reader is encouraged to sketch each of these probabilities and give a repeated sampling interpretation.

It is interesting to note that when ν is "large" there is no appreciable difference between the t and Z values for a given probability. For example, when n = 30, i.e., $\nu = 29$, we see that:

$$P[-2.045 < t_{29} < 2.045] = 0.95$$

and in the limit,

$$P[-1.96 < Z = t_\infty < 1.96] = 0.95$$

The difference in absolute value here is equal to 2.045 - 1.96 = 0.085. This is the reason why our t table stops at n = 30. Hence when n > 30, we use the Z table.

The t distribution is extremely useful when we wish to make probability statements involving the sample mean $\overline{X}$, when the population mean μ_X is given or a value for it is hypothesized, but the population standard deviation σ_X is unknown so that it has to be estimated from a small ($n \leq 30$) random sample. Let us now provide some practical illustrations.

Example 9.2 Suppose that first year unmarried students on the average date four different undergraduate students during their first semester. An interview with nine such randomly selected students at the end of the first semester provides the following results:

Student	No. of dates	$(X_i - \overline{X})^2$
1	1	4
2	0	9
3	6	9
4	3	0
5	4	1
6	6	9
7	2	1
8	3	0
9	2	1

$\Sigma X_i = 27$ $\qquad \Sigma(X_i - \overline{X})^2 = 34$

$\overline{X} = 3$ $\qquad s_X = \sqrt{\frac{34}{8}} = 2.06$

$$s_{\overline{X}} = \frac{2.06}{\sqrt{9}} = 0.69$$

Find the probability that in repeated sampling of nine such students the t value (corresponding to $\overline{X}$ and $s_{\overline{X}}$) will be less than that found from the sample.

To answer this question we make the assumption that the number of undergraduate female students dated by first-year unmarried male undergraduate students, or vice versa, during their first semester, is normally distributed with $\mu_X = 4$. From the above sample calculations, we see that $\overline{X} = 3$ and $s_{\overline{X}} = 0.69$. Hence the value of t from the sample is equal to:

$$\frac{\overline{X} - \mu_X}{s_{\overline{X}}} = \frac{3 - 4}{0.69} = -1.45$$

The required probability is therefore:

$$= P[t_8 < -1.45]$$

$$= P[t_8 > 1.45]$$

$$\doteq 0.09$$

by interpolation of the t table.

Example 9.3 A sociologist claims that dentists have 5 extramarital affairs during their married life. To verify this claim a random sample of 16 dentists are interviewed and the sample calculations show a sample mean of 6.20 and a sample standard deviation of 1.13. What is the probability that in repeated random sampling of 16 dentists the t value (corresponding to $\overline{X}$ and $s_{\overline{X}}$) will be greater than that found in the sample?

We make the assumption that the number of extramarital affairs is a normal variable with mean $\mu_X = 5$. From the sample we have obtained $\overline{X} = 6.20$ and $s_X = 1.13$, so that $s_{\overline{X}} = 1.13/\sqrt{16} = 0.28$. Hence the t value obtained from the sample is:

$$\frac{\overline{X} - \mu_X}{s_{\overline{X}}} = \frac{6.20 - 5}{0.28} = 4.286$$

Therefore the required probability is:

$$P[t_{15} > 4.286] = \text{less than } \tfrac{1}{2}\%$$

by Table A.5.

We will see in Chapter 13, which deals with testing hypotheses, that on the basis of this extremely small probability the sample evidence leads to rejection of the sociologist's claim.

Example 9.4 A sample of 14 IQ scores of grade 6 students gave a mean of 114 and a standard deviation of 5.2. If the population mean is equal to 110, then find the probability that in repeated sampling the absolute value of t will exceed the value found in the sample.

To solve this problem we assume that IQ scores are normally distributed with $\mu_X = 110$. From the sample of 14 scores we have:

$$\overline{X} = 114 \qquad \text{and} \qquad s_{\overline{X}} = \frac{s_X}{\sqrt{n}} = \frac{5.2}{\sqrt{14}} = 1.39$$

Hence the t value from the sample is equal to:

$$\frac{\overline{X} - \mu_X}{s_{\overline{X}}} = \frac{114 - 110}{1.39} = 2.877$$

Therefore the desired probability is:

$$P[|t_{13}| > 2.877] = P[t_{13} > 2.877 \text{ or } t_{13} < -2.877]$$

$$= 2P[t_{13} > 2.877]$$

$$\doteq 2(0.007) = 1.4\%$$

by interpolation from the t table.

Theorem 8.2 dealt with the distribution of the difference $\overline{X} - \overline{Y}$ between two sample means, where the samples $X_1, X_2, \ldots, X_{n_1}$ and $Y_1, Y_2, \ldots, Y_{n_2}$ came from two independent normal populations $N(\mu_X, \sigma_X)$ and $N(\mu_Y, \sigma_Y)$, respectively. To make probability statements concerning $\overline{X} - \overline{Y}$ required the fact that both $\mu_{\overline{X}-\overline{Y}}$ and $\sigma_{\overline{X}-\overline{Y}}$ were specified, because both were needed in the Z standardization:

$$Z = \frac{(\overline{X} - \overline{Y}) - (\mu_{\overline{X}-\overline{Y}})}{\sigma_{\overline{X}-\overline{Y}}}$$

Since $\mu_{\overline{X}-\overline{Y}} = \mu_X - \mu_Y$ and $\sigma_{\overline{X}-\overline{Y}} = \sqrt{\sigma_X^2/n_1 + \sigma_Y^2/n_2}$, we essentially required knowledge of μ_X, μ_Y (or of $\mu_X - \mu_Y$) and σ_X, σ_Y.

Now suppose that μ_X and μ_Y or $\mu_X - \mu_Y$ are specified but both σ_X and σ_Y are not given, and they are estimated from the respective samples $X_1, X_2, \ldots, X_{n_1}$ and $Y_1, Y_2, \ldots, Y_{n_2}$ by s_X and s_Y. If we try to form the analogous t statistic, say,

$$t^* = \frac{(\overline{X} - \overline{Y}) - (\mu_{\overline{X}-\overline{Y}})}{s^*_{\overline{X}-\overline{Y}}}$$

where $s^*_{\overline{X}-\overline{Y}} = (s_X^2/n_1 + s_Y^2/n_2)^{1/2}$, and we try to derive the distribution of t^*, we immediately are confronted with an unsolved problem of great complexity, known as the Behrens-Fisher problem. No one has obtained an explicit formula for the distribution of t^*.

The problem is solved, however, if we can assume $\sigma_X^2 = \sigma_Y^2 = \sigma^2$, say. In this case we do not form separate estimates of σ_X^2 and σ_Y^2. What we do is form a *pooled estimate* of σ, given by

$$s = \sqrt{\frac{\Sigma(X_i - \overline{X})^2 + \Sigma(Y_i - \overline{Y})^2}{n_1 + n_2 - 2}}$$

$$= \sqrt{\frac{(n_1 - 1)s_X^2 + (n_2 - 1)s_Y^2}{n_1 + n_2 - 2}}$$

This estimate is then used to form an estimate of $\sigma_{\overline{X}-\overline{Y}}$; i.e.,

$$s_{\overline{X}-\overline{Y}} = \sqrt{\frac{s^2}{n_1} + \frac{s^2}{n_2}}$$

$$= s\sqrt{\frac{1}{n_1} + \frac{1}{n_2}}$$

$$= s\left(\frac{n_1 + n_2}{n_1 \cdot n_2}\right)^{1/2}$$

And now we are ready for a theorem similar to Theorem 8.2.

THEOREM 9.3 *If random samples* $X_1, X_2, \ldots, X_{n_1}$ *and* $Y_1, Y_2, \ldots Y_2, \ldots, Y_{n_2}$ *are drawn from two independent normal populations with given means* μ_X *and* μ_Y *(or specified difference* $\mu_X - \mu_Y$*) and unknown common finite variance* σ^2*, then the statistic*

$$t = \frac{(\overline{X} - \overline{Y}) - \mu_{\overline{X}-\overline{Y}}}{s_{\overline{X}-\overline{Y}}}$$

where $s_{\overline{X}-\overline{Y}}$ *is given above, follows Student's t distribution with* $n_1 + n_2 - 2$ *degrees of freedom.*

Let us illustrate the use of Theorem 9.3 by an example.

Example 9.5 Suppose we have random samples of IQ scores on two "A" groups of grade 6 students belonging to different elementary schools:

Group I:	124, 115, 117, 124
Group II:	130, 127, 124, 120, 119, 124

Under the assumption that the samples come from two normal populations with equal means and equal standard deviations, find the probability that in repeated sampling the t value (associated with $\bar{X} - \bar{Y}$ and $s_{\bar{X}-\bar{Y}}$) will be less than that found from the data above.

To answer this question we complete the following necessary calculations:

X	Y	$(X - \bar{X})^2$	$(Y - \bar{Y})^2$
124	130	16	36
115	127	25	9
117	124	9	0
124	120	16	16
	119		25
	124		0
$\Sigma X = 480$	$\Sigma Y = 744$	$\Sigma(X - \bar{X})^2 = 66$	$\Sigma(Y - \bar{Y})^2 = 86$
$\bar{X} = 120$	$\bar{Y} = 124$	$\bar{X} - \bar{Y} = -4$	

$$s = \sqrt{\frac{66 + 86}{8}} = \sqrt{19.00} = 4.36$$

$$s_{\bar{X}-\bar{Y}} = 4.36\sqrt{\frac{1}{4} + \frac{1}{6}} = 2.81$$

It follows that the t value from the data is equal to

$$\frac{(\bar{X} - \bar{Y}) - (\mu_{\bar{X}} - \mu_{\bar{Y}})}{s_{\bar{X}-\bar{Y}}} = \frac{(120 - 124) - (\mu_X - \mu_Y)}{2.81}$$

$$= \frac{-4 - 0}{2.81} = -1.42$$

The required probability statement is then equal to:

$$P[t_8 < -1.42] = P[t_8 > 1.42] \doteq 10\%$$

from the t table.

9.3 THE F DISTRIBUTION

To conclude this chapter, let us now proceed with the development of the F distribution. The F distribution is concerned with the F variable, which is essentially the ratio of two independent chi-square variables. Imagine the repeated sampling process of sample sizes n_1 and n_2 associated with two independent normal populations, $N(\mu_X, \sigma_X)$ and $N(\mu_Y, \sigma_Y)$. Assume that σ_X and σ_Y or the ratio σ_X/σ_Y are known and execute the operations shown in Figure 9.8. From the samples form the following ratios:

$$\text{Sample 1:} \quad \frac{[(n_1 - 1)s_{X1}^2/\sigma_X^2]/(n_1 - 1)}{[(n_2 - 1)s_{Y1}^2/\sigma_Y^2]/(n_2 - 1)}$$

$$\text{Sample 2:} \quad \frac{[(n_1 - 1)s_{X2}^2/\sigma_X^2]/(n_1 - 1)}{[(n_2 - 1)s_{Y2}^2/\sigma_Y^2]/(n_2 - 1)}$$

$$\text{Sample 3:} \quad \frac{[(n_1 - 1)s_{X3}^2/\sigma_X^2]/(n_1 - 1)}{[(n_2 - 1)s_{Y3}^2/\sigma_Y^2]/(n_2 - 1)}$$

$$\vdots \qquad\qquad \vdots$$

Note that we have created a new population of ratios and that the random variable is equal to:

$$\frac{[(n_1 - 1)s_X^2/\sigma_X^2]/(n_1 - 1)}{[(n_2 - 1)s_Y^2/\sigma_Y^2]/(n_2 - 1)}$$

$$= \frac{\chi_{\nu_1}^2/\nu_1}{\chi_{\nu_2}^2/\nu_2} \qquad \text{where } \nu_i = n_i - 1, \quad i = 1, 2$$

The following theorem was proved by G. W. Snedecor, who named the resulting distribution the F distribution, to honor R. A. Fisher (1890-1962), unquestionably the most famous statistician of all time!

THEOREM 9.4 *Consider random samples* $X_1, X_2, \ldots, X_{n_1}$ *and* $Y_1, Y_2, \ldots, Y_{n_2}$ *from two independent normal populations* $N(\mu_X, \sigma_X)$ *and* $N(\mu_Y, \sigma_Y)$ *such that* σ_X *and* σ_Y *or the ratio* σ_X/σ_Y *are known. Then the statistic*

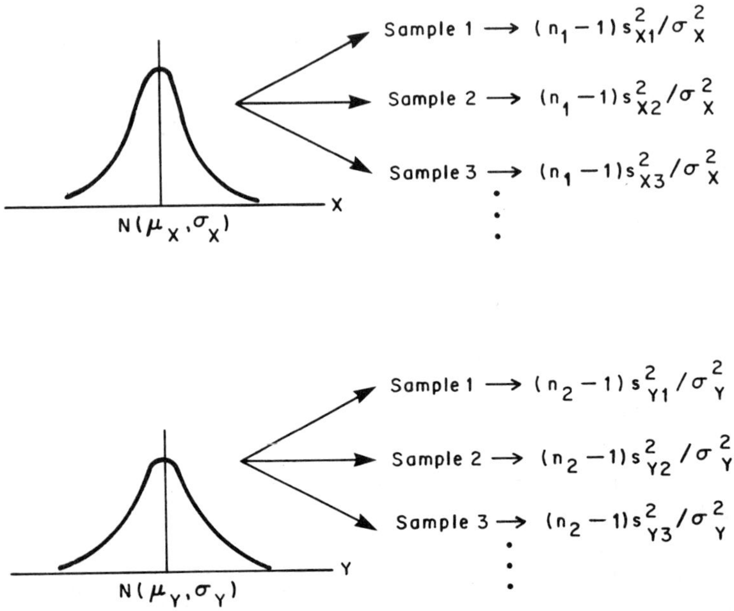

Figure 9.8 Sampling to form the F distribution.

$$F = \frac{[(n_1 - 1)s^2_X/\sigma^2_X]/(n_1 - 1)}{[(n_2 - 1)s^2_Y/\sigma^2_Y]/(n_2 - 1)} = \frac{\chi^2_{\nu_1}/\nu_1}{\chi^2_{\nu_2}/\nu_2}$$

has the F *distribution with* $\nu_1 = n_1 - 1$ *and* $\nu_2 = n_2 - 1$ *degrees of freedom.*

Before giving an example of the use of the F distribution, we do a little bit of elementary algebra:

$$F = \frac{[(n_1 - 1)s^2_X/\sigma^2_X]/(n_1 - 1)}{[(n_2 - 1)s^2_Y/\sigma^2_Y]/(n_2 - 1)}$$

$$= \frac{s^2_X/\sigma^2_X}{s^2_Y/\sigma^2_Y}$$

In this form the F variable is the ratio of the sample variances, divided by their respective population variances. The F statistic can also be written as:

$$F = \frac{s_X^2/\sigma_X^2}{s_Y^2/\sigma_Y^2} = \frac{s_X^2/s_Y^2}{\sigma_X^2/\sigma_Y^2}$$

In this case it is the ratio of the ratio of sample variances and the ratio of the population variances. The F variable assumes nonnegative values, and the explicit formula for drawing the F distribution is given by the horrendous (for mathematics lovers only!) equation:

$$Y = \frac{[(\nu_1 + \nu_2 - 2)/2]!}{[(\nu_1 - 2)/2]!\,[(\nu_2 - 2)/2]!}\left(\frac{\nu_1}{\nu_2}\right)^{\nu_1/2} \cdot \frac{F^{(\nu_1-2)/2}}{[1 + (\nu_1/\nu_2)F]^{(\nu_1+\nu_2)/2}}$$

An F distribution depends on ν_1 and ν_2, which are often referred to as the numerator and denominator degrees of freedom, respectively. It is an asymmetric distribution, with the shapes shown in Figure 9.9 typical for some values of ν_1 and ν_2.

Probability tables of the F distribution are complicated by the fact that you have to worry about two different degrees of freedom ν_1 and ν_2. The most frequently used tabulation is the one which gives the percentage points for 5% and 1% in the right-hand

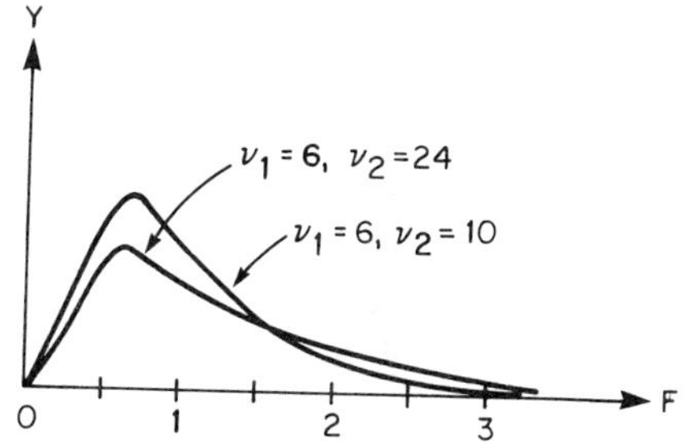

Figure 9.9 F distribution for two pairs of degrees of freedom.

tail of the distribution. This is the form of Table A.6. To understand it correctly, let us give an explicit example.

Suppose that $\nu_1 = 8$ and $\nu_2 = 5$; then from the F table we obtain the following:

$\alpha = 0.05$

$\nu_2 \backslash \nu_1$	... 8 ...
⋮	⋮
5	4.82
⋮	⋮

$\alpha = 0.01$

$\nu_2 \backslash \nu_1$	... 8 ...
⋮	⋮
5	10.29
⋮	⋮

The probability statements which go with the above numbers are:

$$P[F_{8,5} > 4.82] = 0.05$$

and

$$P[F_{8,5} > 10.29] = 0.01$$

That is, in repeated sampling the probability that the F variable with 8 and 5 degrees of freedom exceeds 4.82 is 5% and that it exceeds 10.29 is 1%. Figure 9.10 shows a sketch corresponding to the former statement.

The F distribution is extremely useful in making probability statements about the ratio of the sample standard deviations or

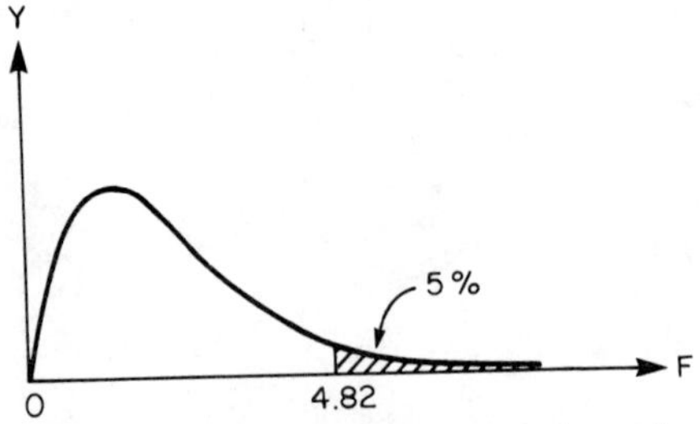

Figure 9.10 Probability that the F variable with 8 and 5 degrees of freedom exceeds 4.82.

sample variances, when random samples are drawn from two independent normal populations.

Example 9.6 In a psychological experiment there were available 11 experimental subjects, who were matched for every conceivable factor to make the group as homogeneous as possible. Five subjects were randomly administered a placebo, and the rest of them were given a stimulant drug (e.g., ethanol). The following times (in seconds) were recorded for completion of an assigned task:

Placebo (X)	4 3 6 3 4	$\overline{X} = 5$, $s_X = 1.66$
Drug (Y)	2 3 3 4 1 5	$\overline{Y} = 3$, $s_Y = 1.41$

Under the assumption that we have random samples from two independent normal populations with equal variances, find the probability that in repeated sampling the ratio of the sample standard deviations exceeds the ratio found in the data above. (By the way, the literature is rich with such experiments, where the range of subjects has included mice, rats, monkeys, and goldfish. A typical example is the paper by MacInnes and Uphouse, 1973.)

In this problem $s_X/s_Y = 1.18$, so that the required probability is equal to:

$$P\left[\frac{s_X}{s_Y} > 1.18\right] = P\left[\frac{s_X^2}{s_Y^2} > 1.39\right]$$

$$= P\left[\frac{s_X^2/\sigma_X^2}{s_Y^2/\sigma_Y^2} > 1.39\right]$$

$$= P[F_{4,5} > 1.39]$$

since $\sigma_X^2 = \sigma_Y^2$. From our limited F table we find that this probability is much greater than 5%. The exact answer can be found from extended F tables.

Our 5% and 1% F tables have a small built-in bonus attached to them, because of the following fact:

$$P[F_{\nu_1,\nu_2} > f_\alpha] = \alpha$$

implies

$$P\left[\frac{1}{F_{\nu_1,\nu_2}} < \frac{1}{f_\alpha}\right] = \alpha$$

which implies

$$P\left[F_{\nu_2,\nu_1} < \frac{1}{f_\alpha}\right] = \alpha$$

(from the definition of the F variable); and this implies

$$P\left[F_{\nu_2,\nu_1} > \frac{1}{f_\alpha}\right] = 1 - \alpha$$

where α could be 5% or 1%, say.

As an illustration consider the 5% F value for $\nu_1 = 6$ and $\nu_2 = 11$; i.e.,

$$P[F_{6,11} > 3.09] = 0.05$$

Hence:

$$P\left[F_{11,6} > \frac{1}{3.09} = 0.32\right] = 0.95$$

We will provide further applications of the F distribution in later chpaters, but for the moment the reader is encouraged to master the repeated sampling process, which leads to the F statistic, and the interpretation of the 5% and 1% F tables.

EXERCISES

9.1 A partial table of the t distribution is reproduced here:

d.f.	$\alpha = 0.10$	$\alpha = 0.05$	$\alpha = 0.025$	$\alpha = 0.01$	$\alpha = 0.005$
3	1.638	2.353	3.182	4.541	5.841
4	1.533	2.132	2.776	3.747	4.604

Complete the following such that the statements are true:

(a) When d.f. = 4 and the α probabilities decrease, then the values of t ____________

(b) When α = 0.05 and d.f. = 3, then $P[|t_3| < 2.353]$ = ______

(c) When α = 0.025 and the degrees of freedom increase, then the values of t __________

9.2 In the pictures shown here, from the t table of Exercise 9.1 find the quantities indicated by question marks.

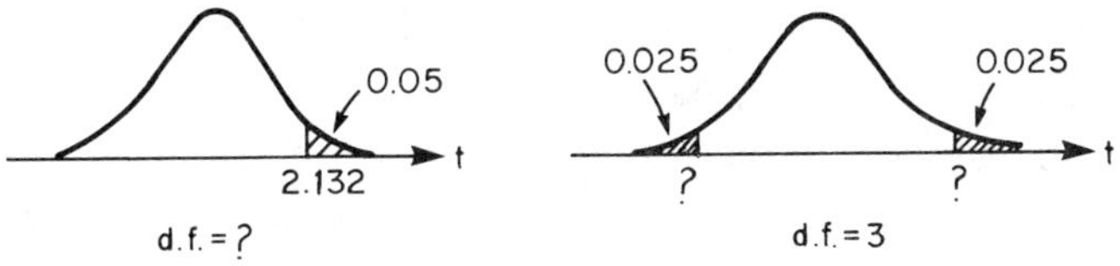

9.3 Make precise probability statements in terms of repeated sampling for the probabilities in Exercise 9.2.

9.4 In the χ^2 table, we find the entry 38.88 for 26 degrees of freedom and 0.05 probability. Make a sketch of the distribution and give a repeated sampling interpretation of the 5% probability.

9.5 In the F distribution we find the entry 3.77 for $\nu_1 = 7$ and $\nu_2 = 19$ and $\alpha = 0.01$. Make a sketch of the distribution and give a repeated sampling interpretation of the 1% probability.

9.6 A random sample of size 10 is extracted from a normal population of scores which has an unknown mean μ_X and hypothesized standard deviation $\sigma_X = 5$. If the sample calculations give $\overline{X} = 69$ and $s_X = 4$, then find the probability that in repeated sampling the sample standard deviation will be less than 4 (i.e., the chance that in repeated sampling the sample standard deviation is off the population standard deviation on the left by more than 1 point).

9.7 A politician claims that the number of murders on Christmas Day in a large city is a normal random variable with mean equal to 5 and unknown standard deviation σ_X. To check on his claim a random sample of 10 years from the records provides $\overline{X} = 6$ and

$s_{\overline{X}} = 0.4$. Calculate the probability that in repeated random sampling the t value (corresponding to $\overline{X}$ and $s_{\overline{X}}$) is greater than that found in the sample.

9.8 A standard intelligence test is given to 12 seventh grade children. The test publisher states that the population mean is equal to 110 and the standard deviation is unknown. If the mean for the 12 children is equal to 112.6 and $s_X = 5$, then find the probability that in repeated sampling of size 12 the absolute value of t (associated with $\overline{X}$ and $s_{\overline{X}}$) will be greater than the value found from the sample.

9.9 An experiment was conducted to investigate blinking rates of individuals relative to readability. For two subjects A and B the data for six 2-minute intervals were:

A	B
24	18
23	17
28	17
30	19
30	18
41	19

Under the assumption that the two underlying normal populations have the same mean and the same standard deviation, find the probability that in repeated sampling the t value (corresponding to $\overline{X}_A - \overline{X}_B$ and $s_{\overline{X}_A - \overline{X}_B}$) will be greater than that found in the data.

9.10 In Exercise 9.9 find the probability that in repeated sampling the ratio of the two sample standard deviations will exceed the ratio found from the data.

9.11 Complete the following probability statements. (*Hint*: Make sketches.)

(a) $P[-1.16 < Z < 1.16] =$ __________

(b) $P[-3.055 < t_{12} < 3.055] =$ __________

(c) $P[\chi^2_7 < 14.07] =$ __________

(d) $P[F_{6,10} >$ __________ $] = 0.05$

(e) $P[Z >$ __________ $] = 0.0054$

(f) $P[\_\_\_\_\_\_\_\_\_\_ < t_{13} < \_\_\_\_\_\_\_\_\_\_] = 0.99$

(g) $P[t_6 < -\_\_\_\_\_\_\_\_\_\_] = 0.025$

9.12 Give some valid arguments why the distributions discussed in this chapter are useful.

9.13 Find journal publications in your own area of interest and cite at least one publication mentioning each of the distributions of this chapter.

9.14 The following experiment can be used to empirically verify the entries in the t table. It is an example of what experimentalists call *simulation*. It requires that a bowl (or some container) contain many (say 300) numbered tags (say 1, 2, ..., 20) such that the frequency histogram is approximately normal, with mean μ_X equal to 10, say. This should be constructed before the experiment is actually done. The steps in the experiment are:

1. Each student successively draws 12 tags and records draw and tag number. [*Note:* (a) The first and second numbers drawn form a sample of size 2, the third and fourth form another sample of size 2. (b) Similarly, the first three numbers form a sample of size 3, the next three form another sample of size 3, and so on. Each student thus has 4 samples of size 3.]
2. Each student now computes (a) the 6 observed values of t_1 and (b) the 4 observed values of t_2.
3. Someone (a laboratory instructor) collects all observed values of t_1 and t_2. The two sample frequency tables and histograms are constructed for the t_1 and t_2 values.
4. Each student counts how many of the collected values satisfy the following inequalities:

(a) $|t_1| \geq 1.0000$ $\quad |t_1| \geq 2.4142$ $\quad |t_1| \geq 6.3138$

(b) $|t_2| \geq 0.8165$ $\quad |t_2| \geq 1.6036$ $\quad |t_2| \geq 2.9200$

9.15 Simulate the χ^2 and F distributions using the equipment in Exercise 9.14.

9.16 The following data on birth rates (live births per 1000 population) are of a random sample of 15 countries in which a majority of the population is literate (labeled H) and 15 countries in which a majority of the population is illiterate (labeled L). (These samples were drawn from all those countries for which data were available; therefore, the population of countries is limited to those which keep records!) The data were taken from two publications: United Nations (1962, 1963). The data on Y, the percentage of Roman Catholics of total population in 1958, were taken from Russet (1964).

H	X_H	Y_H	L	X_L	Y_L
Sweden	13.7	0.3	Sarawak	33.0	3.7
New Zealand	26.3	12.3	Peru	45.0	95.0
Puerto Rico	33.7	91.9	Iran	45.0	0.1
Ceylon	44.0	7.5	Haiti	46.5	69.9
Greece	16.3	0.8	India	43.2	1.4
Hong Kong	38.3	4.3	Brazil	45.0	93.0
Albania	35.3	6.8	Egypt	50.8	0.8
Hungary	17.8	60.9	Jordan	47.4	2.4
W. Germany	16.8	48.4	Malaya	44.4	2.3
USSR	25.3	5.0	Madagascar	45.0	18.7
E. Germany	13.9	11.1	Cambodia	41.4	1.2
Portugal	23.6	92.1	Ghana	54.0	10.4
Austria	16.8	89.4	Sudan	51.0	1.7
Netherlands	21.2	40.4	Guatemala	50.0	92.2
Israel	27.9	1.9	Angola	45.0	26.4

(a) Make the proper assumptions and find the probability that:

(i) t (associated with $\overline{X}_H - \overline{X}_L$ and $s_{\overline{X}_H - \overline{X}_L}$) is greater than that found in the sample.

(ii) F (associated with $s^2_{X_H}$ and $s^2_{X_L}$) is greater than that found in the sample.

(iii) t (associated with $\overline{Y}_H - \overline{Y}_L$ and $s_{\overline{Y}_H - \overline{Y}_L}$) is greater than that found in the sample.

(iv) F (associated with $s^2_{Y_H}$ and $s^2_{Y_L}$) is greater than that found in the sample.

(b) What do these probabilities suggest?

9.17 In Exercise 9.14, step 2(a) of the experiment, verify that

$$t_1 = \frac{(X_1 + X_2 - 2\mu_X)}{|X_1 - X_2|}$$

10

POINT ESTIMATION OF PARAMETERS AND CONFIDENCE INTERVAL ESTIMATION OF THE MEAN AND THE DIFFERENCE OF TWO MEANS BASED ON THE STANDARD NORMAL DISTRIBUTION

In Chapter 8, "Measures of Central Tendency and Variability," we have presented the calculation of the mean, median, standard deviation, etc. We did not distinguish then between parameters and estimators. It is now time to do so and in as rigorous a manner as possible. For this reason, let us at the outset state that we will be dealing with *infinite populations*, which typically arise when we use a probability distribution as our model. As the reader already knows, such populations as the binomial and normal represent infinite populations. The term "infinite" carries with it the connotation of "very large." Since all populations in nature are finite, only "large" ones are approximated by probability distributions. The moment this approximation is accepted we are in essence dealing with an infinite population. The parameters of the population are the parameters of the probability distribution, which are then estimated by sample quantities, called *estimators*. The development for finite populations is different and we will later devote a separate chapter to sample surveys of finite populations.

There are several problems associated with estimation of a parameter of a population. First of all, for any given parameter there is usually a whole class of estimators. For example, to estimate the mean μ_X of a normal population, one may use the sample mean, the sample median, the sample mode, or any other estimator the reader can think of. How do we go about choosing among estimators of a parameter? Well, as rational human beings we introduce certain *criteria of goodness*. Before going into the definitions of such criteria, let us recall the fact that any estimator based on the sample will produce different results under repeated sampling. It seems natural that we should pick an estimator which produces results such that the average under repeated sampling is equal to the value of the parameter in question. If an estimator has this property, then one speaks of an *unbiased estimator*; otherwise it is called a *biased estimator*. It also seems natural that we should pick an estimator which shows the smallest variability under repeated sampling, i.e., the smallest standard deviation. The standard deviation of an estimator is often referred to as the *standard error of the estimator*. It is a measure of *precision* or of *reliability*. Using the criteria of "bias" and "precision" one should pick an estimator which has the smallest bias and highest precision.

There are basically two types of estimators, namely, *point estimators* and *confidence interval estimators*. The first type provides a single value from a sample, while the second one provides two values with the assertion that the parameter lies between them with a specified probability in repeated sampling. Let us now introduce certain basic definitions and become somewhat formal.

10.1 POINT ESTIMATION

DEFINITION 10.1 *A sample estimator, or briefly, an estimator, is a procedure or formula which provides an estimate of a population parameter from a random sample.*

This definition implies that an estimator is a statistic, and hence a random variable, since it depends strictly on the sample. For example, if $X_1, X_2, \ldots, X_n$ is a random sample from any population with a distribution having mean μ_X and standard deviation σ_X, then the sample mean $\overline{X}$ is an estimator of μ_X and the sample standard deviation s_X is an estimator of σ_X. For a given random sample, $\overline{X}$ and s_X will produce particular estimates of μ_X and σ_X. Notice the difference between *estimator* and *estimate*; e.g., $\overline{X} = (1/n)\Sigma X_i$ is the formula, while $\overline{X} = 10$ cm is an estimate if we are dealing with a sample of measurements in centimeters. In practice (and in many books) this distinction is not made, and one refers to both the formula and a particular value from a sample as an estimate. We shall try to distinguish between them.

DEFINITION 10.2 *An estimator is called a point estimator if it provides a single value as an estimate from a random sample.*

For example, $\overline{X}$ is a point estimator of the mean $\mu_X = p$ of the Bernoulli population, because from a sample one calculates a single value as an estimate of μ_X. To illustrate this notion, consider interviewing 10 people regarding a "yes-no" issue. This is a random sample of size 10 from a Bernoulli population. If the 10 responses are 0, 1, 0, 0, 0, 1, 1, 0, 0, 0, where 0 = no and 1 = yes, then the estimate of the population mean μ_X (= p = population proportion saying "yes") obtained from the sample is equal to:

$$\overline{X} = \frac{1}{10}(0 + 1 + 0 + 0 + 0 + 1 + 1 + 0 + 0 + 0) = 0.3$$

The reader is invited to find a point estimate for the standard deviation $\sigma_X = \sqrt{pq}$ from the same 10 interviews.

DEFINITION 10.3 *An estimator is said to be an unbiased estimator of the parameter* θ *if the mean of the estimates over all possible samples from the distribution is equal to* θ*; if the mean over all possible samples is not equal to* θ*, then the estimator is said to be biased.*

It can be shown via a mathematical argument (i.e., without drawing all possible samples) that the sample mean $\overline{X}$ is an unbiased estimator of the population mean μ_X, no matter what the underlying distribution. In the same way, it can be shown that in general the sample median and sample mode are biased estimators of the population mean μ_X. The sample variance s_X^2 is an unbiased estimator of the population variance σ_X^2, but s_X is a biased estimator of the population standard deviation σ_X. If in s_X^2 the divisor (n - 1) is replaced by n, then the resulting estimator will be biased. This is the reason why we divide by n - 1 and not n.

There is a nice interpretation of the concept of unbiasedness. Suppose that all possible samples are drawn from a population and for each sample an estimate is obtained from the estimator $\hat{\theta}$ of θ. *Suppose* that when the probability distribution of the estimator is plotted, it is symmetric and centered at θ; i.e., the picture looks like Figure 10.1. Then it follows that $\hat{\theta}$ is an unbiased estimator of θ. Such a picture actually arises, when we consider the sample mean $\overline{X}$ as an estimator of the mean μ_X of a normal population. The reader knows this from a previous chapter.

Now *assume* that the probability distribution is still symmetric but is not centered at θ; i.e., we have the graph shown in Figure 10.2. Therefore, $\hat{\theta}$ is a biased estimator of θ and the amount of bias in absolute value is given by the distance β, i.e., β is equal to the absolute difference of the mean of the estimates over all possible samples (= mean of the variable $\hat{\theta}$) and the parameter θ. Such a picture can arise in the case that $\overline{X}$ is used as an estimator of a parameter other than μ_X in the normal case, and the reader is invited to give an example of this.

Finally, note that the mean of the estimates as provided by the estimator $\hat{\theta}$ over all possible samples is in fact equal to the mean of the distribution of the estimator $\hat{\theta}$. This mean is known in the statistical language as the *expected value* of $\hat{\theta}$. Thus the bias in

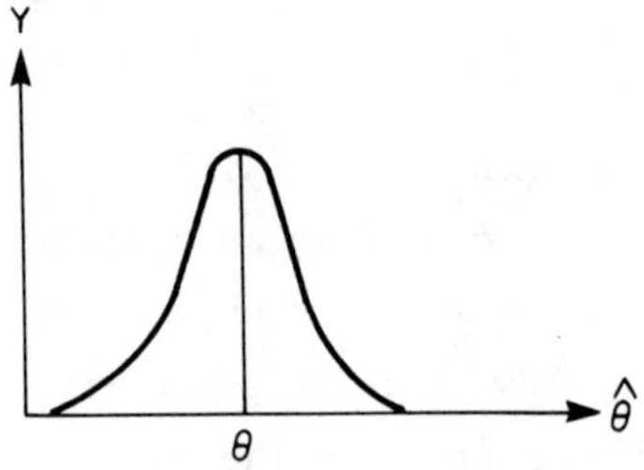

Figure 10.1 Probability distribution of the estimator $\hat{\theta}$ of θ.

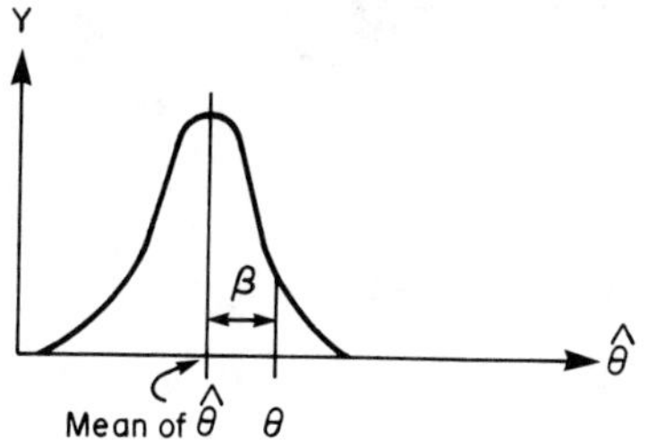

Figure 10.2 Probability distribution of the estimator $\hat{\theta}$, not centered at θ.

absolute value is equal to the absolute difference between the expected value of $\hat{\theta}$ and the parameter θ. This is really a nice way of putting it. In the statistical literature the term bias is often referred to as *systematic error*.

The standard deviation of $\hat{\theta}$ measures the closeness of the estimates around their mean (i.e., around the expected value of $\hat{\theta}$). This is the reason why it is called a *measure of precision*.

DEFINITION 10.4 *The standard deviation* $\sigma_{\hat{\theta}}$ *of an estimator* $\hat{\theta}$ *is called the standard error of* $\hat{\theta}$.

Through some mathematical arguments it can be shown that the sample mean $\overline{X}$, when sampling from an infinite population with finite mean μ_X and finite standard deviation σ_X, has a standard deviation equal to $\sigma_{\overline{X}} = \sigma_X/\sqrt{n}$. The reader knows this fact already when sampling from a normal population.

Since $\sigma_{\hat{\theta}}$ is a measure of closeness of the estimates around their mean in repeated sampling and not necessarily around the parameter θ, unless $\hat{\theta}$ is unbiased, it follows that an ultimate judgment on the goodness of an estimator should be based both on the standard error and the bias. This is so because an estimator may have a high *precision* and yet a large bias, so that its ultimate *accuracy* will be low. Accuracy of an estimator is thus determined by both precision and bias.

DEFINITION 10.5 *The square root of the sum of the variance and the square of the bias of an estimator, i.e.,* $\sqrt{\sigma^2_{\hat{\theta}} + \beta^2_{\hat{\theta}}}$, *is called the root mean square error of* $\hat{\theta}$.

The root mean square error of $\hat{\theta}$ is abbreviated in the literature as RMSE($\hat{\theta}$); i.e.,

$$\text{RMSE}(\hat{\theta}) = \sqrt{\sigma^2_{\hat{\theta}} + \beta^2_{\hat{\theta}}}$$

Here $\beta_{\hat{\theta}}$ is the bias of $\hat{\theta}$, i.e., the expected value of $\hat{\theta}$ minus θ, and it is not necessary to take the absolute difference, because in RMSE($\hat{\theta}$) the square of the bias, which is always positive, is taken.

From what has been said above, it should be clear that when we have to select an estimator from among competing ones, we should select one which has maximum accuracy, i.e., the least RMSE. Such an estimator is called the *minimum RMSE estimator* or *best estimator*. It is also clear that when an estimator has to be selected from among unbiased estimators it is sufficient to pick one which has maximum precision, i.e., *minimum standard error* or *minimum variance*.

There is a popular pictorial presentation of the concept of *accuracy*. Imagine throwing darts repeatedly at a board with the familiar bull's-eye on it. The bull's-eye is the parameter θ and the throws of the dart are likened to drawing a random sample, while the point of impact of the dart is an estimate produced by the thrower, who is likened to the estimator. The four pictures of Figure 10.3 represent four different estimators, $\hat{\theta}_1$, $\hat{\theta}_2$, $\hat{\theta}_3$, and $\hat{\theta}_4$, of the parameter θ.

The following theorem can be proven by those who know some algebra and calculus. We are stating it because it has something important to say about the goodness of the sample mean $\overline{X}$ as an estimator of the population mean μ_X.

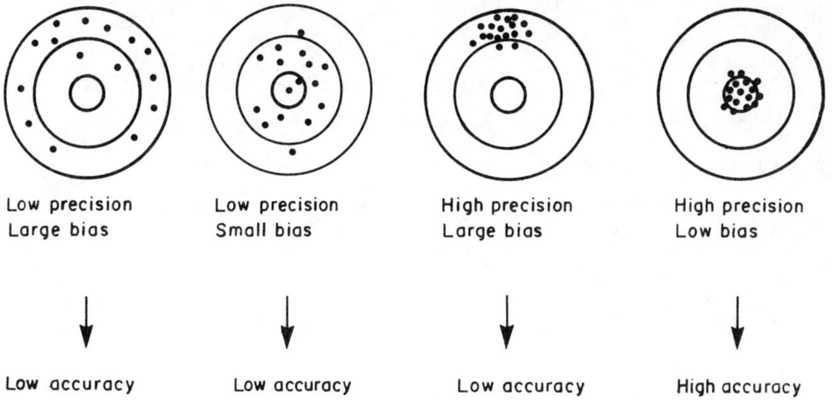

Figure 10.3 Four estimators of θ.

THEOREM 10.1 *Let* $X_1, X_2, X_3, \ldots, X_n$ *be a random sample from a population, whose underlying probability distribution has mean* μ_X *and standard deviation* σ_X. *Then among all possible unbiased estimators of the form* $a_0 + \sum_{i=1}^n a_i X_i$, *the sample mean* $\overline{X}$ *has minimum variance.*

Estimators which are written as $a_0 + \sum_{i=1}^n a_i X_i$, where a_0 and a_i are real numbers, are called *linear estimators*. For the sample mean $\overline{X} = (1/n) \sum_{i=1}^n X_i$, we observe that $a_0 = 0$ and $a_i = 1/n$. Thus Theorem 10.1 tells us that $\overline{X}$ is the best linear unbiased estimator of μ_X. In the literature such estimators are often referred to as *BLUEs*.

We now illustrate point estimation of parameters of three popular probability distributions, namely, the Bernoulli, the binomial, and the normal distributions. Let us repeat that these distributions represent infinite populations.

Example 10.1 A random sample of 10 Canadian female voters were asked the "yes-no" question, "Would you like to see a woman as prime minister of Canada?" The responses obtained were as follows: 1, 0, 0, 1, 1, 1, 0, 1, 1, 1, where 0 = no and 1 = yes. The population of female voters can be seen as a Bernoulli population with mean and standard deviation equal to:

$$\mu_X = p \qquad \sigma_X = \sqrt{pq}$$

The mean μ_X is simply the population proportion answering "yes." The estimators of the two parameters are $\overline{X}$ and s_X, respectively, and from the sample we obtain the point estimates:

$$\overline{X} = \frac{1}{10}\,\Sigma X_i = \frac{1}{10}\,(1 + 0 + 0 + 1 + 1 + 1 + 0 + 1 + 1 + 1) = 0.7$$

$$s_X^2 = \frac{\Sigma X_i^2 - (1/n)(\Sigma X_i)^2}{n - 1} = \frac{7 - (1/10)(49)}{9} = 0.2333$$

$$s_X = \sqrt{0.2333} = 0.48$$

For the standard error of the mean, we find the point estimate:

$$s_{\overline{X}} = \frac{s_X}{\sqrt{n}} = \frac{0.48}{\sqrt{10}} = 0.15$$

Note that 0.7 is an unbiased estimate of μ_X, 0.23 is an unbiased estimate of the variance σ_X^2, and that 0.48 and 0.15 are biased estimates of the standard deviation σ_X and standard error $\sigma_{\overline{X}}$, respectively. The interpretation of the sample estimate $s_{\overline{X}} = 0.15$ is that in repeated sampling the accuracy or precision of $\overline{X}$ is estimated as 0.15.

It is interesting to point out that the sample variance s_X^2 in the Bernoulli case can be simplified to the formula:

$$s_X^2 = \frac{n}{n - 1}\,\overline{X}(1 - \overline{X})$$

This formula conclusively shows that the estimator $\overline{X}(1 - \overline{X})$ will be a biased estimator of σ_X^2. For large sample sizes (i.e., $n > 30$), one often uses this estimator, because $n/(n - 1)$ is then fairly close to 1.

Example 10.2 Five randomly selected clerks in each of 90 randomly selected department stores were asked the "yes-no" question, "Should old age security pensions be increased by the federal government?" The following frequency distribution was obtained from the responses:

No. of yes answers per five interviews	Observed frequencies of stores
0	1
1	2
2	9
3	28
4	40
5	10
Total	90

The population of department stores is considered a binomial population, and we have essentially a random sample of 90 stores, each store being considered a Bernoulli population of clerks, from which a random sample of five is taken.

We know that the mean, standard deviation, and standard error of the sample mean for a binomial population are equal to:

$$\mu_X = 5p \qquad \sigma_X = \sqrt{5pq} \qquad \sigma_{\overline{X}} = \frac{\sigma_X}{\sqrt{n}} = \sqrt{pq}$$

The point estimates of these parameters are obtained from the sample data by utilizing some of the results of Chapter 4:

$$\overline{X} = \frac{1}{90}\Sigma X_i f_i = \frac{314}{90} = 3.49$$

$$s_X^2 = \frac{\Sigma X_i^2 f_i - (\Sigma X_i f_i)^2/90}{89}$$

$$= \frac{1160 - (98{,}596/90)}{89} = 0.72$$

$$s_X = \sqrt{0.72} = 0.85$$

$$s_{\overline{X}} = \frac{s_X}{\sqrt{5}} = \frac{0.85}{2.24} = 0.38$$

Hence it is estimated unbiasedly that the mean number of clerks saying "yes" is equal to 3.49 when five interviews are taken in 90 stores. Note that this gives an unbiased point estimate of p equal to 3.49/5 = 0.70; i.e., 70% are in favor of pensions in this survey.

The variability in the response is estimated to be equal to 0.85, and in repeated sampling the estimated variability of the sample mean is 0.38.

Example 10.3 A random sample of 15 undergraduates obtained the following scores from the Rotter Interpersonal Trust Scale (see Chun and Campbell, 1974):

117, 95, 120, 85, 114, 90, 89, 104, 115, 113, 110, 81,
98, 125, 117

Assuming that such scores of undergraduates follow a normal distribution, with mean μ_X and standard deviation σ_X, we obtain the following point estimates from the sample:

$$\overline{X} = \frac{1}{15}\Sigma X_i = \frac{1573}{15} = 104.87$$

$$s_X^2 = \frac{1}{14}[\Sigma X_i^2 - \frac{1}{15}(\Sigma X_i)^2] = \frac{1}{14}\left(167{,}745 - \frac{2{,}474{,}329}{15}\right)$$

$$= 199.27$$

$$s_X = \sqrt{199.27} = 14.12$$

$$s_{\overline{X}} = \frac{14.12}{\sqrt{15}} = 3.65.$$

This time the reader is invited to interpret the numbers 104.87, 14.12, and 3.65.

10.2 CONFIDENCE INTERVAL ESTIMATION

We now move on to the topic of *confidence interval estimation.* As stated at the beginning of this chapter, we have to provide two estimators based on the sample, such that in repeated sampling we can assert that the parameter in question will be between them with a certain specified probability. Formally, we define confidence interval estimation as follows:

DEFINITION 10.6 *Let* $X_1, X_2, \ldots, X_n$ *be a random sample from a population with a specified probability distribution and a specified* *parameter θ of interest. If two point estimators based on the sample, called the lower limit* L *and the upper limit* U, *can be found for θ such that in repeated sampling we can assert that θ will be between* L *and* U *with a specified probability γ, then the pair* (L, U) *is called a confidence interval estimator for θ.*

The probability γ in Definition 10.6 is called the *confidence coefficient*. It is often referred to as the *coverage probability*, because in repeated sampling we expect the confidence interval estimates obtained from (L, U) to cover the parameter $\gamma \times 100\%$ of the time. The difference between U and L is called the *length* of the confidence interval and the difference between the mean of the estimator U and the mean of the estimator L is called the *expected length* of the confidence interval estimator (L, U). In general we will be working with confidence interval estimators which have *minimum expected length*, i.e., the shortest expected length. There are other criteria, such as unbiasedness, associated with confidence interval estimation; however, we will not go into these.

10.3 CONFIDENCE INTERVAL FOR μ_X

Our basic result is:

THEOREM 10.2 *If $X_1, X_2, \ldots, X_n$ is a random sample from a normal population with unknown mean μ_X and known standard deviation σ_X, then a $\gamma \cdot 100\%$ $[= (1 - \alpha) \cdot 100\%]$ confidence interval estimator for the population mean is equal to* $(L, U) = (\overline{X} - Z_{\alpha/2} \cdot \sigma_{\overline{X}}, \overline{X} + Z_{\alpha/2} \cdot \sigma_{\overline{X}})$, *where $Z_{\alpha/2}$ is the value of Z such that the area under the Z distribution to the right of $Z_{\alpha/2}$ is equal to $\alpha/2$.*

Let us derive (for fun and for a little algebra practice) the theorem for the case $\gamma = 0.95$; i.e., $\alpha = 0.05$. The reader knows from earlier developments that:

1. $\overline{X}$ is $N\left(\mu_X, \sigma_{\overline{X}} = \dfrac{\sigma_X}{\sqrt{n}}\right)$

2. $$Z = \frac{\overline{X} - \mu_X}{\sigma_{\overline{X}}} \text{ is } N(0, 1)$$

Hence:

$$P[-1.96 < Z < 1.96] = 0.95$$

i.e.,

$$P\left[-1.96 < \frac{\overline{X} - \mu_X}{\sigma_{\overline{X}}} < 1.96\right] = 0.95$$

$$P[-1.96\sigma_{\overline{X}} < \overline{X} - \mu_X < 1.96\sigma_{\overline{X}}] = 0.95$$

$$P[-\overline{X} - 1.96\sigma_{\overline{X}} < -\mu_X < -\overline{X} + 1.96\sigma_{\overline{X}}] = 0.95$$

$$P[\overline{X} - 1.96\sigma_{\overline{X}} < \mu_X < \overline{X} + 1.96\sigma_{\overline{X}}] = 0.95$$

Therefore:

$$(L, U) = (\overline{X} - 1.96\sigma_{\overline{X}}, \overline{X} + 1.96\sigma_{\overline{X}})$$

The first probability statement from which the two limits were derived is illustrated in Figure 10.4.

Example 10.4 Assume that the dollar values of houses owned by faculty members of universities are normally distributed with unknown mean μ_X and known standard deviation σ_X = \$7000. A random

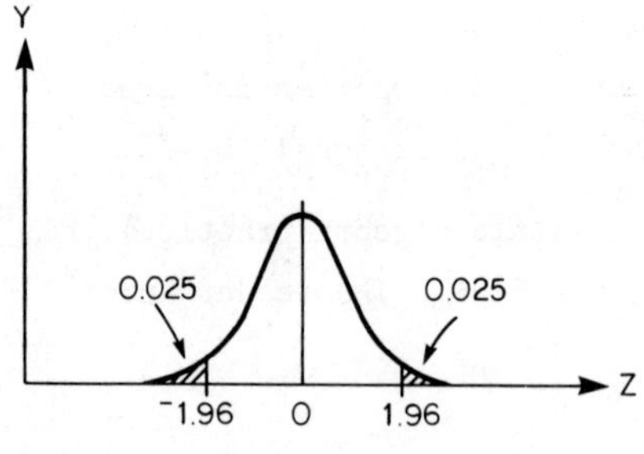

Figure 10.4 Probability that $|Z| > 1.96$

sample of eight faculty members leads to the following information regarding the values of their houses:

\$55,000	\$75,000
\$37,000	\$60,000
\$42,000	\$48,000
\$39,000	\$57,000

From these data we obtain the following 95% confidence interval estimates of the mean value of houses owned by faculty members:

$$L = \overline{X} - (1.96)\sigma_{\overline{X}} = 51{,}625 - (1.96)\frac{7000}{\sqrt{8}} = \$46{,}776.94$$

$$U = \overline{X} + (1.96)\sigma_{\overline{X}} = 51{,}625 + (1.96)\frac{7000}{\sqrt{8}} = \$56{,}473.06$$

Notice that a particular confidence interval estimate need not cover μ_X. The interpretation is in terms of repeated sampling; i.e., had we extracted 100 samples of size 8, then we could expect on the average that 95 of the samples would lead to intervals which would cover the mean and 5 of them would not. However, there is a 95% chance that our particular confidence interval belongs to the set covering the mean μ_X. One therefore says: "That we are 95% *confident* that μ_X lies between \$46,776.94 and \$56,473.06."

Remark. The length of the interval given in Theorem 10.2 is equal to $2(1.96)\sigma_{\overline{X}}$, which is not a random variable. It is beyond the scope of this book to show that the confidence interval given in Theorem 10.2 is the shortest one under the given setting. Note that the confidence interval estimator of the mean in this theorem depends on the sample size n, the confidence coefficient γ, and the standard deviation σ_X. In practice one often uses 95% and 99% coefficients and occasionally one sees a 90% confidence coefficient. The larger the γ, the larger the expected length, and the larger n, the smaller the expected length of the confidence interval. The reader is encouraged to construct a 99% confidence interval for Example 10.4 and compare it with the 95% interval.

10.4 CONFIDENCE INTERVAL FOR THE DIFFERENCE $\mu_X - \mu_Y$

The following theorem is an extension of Theorem 10.2 to the case of confidence interval estimation for the difference between means of two normal populations.

THEOREM 10.3 *If* $X_1, X_2, \ldots, X_{n_1}$ *and* $Y_1, Y_2, \ldots, Y_{n_2}$ *are random samples from two independent normal populations* $N(\mu_X, \sigma_X)$ *and* $N(\mu_Y, \sigma_Y)$*, respectively, such that* μ_X *and* μ_Y *are unknown and* σ_X *and* σ_Y *are known, then a* $\gamma \cdot 100\%$ $[= (1 - \alpha) \cdot 100\%]$ *confidence interval estimator for the difference* $\mu_{\overline{X}-\overline{Y}} = \mu_X - \mu_Y$ *is given by* $(L, U) = (\overline{X} - \overline{Y} - Z_{\alpha/2} \cdot \sigma_{\overline{X}-\overline{Y}}, \overline{X} - \overline{Y} + Z_{\alpha/2} \cdot \sigma_{\overline{X}-\overline{Y}})$*, where* $Z_{\alpha/2}$ *is as defined in Theorem 10.2.*

The reader should recall that $\sigma_{\overline{X}-\overline{Y}} = \sqrt{\sigma_X^2/n_1 + \sigma_Y^2/n_2}$, so that its calculation is trivial when σ_X^2 and σ_Y^2 are given. Let us illustrate Theorem 10.3 with an example.

Example 10.5 Suppose that in a mathematics aptitude test for college entrance examinations it is known that the standard deviation for females is equal to 10 and for males is equal to 12. Six randomly selected scores for females and eight for males show the following results:

Females (X)	Males (Y)
78	42
82	67
92	87
69	93
46	69
51	77
	57
	42
$\Sigma X = 418$	$\Sigma Y = 534$
$\overline{X} = 69.67$	$\overline{Y} = 66.75$

A 99% confidence interval for the difference between the means of the female and male population of scores of college entrants obtained from the data is equal to:

$$L = (69.67 - 66.75) - 2.58\sqrt{\frac{10^2}{6} + \frac{12^2}{8}} = 2.92 - 15.19 = -12.27$$

$$U = (69.67 - 66.75) + 2.58\sqrt{\frac{10^2}{6} + \frac{12^2}{8}} = 2.92 + 15.19 = 18.11$$

Note that in this example $Z_{\alpha/2} = Z_{0.005} = 2.58$, from the Z table. The reader should give an interpretation of the confidence interval obtained in this example.

10.5 LARGE SAMPLE CONFIDENCE INTERVAL ESTIMATION

The following theorem is one of the most important theorems in the discipline of statistics. It settles several problems for us; among others, the construction of a confidence interval for the mean of a population when the sample size is large, i.e., when we are tackling *large sample confidence interval estimation.*

THEOREM 10.4 *If* $X_1, X_2, \ldots, X_n$ *is a random sample from a population with a distribution such that the mean* μ_X *and standard deviation* σ_X *are finite, then the sample mean* $\overline{X}$*, for sufficiently large* n*, is approximately normally distributed with mean equal to* μ_X *and standard deviation* $\sigma_{\overline{X}} = \sigma_X/\sqrt{n}$.

This theorem is known as the *central limit theorem* and its importance cannot be over emphasized. By "sufficiently large" one means in practice a sample of size $n > 30$. This theorem is especially noteworthy because it says that no matter how the population distribution looks, as long as σ_X exists and $n > 30$, the sample mean $\overline{X}$ in repeated sampling will be approximately normally distributed. This means that the standardized variable

$$Z = \frac{\overline{X} - \mu_X}{\sigma_X/\sqrt{n}}$$

in such situations will have the standard normal distribution.

Often the central limit theorem is stated for the sample sum rather than the sample mean. For the sample sum it means that ΣX_i, for sufficiently large n (i.e., n > 30), is approximately normally distributed with mean equal to $n\mu_X$ and standard deviation $\sqrt{n}\sigma_X$. The standardization for the sample sum is given by:

$$Z = \frac{\Sigma X_i - n\mu_X}{\sqrt{n}\sigma_X}$$

In the construction of a confidence interval for the mean μ_X (or for $n\mu_X$) one replaces σ_X by s_X, which is obtained from the large sample, unless σ_X is given. Since n is large, one uses $s_X = \sqrt{(1/n)\Sigma(X_i - \bar{X})^2}$ rather than dividing through by n - 1. The reason why the t table stops at $\nu = 29$ degrees of freedom (i.e., n = 30) is also due to this result.

The reader should appreciate the formidable nature of the central limit theorem. It tells us that for large sample sizes the only distribution needed for construction of a confidence interval for a population mean (mind you, any population!) is the standard normal distribution. Hence the only table needed in the large sample case is the Z table, and the confidence interval is simply equal to:

$$L = \bar{X} - Z_{\alpha/2} \cdot \frac{s_X}{\sqrt{n}}$$

$$U = \bar{X} + Z_{\alpha/2} \cdot \frac{s_X}{\sqrt{n}}$$

Example 10.6 On the basis of interviewing 36 randomly selected male undergraduate students on the "yes-no" question, "Should females engage in the sport of boxing?" it is found that the percentage saying "yes" is equal to 30%. On the basis of this result, find a 95% confidence interval estimate for the unknown proportion of males saying "yes."

The sample is from a Bernoulli population with mean equal to $\mu_X = p$ and $\sigma_X = \sqrt{pq}$. The standard error of the sample mean $\bar{X}$ is equal to $\sigma_X/\sqrt{n} = \sqrt{pq/n}$. Since the sample size is large, we replace σ_X by its estimate

$$s_X = \sqrt{(\bar{X})(1 - \bar{X})} = \sqrt{(0.3)(0.7)} = \sqrt{0.21} = 0.46$$

which, for large n, is close to $s_X = \sqrt{[\Sigma X_i^2 - (\Sigma X)^2/n]/(n - 1)}$, as was stated earlier, when we dealt with point estimation in the case of the Bernoulli population. Hence from the central limit theorem we have:

$$L = 0.30 - (1.96)\frac{0.46}{\sqrt{36}} = 0.14 \qquad U = 0.30 + (1.96)\frac{0.46}{\sqrt{36}} = 0.46$$

Example 10.7 Weiskott (1974) has observed that 390 out of 736 telephone calls received at a counseling service occurred during the 2-week span of the new moon period, i.e., an observed proportion of 0.53. The 95% confidence interval for the true proportion is

$$0.530 \pm 1.96\sqrt{\frac{(0.53)(0.47)}{736}} = 0.530 \pm 0.036$$

Notice that this interval contains 1/2.

Example 10.8 A random sample of 49 stay-at-home housewives in a certain city were asked how many hours they watched the morning TV programs between 7 a.m. and 12 noon. The answers resulted in a mean of 2.6 hr and a standard deviation of 0.7 hr.

From these data we may construct a 95% confidence interval for the mean number of hours of watching morning TV programs for the population of housewives, i.e.,

$$L = 2.6 - (1.96)\frac{0.7}{\sqrt{49}} = 2.4 \qquad U = 2.6 + (1.96)\frac{0.7}{\sqrt{49}} = 2.8$$

EXERCISES

10.1 A random sample of 9 undergraduate students who have registered cars was selected in a university to see whether they had any campus traffic violations or not. From the records it was found that 2 had received tickets and were fined.

(a) What is the name of the distribution underlying the population?

(b) Find point estimates of μ_X and σ_X and also find an estimate of the standard error of $\overline{X}$.

10.2 Set up your own "sampling from a binomial population" example and find point estimates of μ_X, σ_X, and $\sigma_{\overline{X}}$.

10.3 Five highly overweight women were randomly selected from among a large group; their weights were: 210, 175, 189, 250, and 195 lb. Make a distribution assumption and find point estimates of μ_X and σ_X and the standard error $\sigma_{\overline{X}}$.

10.4 The absolute bias of the estimator $s_X^2 = (1/n)\Sigma(X_i - \overline{X})^2$ of σ_X^2 is equal to σ_X^2/n. What is the meaning of this?

10.5 Describe verbally the difference between precision and accuracy.

10.6 What is the basic difference between a point estimator and a confidence interval estimator of a parameter θ?

10.7 Suppose that the variance of an estimator of a parameter is equal to 10 and its absolute bias is equal to 2. What measure is proper to judge the goodness of the estimator? Find its value.

10.8 Give an example of a biased estimator of μ_X in the case of a Bernoulli population. Why is it biased?

10.9 Why do you think that $\overline{X}^2$ is a biased estimator of the population variance σ_X^2?

10.10 A random sample of 10 measurements from a normal population with $\sigma_X = 2$ cm included the following: 4, 6, 8, 3, 12, 5, 11, 6, 9, 6. Find a 90% confidence interval for the mean μ_X and give an interpretation.

10.11 A control treatment and a new treatment are tested in a mental institution on patients. These treatments are given to two random groups of sizes 9 and 10, respectively, which have been matched for other factors. Assume that the task completion times (in seconds) were recorded as:

Control	Treatment
2	1
3	3
2	2
4	2
3	2
6	4
2	1
3	2
6	2
	3

(a) Under the assumption that the underlying populations are normal with standard deviations equal to 2 and 1, respectively, find a 95% confidence interval for the difference between the two population means.

(b) Make a sketch of the distribution of the difference between the sample means, showing the estimate under (a).

10.12 Do we need the central limit theorem when the large sample comes from a normal population? Defend your answer.

10.13 Set up a large sample situation for a binomial population (use Example 10.2 as a guide) and find a 95% confidence interval for the mean μ_X.

10.14 A mobility study was conducted among a random sample of 1000 university graduates (B.A.'s) in a particular province over a period of 15 years. The following results were found for the distance between the university and present permanent residence:

$\overline{X} = 1800$ k

$s_X = 520$ k

Find a 99% confidence interval for the mean distance for the population. What is the estimated length of the calculated confidence interval?

10.15 Make a true statement and then a false statement for each of the following concepts:

(a) Bias
(b) Estimator
(c) Standard error of $\overline{X}$
(d) Expected value of an estimator
(e) Accuracy
(f) Root mean square error of an estimator
(g) Confidence coefficient
(h) Confidence interval based on Z for $\mu_X - \mu_Y$
(i) Central limit theorem
(j) Precision

Note: If someone claims that $\mu_X = 10$, say, and then if by a legitimate sampling procedure you derive the confidence interval estimate for μ_X as (12, 15), then there is some evidence against this person's claim because your interval does not contain his claimed value. On the other hand, if your interval had contained μ_X you lack evidence against his claim. An argument similar to the above should be used in Exercises 10.16 to 10.22. Chapter 11 will present another and, in fact, a more rigorous procedure for handling such questions.

10.16 A normal population has mean 20 and standard deviation 2. A sample of 6 items from this population has a mean 18.2. Can the sample be reasonably regarded as a random sample from the population?

10.17 A company that sells frozen shrimp prints "Contents 12 oz" on the package. Suppose it is known from past experience that the population of package weights has a standard deviation of 0.5 oz. A sample of 36 packages yields an average of 11.82 oz. What conclusion would be drawn concerning the standard which the company is trying to achieve?

10.18 A manufacturer of flashlight batteries claims that the average life of this product will exceed 30 hr. A company is willing to buy a very large shipment of batteries if the claim is true. A random sample of 36 batteries is tested, and it is found that the sample mean is 40 hr. If the population of batteries has a standard deviation of 5 hr, is it likely that the batteries will be purchased?

10.19 Incoming freshmen are given entrance examinations in a number of fields, including mathematics. Over a period of years it has been found that the average score in the mathematics examination is 67 and the standard deviation of the scores is 7.5. A mathematics instructor looks up the scores of his class of 36 and finds that their average is 70. Assuming the standard deviation of 7.5 can be used, can the instructor claim that the average is greater than 67?

10.20 A fisherman decides that he needs a line that will test 10 lb if he is to catch the size fish he desires. He tests 16 pieces of Brand X line and finds a sample mean of 10.4. If it is known that $\sigma = 0.5$ lb, what can he conclude about Brand X?

10.21 For the Bernoulli random variable verify that $s_X^2 = [n/(n - 1)]\overline{X}(1 - \overline{X})$.

10.22 A researcher, wishing to determine what proportion of a very large population possesses a particular genetic trait, has decided to choose a random sample of individuals. He has considered three different sample sizes, n equal to (a) 36, (b) 64, and (c) 100. Suppose that (unknown to the investigator) the true proportion is 0.5. Find the probability that the estimated proportion differs by more than 0.1 from the true proportion, $p = 0.5$, for each of the three cases. In the first case, for example, we want

$$P[\hat{p} < 0.4 \text{ or } \hat{p} > 0.6] = P[X < (0.4)(36) \text{ or } X > (0.6)(36)]$$

where X is the number of individuals in the sample possessing the trait (so $\hat{p} = X/n$). What is happening to the probability of a "bad" estimate of p (i.e., $\hat{p} < 0.4$ or $\hat{p} > 0.6$) as n increases? (We say that $\hat{p}$ is a consistent estimator of p if the probability of obtaining a "bad" estimate of p decreases to zero as n increases to infinity.)

11

SMALL SAMPLE CONFIDENCE INTERVAL ESTIMATION BASED ON THE t, χ^2, AND F DISTRIBUTIONS AND DETERMINATION OF SAMPLE SIZE

In Chapter 10 the central limit theorem told us that for sufficiently large samples, there was no need to worry about the nature of the distribution underlying a population from which the sample was selected. When dealing with small samples, the story is somewhat different. The theory of confidence interval estimation in the small sample case is well developed when sampling from a normal population. Throughout this chapter we will be concerned with small ($n < 30$) samples from normal populations, and on the basis of the samples find confidence intervals for specified parameters, such as the mean μ_X, σ_X, and the ratio of two standard deviations σ_X/σ_Y. We will rely heavily on the t, χ^2, and F distributions, which have been developed in Chapter 9.

11.1 CONFIDENCE INTERVAL FOR THE MEAN μ_X BASED ON THE t DISTRIBUTION

We first consider the problem of finding a confidence interval estimator for the mean of a normal population. Let $X_1, X_2, \ldots, X_n$ be a random sample from $N(\mu_X, \sigma_X)$, where both μ_X and σ_X are unknown. We know from Chapter 9 that in repeated sampling the random variable

$$t = \frac{\overline{X} - \mu_X}{s_{\overline{X}}}$$

has Student's t distribution, with $\nu = n - 1$ degrees of freedom. Hence the following probability statement is true:

$$P[-t_{n-1,\alpha/2} < t_{n-1} < t_{n-1,\alpha/2}] = 1 - \alpha = \gamma$$

where $t_{n-1,\alpha/2}$ is the value of t_{n-1} in the table with n - 1 degrees of freedom such that the area to its right under the t curve is equal to $\alpha/2$. Pictorially this probability statement looks like the curve shown in Figure 11.1. For example, when n = 10 and $\gamma = 0.95$, from the t table $t_{9,0.025} = 2.262$; therefore

$$P[-2.262 < t_9 < 2.262] = 0.95$$

Going back to the original statement, we substitute the definition of the t variable and obtain the following result by manipulating the inequalities:

$$P\left[-t_{n-1,\alpha/2} < \frac{\overline{X} - \mu_X}{s_{\overline{X}}} < t_{n-1,\alpha/2}\right] = 1 - \alpha$$

i.e.,

$$P[-t_{n-1,\alpha/2}\, s_{\overline{X}} < \overline{X} - \mu_X < t_{n-1,\alpha/2}\, s_{\overline{X}}] = 1 - \alpha$$

$$P[-\overline{X} - t_{n-1,\alpha/2}\, s_{\overline{X}} < -\mu_X < -\overline{X} + t_{n-1,\alpha/2}\, s_{\overline{X}}] = 1 - \alpha$$

$$P[\overline{X} - t_{n-1,\alpha/2}\, s_{\overline{X}} < \mu_X < \overline{X} + t_{n-1,\alpha/2}\, s_{\overline{X}}] = 1 - \alpha$$

This derivation leads us to the following:

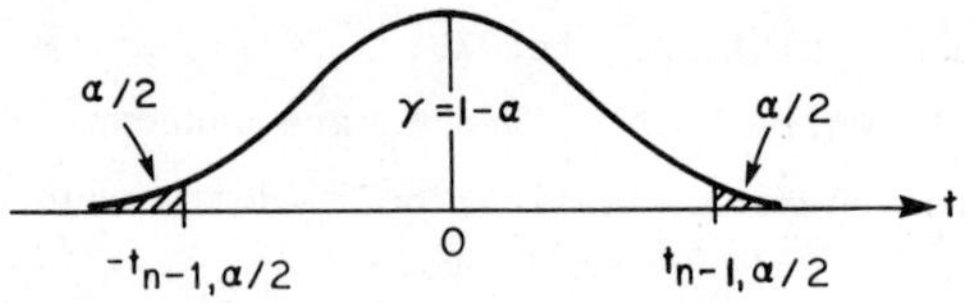

Figure 11.1 Tail probabilities for the t variable.

THEOREM 11.1 *If* $X_1, X_2, \ldots, X_n$ *is a random sample from a normal population with unknown mean* μ_X *and unknown standard deviation* σ_X, *then a* $\gamma = 1 - \alpha$ *confidence interval for the mean* μ_X *is given by:*

$$(L, U) = (\overline{X} - t_{n-1,\alpha/2}\, s_{\overline{X}}, \overline{X} + t_{n-1,\alpha/2}\, s_{\overline{X}})$$

Example 11.1 To get a quick idea about the mean grade in a final examination given to students in a large statistics course, the professor in charge of the course randomly selected 9 students and found from the sample an average of 67 and a standard deviation of 4. Find a 95% confidence interval estimate for the mean of the population of students.

To solve this problem, we assume that the final examination grades form a normal population, with unknown mean μ_X and unknown standard deviation σ_X. Hence from Theorem 11.1 we have:

$$L = \overline{X} - (t_{8,0.025})\frac{s_X}{\sqrt{n}} = 67 - (2.306)\frac{4}{3} = 63.9$$

$$U = \overline{X} + (t_{8,0.025})\frac{s_X}{\sqrt{n}} = 67 + (2.306)\frac{4}{3} = 70.1$$

The interpretation is that in a repeated sampling of size 9 from the above population, we should expect that 95% of the calculated confidence intervals will cover the mean and 5% of the intervals will not; i.e., we are 95% confident that μ_X is in (63.9, 70.1).

Example 11.2 One aspect of an undergraduate thesis, "Leisure Time of Undergraduate Students at a University," was concerned with interviewing a random sample of 25 B.A. students and asking them how many hours per week were spent in nonuniversity leisure activities. From the data it was found that $\overline{X} = 9.2$ hr and $s_X = 2.3$ hr. Find a 90% confidence interval for the mean of the population of B.A. students.

Again, we assume that the number of leisure hours is normally distributed with unknown mean μ_X and unknown standard deviation σ_X. From Theorem 11.1 we find:

$$L = 9.2 - (t_{24,0.05})\frac{2.3}{5} = 9.2 - 0.7870 = 8.413$$

$$U = 9.2 + (t_{24,0.05})\frac{2.3}{5} = 9.2 + 0.7870 = 9.987$$

where $t_{24,0.05} = 1.711$.

11.2 CONFIDENCE INTERVALS FOR THE DIFFERENCE $\mu_X - \mu_Y$ BASED ON THE t DISTRIBUTION

An extension of Theorem 11.1 to the two-population case is given in the next theorem.

THEOREM 11.2 *If* $X_1, X_2, \ldots, X_{n_1}$ *and* $Y_1, Y_2, \ldots, Y_{n_2}$ *are random samples from the two independent populations* $N(\mu_X, \sigma_X)$ *and* $N(\mu_Y, \sigma_Y)$, *respectively, with unknown means and unknown but equal standard deviations, then a* $\gamma = 1 - \alpha$ *confidence interval for the difference* $\mu_X - \mu_Y$ *between the population means is given by*

$$(L, U) = ((\overline{X} - \overline{Y}) - t_{n_1+n_2-2,\alpha/2}\, s_{\overline{X}-\overline{Y}},\ (\overline{X} - \overline{Y}) + t_{n_1+n_2-2,\alpha/2}\, s_{\overline{X}-\overline{Y}})$$

The reader should recognize from Chapter 9 that this result can be derived by noting the fact that

$$t_{n_1+n_2-2} = \frac{(\overline{X} - \overline{Y}) - (\mu_X - \mu_Y)}{s_{\overline{X}-\overline{Y}}}$$

and going through the same procedure as in Theorem 11.1. Also, the pooled standard deviation $s_{\overline{X}-\overline{Y}}$ has already been introduced in that same chapter, and its calculation formula is explicitly equal to:

$$s_{\overline{X}-\overline{Y}} = s\sqrt{\frac{1}{n_1} + \frac{1}{n_2}}$$

where

$$s = \sqrt{\frac{[\Sigma X_i^2 - (1/n_1)(\Sigma X_i)^2] + [\Sigma Y_i^2 - (1/n_2)(\Sigma Y_i)^2]}{n_1 + n_2 - 2}}$$

or

$$s = \sqrt{\frac{(n_1 - 1)s_X^2 + (n_2 - 1)s_Y^2}{n_1 + n_2 - 2}}$$

Example 11.3 Two groups of schizophrenic patients, 9 paranoid and 9 nonparanoid, were tested for "auditory thresholds" (a subject's ability to discriminate speech sounds). The results were as follows (data slightly abridged from Bull and Venables, 1974):

Paranoid	Nonparanoid
$\overline{X} = 60.2$	$\overline{Y} = 64.7$
$s_X^2 = 237$	$s_Y^2 = 256$
$n_1 = 9$	$n_2 = 9$

Assuming that these are random samples from two independent normal populations with unknown means and equal standard deviations, we find the following 99% confidence interval for the difference $\mu_X - \mu_Y$:

$$\overline{X} - \overline{Y} = 60.2 - 64.7 = -4.5$$

$$s = \sqrt{\frac{(8)(237) + (8)(256)}{9 + 9 - 2}} = 15.69$$

$$s_{\overline{X}-\overline{Y}} = (15.69)\sqrt{\frac{1}{9} + \frac{1}{9}} = 7.401$$

$$t_{16,0.005} = 2.921$$

so that:

$$L = -4.5 - (2.921)(7.401) = -26.12$$

$$U = -4.5 + (2.921)(7.401) = 17.12$$

Note that this confidence interval estimate contains 0 in it, since it ranges from a negative number to a positive number. This means essentially that the two means can be equal (in the probabilistic sense) which, as will be seen later, is related to the topic of hypothesis testing.

11.3 CONFIDENCE INTERVALS FOR σ_X AND σ_X^2 BASED ON THE χ^2 DISTRIBUTION

We now derive confidence intervals for the other parameter of the normal distribution. In Chapter 9, we discussed the χ^2 distribution and we showed how this distribution can be used to make probability statements regarding the sample variance s_X^2, when a random sample $X_1, X_2, \ldots, X_n$ is selected from a normal population. Assume that the standard deviation σ_X of the normal population is unknown, and our problem is to find a confidence interval estimator for this parameter.

Recall that the random variable

$$\chi_{n-1}^2 = \frac{(n - 1)s_X^2}{\sigma_X^2}$$

has the chi-square distribution with n - 1 degrees of freedom, whatever the value of σ_X may be. Now consider the following algebraic manipulations:

$$P[\chi_{n-1,1-\alpha/2}^2 < \chi_{n-1}^2 < \chi_{n-1,\alpha/2}^2] = 1 - \alpha$$

$$P\left[\chi_{n-1,1-\alpha/2}^2 < \frac{(n - 1)s_X^2}{\sigma_X^2} < \chi_{n-1,\alpha/2}^2\right] = 1 - \alpha$$

$$P\left[\frac{\chi_{n-1,1-\alpha/2}^2}{(n - 1)s_X^2} < \frac{1}{\sigma_X^2} < \frac{\chi_{n-1,\alpha/2}^2}{(n - 1)s_X^2}\right] = 1 - \alpha$$

$$P\left[\frac{(n - 1)s_X^2}{\chi_{n-1,\alpha/2}^2} < \sigma_X^2 < \frac{(n - 1)s_X^2}{\chi_{n-1,1-\alpha/2}^2}\right] = 1 - \alpha$$

$$P\left[\sqrt{\frac{(n-1)s_X^2}{\chi^2_{n-1,\alpha/2}}} < \sigma_X < \sqrt{\frac{(n-1)s_X^2}{\chi^2_{n-1,1-\alpha/2}}}\right] = 1 - \alpha$$

The notation $\chi^2_{n-1,\alpha/2}$ means the value of χ^2_{n-1} in the table such that the area to the right of it under the chi-square curve is equal to $\alpha/2$. There is a similar interpretation for $\chi^2_{n-1,1-\alpha/2}$. Graphically the first probability statement looks like the curve shown in Figure 11.2. This work leads us to the following theorem:

THEOREM 11.3 *If* $X_1, X_2, \ldots, X_n$ *is a random sample from a normal population with unknown mean* μ_X *and unknown standard deviation* σ_X, *then a* $\gamma = 1 - \alpha$ *confidence interval estimator of the standard deviation* σ_X *is equal to*

$$(L, U) = \left(\sqrt{\frac{(n-1)s_X^2}{\chi^2_{n-1,\alpha/2}}},\ \sqrt{\frac{(n-1)s_X^2}{\chi^2_{n-1,1-\alpha/2}}}\right)$$

Note that in the expression for (L, U), if the square root sign is left out, then we have a $\gamma = 1 - \alpha$ confidence interval for the variance σ_X^2.

Example 11.4 From the data in Example 11.3, find a 90% confidence interval estimate for the standard deviation σ_X of the first group. The solution is as follows:

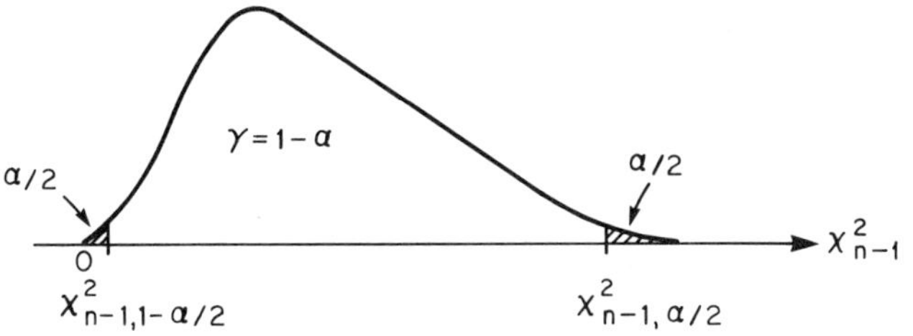

Figure 11.2 Tail probabilities for the χ^2 variable.

$s_X^2 = 60.2$

$n - 1 = 8$

$\chi^2_{8,0.05} = 15.51 \qquad \chi^2_{8,0.95} = 2.73$

Hence:

$$L = \sqrt{\frac{(8)(60.2)}{15.51}} = 5.6$$

$$U = \sqrt{\frac{(8)(60.2)}{2.73}} = 13.3$$

This means that in repeated sampling of size 9 from the population, we can expect 90% of the intervals calculated in this way to cover the standard deviation σ_X. Therefore we are 90% confident that the interval (5.6, 13.3) covers σ_X.

11.4 CONFIDENCE INTERVAL FOR THE RATIO σ_X/σ_Y OR σ_X^2/σ_Y^2 BASED ON THE F DISTRIBUTION

We are now ready to utilize one more jewel from Chapter 9, namely, the F distribution. The setup calls for sampling from two normal populations, $N(\mu_X, \sigma_X)$ and $N(\mu_Y, \sigma_Y)$, with unknown means and standard deviations. The problem is to provide a confidence interval estimator for the ratio σ_X/σ_Y.

The random variable that solves this problem is:

$$F_{n_1-1,n_2-1} = \frac{s_X^2/s_Y^2}{\sigma_X^2/\sigma_Y^2}$$

Now consider the following manipulations:

$$P[F_{n_1-1,n_2-1,1-\alpha/2} < F_{n_1-1,n_2-1} < F_{n_1-1,n_2-1,\alpha/2}] = 1 - \alpha$$

$$P\left[F_{n_1-1,n_2-1,1-\alpha/2} < \frac{s_X^2/s_Y^2}{\sigma_X^2/\sigma_Y^2} < F_{n_1-1,n_2-1,\alpha/2}\right] = 1 - \alpha$$

$$P\left[\frac{F_{n_1-1,n_2-1,1-\alpha/2}}{s_X^2/s_Y^2} < \frac{1}{\sigma_X^2/\sigma_Y^2} < \frac{F_{n_1-1,n_2-1,\alpha/2}}{s_X^2/s_Y^2}\right] = 1 - \alpha$$

$$P\left[\frac{s_X^2/s_Y^2}{F_{n_1-1,n_2-1,\alpha/2}} < \frac{\sigma_X^2}{\sigma_Y^2} < \frac{s_X^2/s_Y^2}{F_{n_1-1,n_2-1,1-\alpha/2}}\right] = 1 - \alpha$$

$$P\left[\frac{s_X/s_Y}{\sqrt{F_{n_1-1,n_2-1,\alpha/2}}} < \frac{\sigma_X}{\sigma_Y} < \frac{s_X/s_Y}{\sqrt{F_{n_1-1,n_2-1,1-\alpha/2}}}\right] = 1 - \alpha$$

Recall that the notation $F_{n_1-1,n_2-1,\alpha/2}$ is the value in the F table such that the area to its right under the F distribution with $\nu_1 = n_1 - 1$ and $\nu_2 = n_2 - 1$ degrees of freedom is equal to $\alpha/2$. The number $F_{n_1-1,n_2-1,1-\alpha/2}$ has a similar interpretation. A clarifying picture is given in Figure 11.3.

The derivation above leads us to the following results:

THEOREM 11.4 *If* $X_1, X_2, \ldots, X_{n_1}$ *and* $Y_1, Y_2, \ldots, Y_{n_2}$ *are random samples obtained from two independent normal populations* $N(\mu_X, \sigma_X)$ *and* $N(\mu_Y, \sigma_Y)$, *respectively, such that the means and standard deviations are unknown, then a* $\gamma = 1 - \alpha$ *confidence interval estimator for the ratio* σ_X/σ_Y *is given by:*

$$(L, U) = \left(\frac{s_X/s_Y}{\sqrt{F_{n_1-1,n_2-1,\alpha/2}}}, \frac{s_X/s_Y}{\sqrt{F_{n_1-1,n_2-1,1-\alpha/2}}}\right)$$

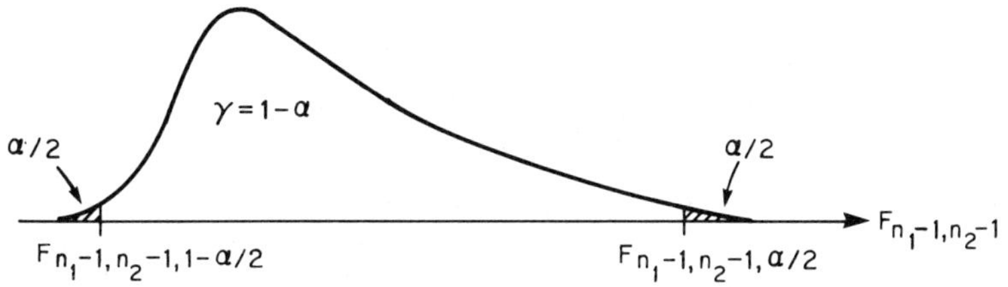

Figure 11.3 Tail probabilities for the F variable.

Since the F table in the Appendix has only 5% and 1% percentage points, we can construct only 90% and 98% confidence intervals for the ratio σ_X/σ_Y. For other confidence coefficients the reader is advised to use extended F tables.

Example 11.5 Again, from the data in Example 11.3, find a 98% confidence interval for the ratio σ_X/σ_Y. The necessary items to do this are:

$$\frac{s_X}{s_Y} = \frac{\sqrt{256}}{\sqrt{237}} = 1.04$$

$$\sqrt{F_{8,8,0.01}} = \sqrt{6.03} = 2.46$$

and

$$\sqrt{F_{8,8,0.99}} = \sqrt{\frac{1}{F_{8,8,0.01}}} = \sqrt{\frac{1}{6.03}} = 0.41$$

Hence from Theorem 11.4 we have:

$$L = \frac{1.04}{2.46} = 0.42$$

$$U = \frac{1.04}{0.41} = 2.54$$

A 98% confidence interval estimate for the ratio of the two variances is obtained by squaring these two limits. As an exercise for the reader, find 98% confidence limits for σ_Y/σ_X.

11.5 SAMPLE SIZE DETERMINATION

In our presentation so far, we have assumed that the sample size n is given in advance. We have learned that the larger the sample size, the higher the accuracy of an estimator of any arbitrary parameter θ. However, we see that the more observations in the sample, the costlier the investigation will be. Hence sample size depends on the degree of accuracy one is willing to attain and on the dollar amount of the budget. Assume that cost is of no consideration and that we are dealing strictly with an unbiased estimator of θ. This means that the sample size will depend on the standard error of $\hat{\theta}$. In this approach we need the following definition:

DEFINITION 11.1 *The coefficient of variation* $CV(\hat{\theta})$ *of an unbiased estimator* $\hat{\theta}$ *of a parameter* θ *is equal to the standard error of the estimator divided by the parameter; i.e.,* $CV(\hat{\theta}) = \sigma_{\hat{\theta}}/\theta$.

The CV is often expressed as a percentage by multiplying the ratio $\sigma_{\hat{\theta}}/\theta$ by 100%. Note that the CV measures the standard error or precision of the estimator in terms of θ. For example, the CV of the sample mean $\overline{X}$ as a percentage is equal to

$$\frac{\sigma_{\overline{X}}}{\mu_X} \cdot 100\% = \frac{\sigma_X/\sqrt{n}}{\mu_X} \cdot 100\%$$

If we specify a CV of k% for the sample mean $\overline{X}$, then by elementary algebra the following steps should be clear:

$$\frac{\sigma_{\overline{X}}}{\mu_X} \cdot 100 = k$$

$$\frac{\sigma_{\overline{X}}}{\mu_X} = \frac{k}{100}$$

$$\frac{\sigma_X/\sqrt{n}}{\mu_X} = \frac{k}{100}$$

$$\frac{1}{\sqrt{n}} = \frac{k/100}{\sigma_X/\mu_X}$$

$$\sqrt{n} = \frac{\sigma_X/\mu_X}{k/100}$$

$$n = \left(\frac{\sigma_X/\mu_X}{k/100}\right)^2$$

Since μ_X and σ_X are unknown parameters, the application of the formula requires advance estimates of σ_X and μ_X or of the CV(X). Note that the sample size above is the square of the ratio of the CV(X) and the specified CV(X). It is known as the *sample size for fixed* CV(X). Let us state it as a theorem.

THEOREM 11.5 *If the coefficient of variation of the sample mean for estimating the population mean is specified to be* k%, *then the required sample size is given by*

$$n = \left(\frac{\sigma_X/\mu_X}{k/100}\right)^2$$

As you may have realized, σ_X/μ_X is usually unknown. What one does in practice is to use information from previous studies, or a pilot study is carried out to estimate σ_X and μ_X. The estimates are substituted into the formula to give us the sample size. The better the estimate of σ_X/μ_X, the better the determination of the sample size.

Example 11.6 Suppose that the objective of a sample survey is to estimate the mean amount of hard liquor consumed per week by blue-collar workers in a city. This is to be done with a relative precision of 10%. If from a previous study the mean was found to be equal to 1.24 liters and the standard deviation equal to 0.8 liters, then find the required sample size for the forthcoming survey.

Using Theorem 11.5 we obtain:

$$n = \left(\frac{0.8/1.24}{10/100}\right)^2 = \left(\frac{0.65}{0.10}\right)^2 = 42.25$$

Hence the sample size is 43 (to be on the safe side) for estimating the mean amount of liquor per week with a CV of 10%.

Now the reader may question why no confidence interval approach was used to determine the sample size. After all, this chapter is basically dealing with confidence interval estimation. Well, we will not disappoint the reader, and we now proceed with an approach known as the *sample size for fixed confidence tolerance*.

In many investigations it is desirable to specify within how many units one wishes to estimate a parameter θ by the estimator $\hat{\theta}$, i.e., a *specification of the error of estimate* or *tolerance*. This leads us to the following definition.

DEFINITION 11.2 *The error of estimate of an estimator* $\hat{\theta}$ *of a parameter* θ *is equal to the absolute difference between the estimator and the parameter; i.e., error of estimate* = $|\hat{\theta} - \theta|$.

Specifying the error of estimate by k units, also called the *error bound*, means setting the tolerance at k units, which implies that one wishes to be within k units of the parameter. Mathematically this is written as:

$$|\hat{\theta} - \theta| \le k$$

For example, if a problem requires us to estimate the population mean by the sample mean, and if the tolerance is k units, then what we are saying is that:

$$|\overline{X} - \mu_X| \le k$$

Of course, in repeated sampling it is desirable that the calculated estimates remain within k units of θ with a certain probability, say γ. This probability is referred to as the *confidence level*, and we write:

$$P[|\hat{\theta} - \theta| \le k] = \gamma$$

In the example of the sample mean, this becomes:

$$P[|\overline{X} - \mu_X| \le k] = \gamma$$

Let us now do a little more elementary algebra. Assuming that the random sample $X_1, X_2, \ldots, X_n$ comes from a normal population with mean μ_X and standard deviation σ_X, and if the problem is to estimate μ_X within k units by $\overline{X}$ with a confidence level of $\gamma \cdot 100\% = (1 - \alpha) \cdot 100\%$, say, then the steps of the argument are:

$$P[|\overline{X} - \mu_X| \le k] = 1 - \alpha$$

$$P\left[\left|\frac{\overline{X} - \mu_X}{\sigma_{\overline{X}}}\right| \le \frac{k}{\sigma_{\overline{X}}}\right] = 1 - \alpha$$

$$P\left[|Z| \le \frac{k}{\sigma_{\overline{X}}}\right] = 1 - \alpha$$

$$P\left[\frac{-k}{\sigma_{\overline{X}}} < Z < \frac{k}{\sigma_{\overline{X}}}\right] = 1 - \alpha$$

$$P\left[Z > \frac{k}{\sigma_{\overline{X}}} \text{ or } Z < \frac{-k}{\sigma_{\overline{X}}}\right] = \alpha$$

$$2P\left[Z > \frac{k}{\sigma_{\overline{X}}}\right] = \alpha$$

$$P\left[Z > \frac{k}{\sigma_{\overline{X}}}\right] = \frac{\alpha}{2}$$

As before, $Z_{\alpha/2}$ is the value in the Z table such that the area to its right under the Z curve is equal to $\alpha/2$. Thus,

$$P[Z > Z_{\alpha/2}] = \frac{\alpha}{2}$$

i.e.,

$$\frac{k}{\sigma_{\overline{X}}} = Z_{\alpha/2}$$

Hence:

$$\frac{k}{\sigma_X/\sqrt{n}} = Z_{\alpha/2}$$

i.e.,

$$\sqrt{n} = \frac{Z_{\alpha/2}\sigma_X}{k}$$

i.e.,

$$n = \left(\frac{Z_{\alpha/2}\sigma_X}{k}\right)^2$$

We have thus obtained the following theorem:

THEOREM 11.6 *If* $X_1, X_2, \ldots, X_n$ *is a random sample from* $N(\mu_X, \sigma_X)$, *then the required sample size* n *to estimate the population mean* μ_X *with a tolerance of* k *units and a confidence level of* $\gamma \cdot 100\% = (1 - \alpha) \cdot 100\%$ *is equal to*

$$n = \left(\frac{Z_{\alpha/2}\sigma_X}{k}\right)^2$$

Since σ_X is unknown, the application of Theorem 11.6 requires an advance estimate of it, either from previous studies or pilot studies. From our experience in confidence interval estimation, we know that when $\gamma = 95\%$, then $Z_{0.025} = 1.96$, so that:

$$n = \left(\frac{1.96\sigma_X}{k}\right)^2$$

Example 11.7 Suppose that in Example 11.6 we wish to estimate the mean consumption of hard liquor within 0.2 liters with a confidence of 95%. From Theorem 11.6 we obtain the following sample size (under the assumption of normality):

$$n = \left(\frac{1.96\sigma_X}{k}\right)^2 = \left(\frac{(1.96)(0.8)}{0.2}\right)^2 = (7.84)^2 = 61.47$$

i.e., a sample of size 62 would be required.

EXERCISES

11.1 A sample of 12 measurements from a normal population yielded a mean of 15 cm and a standard deviation of 2 cm. Construct a 95% confidence interval for the mean of the population.

11.2 In your own area of interest, draw a random sample of measurements from a population and estimate its mean by a 99% confidence interval. (Do not forget to state the assumptions.)

11.3 A standard intelligence test is administered to 15 randomly selected grade 6 students, and the results are: 114, 106, 118, 130, 99, 104, 102, 89, 113, 95, 113, 100, 112, 86, 90. State the necessary assumptions and find a 99% confidence interval for the mean. Give a sampling interpretation. necessary assumptions and find a 99% confidence interval for the mean. Give a sampling interpretation.

11.4 Find a 90% confidence interval for the standard deviation in Exercise 11.1 and give a repeated sampling interpretation.

11.5 Find a 90% confidence interval for the standard deviation from the data in Exercise 11.3.

11.6 Two groups of subjects, called "normals" and "schizophrenics," were observed and scored on the Ammons Full-Range Picture Vocabulary Test (Ecker *et al.*, 1973):

Normals	141	109	58	83	114	150	63	72	120	73
Schizophrenics	28	12	10	34	22	36	19			

State the necessary assumptions and:

(a) Find a 95% confidence interval for the difference of the means of the underlying populations.

(b) Find a 90% confidence interval for the ratio of the standard deviations.

(c) Give a repeated sampling interpretation to the confidence limits obtained in (a) and (b).

11.7 True or false:

(a) A confidence interval for μ_X based on the t distribution requires advance knowledge of the standard deviation σ_X.

(b) The higher the confidence coefficient of a confidence interval estimate for the variance σ_X^2 of a normal population, the smaller the resulting length of the confidence interval.

(c) The F distribution can be used in setting confidence intervals on the ratio of variances of two independent normal populations.

(d) In the confidence-tolerance approach toward calculation of the sample size for the estimation of the mean of a normal population, it follows that the higher the value of γ, the larger the sample size n.

(e) The coefficient of variation of s_X is equal to σ_X/μ_X.

11.8 Make true statements (not formulas!) about the following:

(a) The confidence coefficient when using the t distribution.

(b) The usefulness of the χ^2 distribution.

(c) The t distribution when the sample size is large in relation to confidence interval estimation.

(d) Sampling structure which leads to an F distribution in relation to confidence interval estimation.

(e) The sample size for fixed CV.

11.9 In a forthcoming survey on the smoking of marijuana among undergraduate students at a university, the problem arose of how many students to take in a sample in order to estimate the proportion of students who had smoked the herb at least once. It was decided to set the CV at 5%. From a previous study, the percentage of "at-least-once-smokers" was known to be equal to 30%. Find the required sample size by the CV method. (*Hint*: $\mu_X = p$, $\sigma_X = \sqrt{pq}$.)

11.10 Explore the scientific literature in your area of interest and report one study where confidence interval estimation was used. You must be specific.

11.11 In Exercise 11.6, if the confidence limits for $\mu_X - \mu_Y$ contain 0, then we have some evidence to suggest that $\mu_X = \mu_Y$, at least at the 95% confidence level. If the interval is to the right of zero, then $\mu_X > \mu_Y$ with 0.95 confidence, and if the interval is to the left of zero then $\mu_X < \mu_Y$, with confidence 0.95. On the average, do the data in Exercise 11.6 suggest that $\mu_X > \mu_Y$?

Note: The following questions utilize the argument in Exercise 11.11.

11.12 A standard variety of wheat yields, on the average, 30 bushels per acre in a certain region. A new improved variety is planted on 9 randomly selected 1-acre plots. The average yield from the new variety is 33.4 bushels. If the sample standard deviation $s_X = 5.1$ bushels, what conclusion can be drawn by using $\alpha = 0.05$?

11.13 Samples of two types of electric light bulbs were tested for length of life, and the following data were recorded:

Type 1	Type 2
$n_1 = 5$	$n_2 = 7$
$\bar{X}_1 = 1224$ hr	$\bar{X}_2 = 1036$ hr
$s_1 = 36$ hr	$s_2 = 40$ hr

Is the difference in means sufficient to warrant the conclusion that Type 1 is superior to Type 2?

11.14 Two distinct groups of rats, X (normal) and Y (adrenalectomized), were tested for blood viscosity. The sample sizes and the means and standard deviations of the viscosity readings were:

	n	Mean	SD
X	11	3.921	0.527
Y	9	4.111	0.601

Find the 90% confidence limits for σ_X^2/σ_Y^2 and $\mu_X - \mu_Y$. What do these limits suggest?

11.15 Two populations are said to be homoscedastic if their variances are equal; i.e., if $\sigma_X^2/\sigma_Y^2 = 1$. The densities of sulfuric acid in two containers were measured, four determinations being made on one and six on the other. The results were:

X	1.842	1.846	1.843	1.843		
Y	1.848	1.843	1.846	1.847	1.847	1.845

By finding the 98% confidence limits for σ_X^2/σ_Y^2, does this suggest that the two containers are homoscedastic?

11.16 A company is trying to decide which of two types of tires to buy for their trucks. They would like to adopt Brand X, unless there is some evidence that Brand Y is better. An experiment is conducted in which 15 tires from each brand are used. The tires are run under similar conditions until they wear out. The results are:

Brand X: $\overline{X} = 26{,}000$ miles $s_X = 4200$ miles

Brand Y: $\overline{Y} = 25{,}000$ miles $s_Y = 2800$ miles

What conclusions can be drawn? (Use $\alpha = 0.05$ and state all necessary assumptions.)

11.17 Two methods of teaching statistics are being tried by a professor. A class of 14 students is taught by method 1 and a class of 15 by method 2. The two classes are given the same final examination. These scores yield $\overline{X}_1 = 67$, $\overline{X}_2 = 63$, $s_1 = 10$, $s_2 = 8$. Using $\alpha = 0.05$, can we conclude that the mean scores μ_X and μ_Y of the two methods are different?

11.18 Milicer (1968) has reported that in a sample of 332 girls, from a poor rural area of Poland, the mean menarcheal age was 14.40 years with a standard deviation of 1.44 years. From an urban area of Poland, a sample of 464 girls had a mean of 12.87 years with a standard deviation of 1.24 years. She claims that this difference is not "significant." After stating the necessary assumptions, verify this claim.

12

HYPOTHESIS TESTING AND TESTS OF HYPOTHESES BASED ON THE Z DISTRIBUTION

In Chapters 10 and 11 we dealt with the problem of providing point and confidence interval estimators of parameters. This chapter is concerned with the development of tests of hypotheses. Estimation of parameters and testing hypotheses are the two most important aspects of statistical inference.

Let us again state the framework in which we will be approaching the hypothesis testing problem. The populations we will be dealing with are infinite populations associated with a probability distribution, e.g., the Bernoulli population, the binomial population, and the normal population. A hypothesis is basically a statement about a parameter of a population, and by testing we mean the use of a procedure or decision rule which on the basis of the sample results leads to either its support or its rejection.

There is a certain amount of technical terminology associated with hypothesis testing; let us master it first.

12.1 CONCEPTS OF HYPOTHESIS TESTING

DEFINITION 12.1 *A null hypothesis* H_0 *is a statement about a parameter of a population, which can be tested on the basis of a sample.*

This definition concerns a hypothesis about a specific parameter. Often this is referred to as a *parametric* hypothesis. If the null hypothesis states, for example, that a distribution has a specific structure, like normality, in which no parameter is explicitly involved, then we speak of a *nonparametric* hypothesis.

Definition 12.1 also implies that null hypotheses are testable hypotheses. Examples of such hypotheses are:

(a) The percentage of voters in Canada approving the Prime Minister's performance today is 40%; i.e., H_0: $p = 0.4$.

(b) The mean income of parents of undergraduate students at the 15 Ontario universities is equal to \$20,000; i.e., H_0: $\mu_X = \$20{,}000$.

(c) There is no difference between the mean IQs of male and female undergraduate students in a given province; i.e., H_0: $\mu_X = \mu_Y$ or H_0: $\mu_X - \mu_Y = 0$.

(d) The standard deviation of heights of females in a given ethnic group in a certain country is equal to 10 cm; i.e., H_0: $\sigma_X = 10$ cm.

(e) The distribution of IQs for male and female university students is homoscedastic (equally variable); i.e., H_0: $\sigma_X^2 = \sigma_Y^2$ (or H_0: $\sigma_X^2/\sigma_Y^2 = 1$).

In each of these five examples a sample is taken from the underlying population (which is assumed to be infinite and has a probability distribution) which will contain evidence for or against the null hypothesis. In the first example, we could calculate the sample mean, i.e., the sample percentage of voters, and if a "large" discrepancy shows up between this and 0.40, then the decision should be to *reject* the null hypothesis. If the discrepancy between the sample percentage and 0.40 is not "large" enough, then we should not reject, i.e., we should retain the null hypothesis. The question to be resolved is, what is a large discrepancy?

Undoubtedly the reader can find, for the other examples, the relevant statistics which will lead to discrepancies relative to the stated null hypothesis, and again the question will arise about what a large discrepancy is. This is settled in the development below.

DEFINITION 12.2 *An alternative hypothesis* H_A *(or* H_1*) is a statement about a parameter of a population, which is an alternative to a stated null hypothesis, and which competes against it.*

An alternative hypothesis is stated to counter or to compete against a null hypothesis. For the five examples of null hypotheses above we can have, for example, the following alternatives:

(a') The percentage of voters in Canada approving the Prime Minister's performance today is, say 30%; i.e., H_A: $\mu_X = p = 0.3$.

(b') The mean income of parents of undergraduate students at the 15 Ontario universities is more than \$20,000; i.e., H_A: $\mu_X > \$20,000$.

(c') There is a difference between the mean IQs of male and female undergraduate students in a given province; i.e., H_A: $\mu_X \neq \mu_Y$ or H_A: $\mu_X - \mu_Y \neq 0$.

(d') The standard deviation of heights of females of a given ethnic group in a certain country is greater than 10 cm; i.e., H_A: $\sigma_X > 10$ cm.

(e') The distribution of IQs for male and female university students is not homoscedastic; i.e., H_A: $\sigma_X^2 \neq \sigma_Y^2$ (or H_A: $\sigma_X^2/\sigma_Y^2 \neq 1$).

It follows that when a null hypothesis is rejected, the evidence in the sample is in favor of the alternative hypothesis, so that it must be accepted. Likewise, when we accept a null hypothesis, it implies that we reject the alternative to it. Often the roles of a null hypothesis and its alternative can be interchanged.

DEFINITION 12.3 *A hypothesis (either null or alternative) is called simple if it completely specifies the probability distribution underlying the population, and if this is not the case, then it is called composite.*

In the examples stated above, note that both (a) and (a') are simple, because the probability distribution for a Bernoulli population is completely specified once p is stated. If we make the assumption that income is normally distributed with unknown mean μ_X and unknown standard deviation σ_X, then both (b) and (b') are composite, because neither completely specifies the distribution.

If σ_X is known, then (b) is simple, but (b') is composite. The reader is invited to comment regarding (c) and (c'), (d) and (d'), and (e) and (e').

In Figure 12.1 we sketch the distributions for (a) and (a') and (b) and (b'). The pictures for (a) and (a') are self-explanatory. For (b) there are an infinite number of normal distributions, which are all centered at \$20,000 but have different standard deviations. For (b') there are also an infinite number of normal distributions, centered at means which are greater than \$20,000 and having different standard deviations.

To keep the development of this topic as simple as possible, we consider a normal population of IQ scores for grade 7 students with

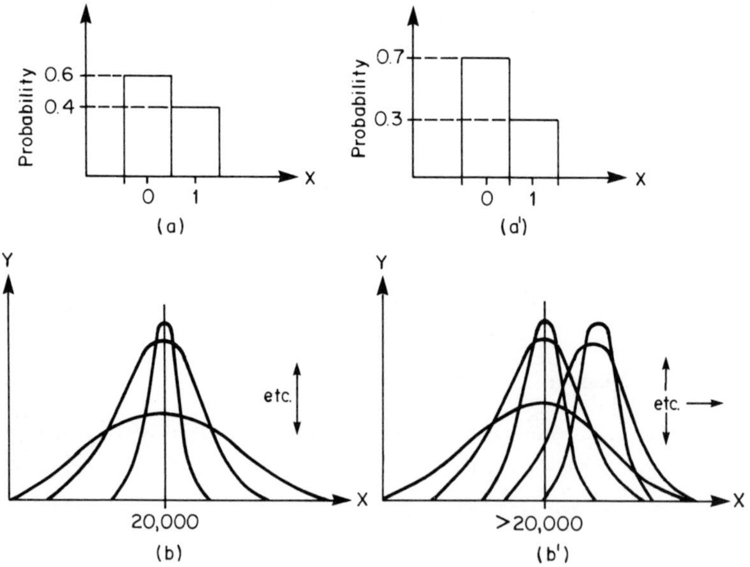

Figure 12.1 Probability distributions for null and alternative hypotheses (a), (a') and (b), (b').

known standard deviation $\sigma_X = 10$, and consider the following simple null and alternative hypotheses:

$$H_0: \quad \mu_X = 115$$

$$H_A: \quad \mu_X = 120$$

The normal distributions specified by these hypotheses are N(115, 10) and N(120, 10). Since the evidence for making a decision regarding the null hypothesis is contained in the statistic $\overline{X}$, we have to worry about its behavior in repeated sampling, i.e., in the distribution of $\overline{X}$. But from H_0 and H_A we immediately have the following:

$$H_0 \longrightarrow \overline{X} \text{ is } N\left(115, \frac{10}{\sqrt{n}}\right)$$

$$H_A \longrightarrow \overline{X} \text{ is } N\left(120, \frac{10}{\sqrt{n}}\right)$$

Pictorially the situation is presented in Figure 12.2.

Recall that for rejection of H_0 we need a "large" discrepancy between $\overline{X}$ and 115. Suppose that we use the following decision rule: Reject H_0 if $\overline{X}$ exceeds 115 by more than 1.64 standard deviations, i.e., by $1.64\sigma_{\overline{X}}$, and accept H_0 if the difference is less than or equal to $1.64\sigma_{\overline{X}}$. Since $\sigma_{\overline{X}} = 10/\sqrt{n}$ depends on the sample size n, suppose that n = 32, so that H_0 will be rejected if $\overline{X}$ differs from

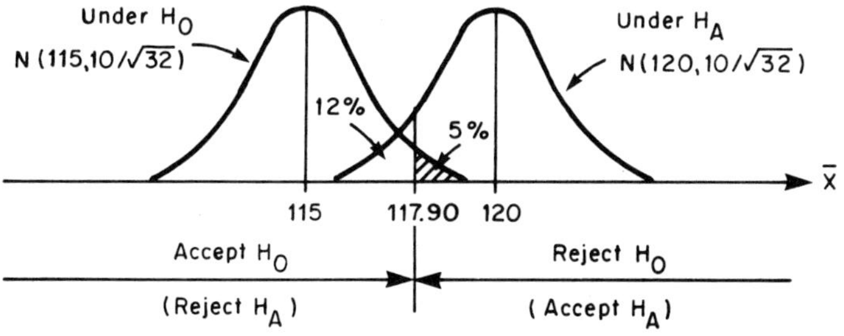

Figure 12.2 Simple null hypothesis against simple alternative hypothesis.

115 by more than $1.64(10/\sqrt{32}) = 2.90$. This means that in repeated sampling all those sample means which are greater than 117.90 will lead to rejection of H_0. The percentage of such means is found by calculating the probability

$$P[\overline{X} > 117.90] = P[\overline{X} > 115 + (1.64)(\sigma_{\overline{X}})]$$
$$= 0.05$$

from our knowledge of the normal curve. Hence, 5% of the sample means in repeated sampling will lead to the rejection of H_0 (and therefore acceptance of H_A that the mean is 120) and 95% of them will lead to acceptance of H_0 (and hence rejection of H_A). What we are saying, in fact, is that in repeated sampling there is a 5% chance of rejecting H_0 (and hence a 95% chance of accepting it), when H_0 is true, if the stated decision rule is used.

In this reasoning we can make two errors. The *first type of error* is committed if those 5% of the means which were said to belong to $N(120, 10/\sqrt{32})$ in fact may have come from $N(115, 10/\sqrt{32})$. Hence there is a 5% chance of committing this first type of error, i.e., of rejecting H_0 when in fact it is true. This is the area to the right of 117.90 for the normal curve with mean 115. The *second type of error* is committed when a sample mean is said to belong to $N(115, 10/\sqrt{32})$, when indeed it might have come from $N(120, 10/\sqrt{32})$. The chance of committing this second type of error, i.e., of accepting H_0 when in fact it is false, is given by the area to the left of 117.90 under the curve with mean 120. We can calculate this probability in the following way:

$$P[\text{second type of error}] = P[\overline{X} < 117.90, \text{ when } H_A \text{ is true}]$$
$$= P\left[\frac{\overline{X} - 120}{10/\sqrt{32}} < \frac{117.90 - 120}{10/\sqrt{32}}\right]$$
$$= P[Z < -1.19] = 12\%$$

So far so good, but who dictates that we use the stated decision rule? Well, no one! For the rule used, notice that the rejection region, also called the *critical region* for rejecting H_0, consisted of all those sample means in $N(115, 10/\sqrt{32})$, which were greater than $115 + 1.64\sigma_{\overline{X}} = 115 + 1.64(10/\sqrt{32}) = 117.90$. This particular rule was based on the statistic $\overline{X}$ and it gave probabilities of the two types of errors of 5% and 12%, respectively. Note further that the probability of rejecting H_0 when it is false is equal to one minus the probability of the error of the second type, i.e., $1 - 0.12 = 0.88$. One says that the *power* of the rule is equal to 0.88. Hence, note that power equals the probability of rejecting H_0 when in fact it is false. In general, one uses the notation α and β to indicate the probabilities of the errors of the first and second type, respectively. Thus power of a decision rule is equal to $P = 1 - \beta$. All these notions will be defined later in a more rigorous framework.

One may conceive of a situation where there are many competing decision rules, based on different statistics, whose distributions are known when the null hypothesis is true. Each of these rules will have corresponding errors of the two types and power. Schematically, we have:

Decision rule	Critical region under H_0	α	β	Power = $1 - \beta$
1	C_1	α_1	β_1	$P_1 = 1 - \beta_1$
2	C_2	α_2	β_2	$P_2 = 1 - \beta_2$
⋮	⋮	⋮	⋮	⋮
Our rule	$\overline{X} > 117.90$	5%	12%	88%
⋮	⋮	⋮	⋮	⋮

Ideally one should select that decision rule which minimizes the probabilities of both types of errors. But from Figure 12.2 one clearly sees that this is not possible. If we use our rule based on the sample mean $\overline{X}$ to make the probability α of the first type of error less than 5% (e.g., by picking a number larger than 1.64), there is a corresponding increase for the probability β of the error of the second type. So the two types of errors work against each other. What do we do? In practice, one sets the probability of the first type error at the fixed level for each rule. One then selects that rule which minimizes the probability β of the error of the second type, i.e., which maximizes the power.

In our example, we have set α at 5% by picking 1.64 in our rule. It can be shown (by mathematical means) that no other decision rule with fixed $\alpha = 5\%$ will have a larger power than 0.88, for this particular example. Hence ours is the best $\alpha = 5\%$ decision rule. In fact, for any other fixed α our rule is the best!

There are a couple of things remaining, which should be brought to the forefront.

1. Our decision rule relied on the test statistic $\overline{X}$. We could have equivalently stated it in terms of the Z variable in the following way: Reject H_0 if $Z = (\overline{X} - 115)/\sigma_{\overline{X}} > 1.64$. Thus, if in a particular sample of size 32 we find that $\overline{X} = 119$, then the corresponding $Z = (119 - 115)/(10/\sqrt{32}) = 4/1.77 = 2.26$ is greater than 1.64, so that H_0 is rejected.
2. Since a Z value of 1.64 corresponds to a probability of the error of the first type of 5%, it follows that any calculated Z value which is greater than 1.64 will lead to calculated probability of less than 5%. Thus if Z for a particular sample was 2.26 then the corresponding calculated probability is equal to: $P[Z > 2.26] \doteq 1\%$, which is much less than 5%, so that H_0 is rejected.

The reader should recognize that the decision rule can be applied for a particular sample either in terms of the calculated Z value or in terms of the calculated probability.

We are now ready to state formally the concepts used in the above development.

DEFINITION 12.4 *A test statistic is a statistic such that its probability distribution is known if the stated null hypothesis* H_0 *is true.*

In the example, H_0: $\mu_X = 115$ against H_A: $\mu_X = 120$ for IQ scores of grade 7 students, the test statistic was $\overline{X}$ and its distribution was $N(115, 10/\sqrt{32})$. This has been graphed in Figure 12.2. Equivalently, we could have used the test statistic $Z = (\overline{X} - 115)/(10/\sqrt{32})$, which has the standard normal distribution. For hypotheses concerning the standard deviation of a normal population the natural statistic is s_X, which can be transformed to a chi-square statistic, as we already know from Chapter 9.

DEFINITION 12.5 *A statistical test, or simply test, is a decision rule based on a test statistic, which leads to acceptance or rejection of a stated null hypothesis* H_0 *against a competing alternative* H_A.

Most statistical tests, or simply tests, are decision rules with the spirit of rejecting a null hypothesis if the test statistic shows a large discrepancy relative to the null hypothesis. In our example about IQ scores, the statistical test was based on $\overline{X}$ or Z and led to the decision rule: Reject H_0 if $Z > 1.64$ and accept H_0 otherwise.

DEFINITION 12.6 *The critical region of a statistical test is the set of all values of the underlying test statistic which leads to rejection of a stated null hypothesis against an alternative.*

In the example about the IQ scores, the critical region of the statistical test consisted of values of $\overline{X}$, which in repeated sampling led to values greater than 117.90. In terms of Z the critical region is given by those Z values which in repeated sampling are greater than 1.64.

The types of errors which can be committed when testing a null hypothesis against an alternative are known as the *Type I* and *Type II errors.* We have met these errors extensively in our example of the IQ scores. They are summarized and formally defined here:

		H_0 is	
		True	False
Conclusion from test	Reject H_0	Type I error	No error
	Accept H_0	No error	Type II error

DEFINITION 12.7 *The probability* α *of a Type I error of a test is the probability of rejecting a stated null hypothesis* H_0 *(i.e., accepting* H_A*) if in fact* H_0 *is true (i.e.,* H_A *is not true).*

The probability α of a Type I error is also known as the *significance level* of the test. In the example on IQ scores, the significance level α was equal to 5%.

DEFINITION 12.8 *The probability* β *of a Type II error of a test is the probability of accepting a stated null hypothesis* H_0 *(i.e., rejecting* H_A*), if in fact* H_0 *is false (i.e.,* H_A *is true).*

Again, in our example about the IQ scores, we calculated the probability of a Type II error and we found that it was equal to 12%; the *power* of the test was equal to 88%. The reader should observe that we can calculate for any given alternative value a corresponding power.

DEFINITION 12.9 *A test is said to be significant for a given sample if the corresponding calculated test statistic leads to rejection of a stated null hypothesis at a given significance level* α.

The term "significant" is derived from the fact that events with a "small" probability are called *significant events*. Since in practice one always chooses a small α (e.g., 5% or 1%), this means that when a particular sample leads to rejection of H_0, the corresponding calculated probability will be smaller than α, i.e., a significant event and hence a significant result. In our example, a particular sample of 36 scores with a mean of $\overline{X} = 119$ led to a probability of 1%, which was smaller than $\alpha = 5\%$, and hence the test was significant. A significant test thus leads to a large departure of the test statistic evaluated from a particular sample and the stated null hypothesis.

12.2 Z TESTS FOR THE MEAN μ_X

It is now time to derive tests of hypotheses using the Z distribution. Consider the following three types of hypotheses concerning the mean of a normal population, whose standard deviation σ_X is *known*:

(a) H_0: $\mu_X = \mu_0$

H_A: $\mu_X > \mu_0$ (i.e., $\mu_X = \mu_1$, such that $\mu_1 > \mu_0$)

(b) H_0: $\mu_X = \mu_0$

H_A: $\mu_X < \mu_0$ (i.e., $\mu_X = \mu_1$, such that $\mu_1 < \mu_0$)

(c) H_0: $\mu_X = \mu_0$

H_A: $\mu_X \neq \mu_0$ (i.e., $\mu_X = \mu_1$, such that $\mu_1 < \mu_0$ or $\mu_1 > \mu_0$)

The null hypothesis in each of these three cases is the same, but the alternatives are different. The first one has a *right-sided* alternative, the second one a *left-sided* alternative, and the last one a *two-sided* alternative. From that which has been developed earlier we are led to the following theorem:

THEOREM 12.1 *Statistical tests at the significance level* α *for the hypotheses* (a), (b), and (c) *above are, respectively:*

(a) *Reject* H_0 *if the sample* $Z = (\bar{X} - \mu_0)/\sigma_{\bar{X}} > Z_\alpha$, *where* $\sigma_{\bar{X}} = \sigma_X/\sqrt{n}$.

(b) *Reject* H_0 *if the sample* $Z = (\bar{X} - \mu_0)/\sigma_{\bar{X}} < -Z_\alpha$.

(c) *Reject* H_0 *if the sample* $|Z| = |(\bar{X} - \mu_0)/\sigma_{\bar{X}}| > Z_{\alpha/2}$, *i.e., if* $(\bar{X} - \mu_0)/\sigma_{\bar{X}} < -Z_{\alpha/2}$ *or* $> Z_{\alpha/2}$.

In each case H_0 *is otherwise accepted.*

For the given normal setting these tests are the most powerful, in the sense that no other test will have a higher power. These tests are depicted in Figure 12.3.

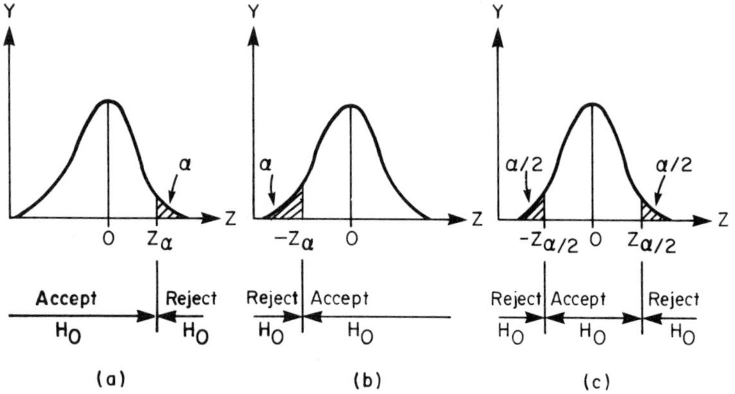

Figure 12.3 Z tests for hypotheses on the mean of a normal population with given standard deviation.

Tests (a) and (b) are known as *one-tail Z tests*, because the alternative hypotheses are one sided, so that rejection of H_0 takes place for extremal values in the right-hand or left-hand tail. Test (c) is a *two-tailed Z test* because the alternative is two sided, and hence rejection takes place for extremal values in both tails.

The power of each of these tests can be sketched by calculating the $1 - \beta$ probabilities for specific alternatives. They are sketched in Figure 12.4.

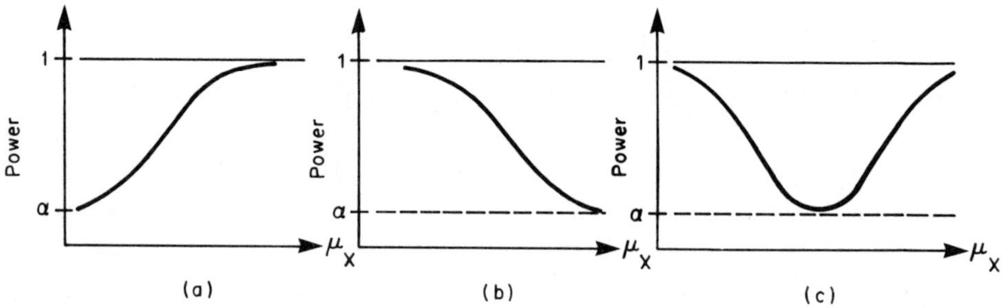

Figure 12.4 Power curves for the three tests in Theorem 12.1.

Example 12.1 In Section 12.1 we considered the normal population of IQ scores of grade 7 students, where we stated:

H_0: $\mu_X = 115$

H_A: $\mu_X = 120$

This is a special case of

H_0: $\mu_X = 115$

H_A: $\mu_X > 115$

So, if a random sample of n students gives a calculated $Z = (\overline{X} - 115)/(10/\sqrt{n}) > 1.64$, then we reject H_0 at the 5% significance level and otherwise we accept H_0. This has already been illustrated earlier, when a sample of 32 gave an $\overline{X}$ of 119, which led to $Z = (119 - 115)/(10/\sqrt{32}) = 2.26 > 1.64$, so that H_0 was rejected.

If we wish to test

H_0: $\mu_X = 115$

H_A: $\mu_X < 115$

then it follows that we reject H_0 if a sample produces a calculated $Z = (\overline{X} - 115)/(10/\sqrt{n}) < -1.64$ and we accept otherwise.

Finally, for testing:

H_0: $\mu_X = 115$

H_A: $\mu_X \neq 115$

Theorem 12.1 tells us to reject H_0 if a sample results in a calculated $Z = (\overline{X} - 115)/(10/\sqrt{n})$ which is less than -1.96 or greater than 1.96, i.e., the absolute value of the calculated $Z > 1.96$.

The calculation of the value of the test statistic Z in each of the three tests is accomplished by substituting the value of $\overline{X}$ found in the sample and the given size n of the sample.

It is interesting to note that the same tests are effective when, in (a) and (b), $\mu_X = \mu_0$ is replaced by $\mu_X \leq \mu_0$ and $\mu_X \geq \mu_0$, respectively.

12.3 Z TESTS FOR EQUALITY OF TWO MEANS

Let us now extend the Z tests from the one-population case to the *two-population case*. For this reason, consider two independent normal populations $N(\mu_X, \sigma_X)$ and $N(\mu_Y, \sigma_Y)$, where the standard deviations are *known* or *given*. We then have the following three types of hypotheses:

1. H_0: $\mu_X = \mu_Y$
   H_A: $\mu_X > \mu_Y$
2. H_0: $\mu_X = \mu_Y$
   H_A: $\mu_X < \mu_Y$
3. H_0: $\mu_X = \mu_Y$
   H_A: $\mu_X \neq \mu_Y$

These are tested by the following theorem.

THEOREM 12.2 *Statistical tests at the significance level* α *for the hypotheses* 1, 2, *and* 3 *are, respectively:*

1. *Reject* H_0 *if* $Z > Z_\alpha$
2. *Reject* H_0 *if* $Z < -Z_\alpha$
3. *Reject* H_0 *if* $|Z| > Z_{\alpha/2}$

The test statistic Z *in each case is defined by*

$$Z = \frac{\overline{X} - \overline{Y}}{\sigma_{\overline{X}-\overline{Y}}}$$

In test 3, $|Z| > Z_{\alpha/2}$ *means* $Z < -Z_{\alpha/2}$ *or* $Z > Z_{\alpha/2}$. *In each case* H_0 *is otherwise accepted.*

In these three tests the standard deviation of the difference $\overline{X} - \overline{Y}$ is obtained from the given standard deviations by the formula:

$$\sigma_{\overline{X}-\overline{Y}} = \sqrt{\frac{\sigma_X^2}{n_1} + \frac{\sigma_Y^2}{n_2}}$$

Example 12.2 Two middle-class neighborhoods in a large city were compared in a small poll with respect to income. From earlier studies the variances for income in the two neighborhoods were found to be \$9950 and \$7880. Samples of households yielded the following information:

Neighborhood 1	Neighborhood 2
$\overline{X}$ = \$25,312	$\overline{Y}$ = \$25,257
$n_1 = 12$	$n_2 = 13$

On the basis of these sample results can we conclude that the mean income of neighborhood 1 is the same as that of neighborhood 2?

We are interested in testing the hypothesis:

$$H_0: \quad \mu_X = \mu_Y \qquad \text{against} \qquad H_A: \quad \mu_X \neq \mu_Y$$

Now, from Theorem 12.2 we find the value of the test statistic to be

$$|Z| = \left|\frac{25312 - 25257}{\sqrt{9950/12 + 7880/13}}\right| = \left|\frac{55.00}{37.88}\right| = |1.45| = 1.45$$

Using a 5% significance level, we observe that 1.45 is less than $Z_{0.025} = 1.96$. Therefore we have insufficient evidence to reject H_0 at the 5% significance level.

12.4 LARGE SAMPLE Z TESTS

So far we have assumed that the sample came from a normal population with a known or given standard deviation. In many settings we do not know the underlying distribution, nor do we know the standard deviation. In such cases we appeal to the *central limit theorem* (see Theorem 10.4), which was so successfully used in confidence interval estimation. We know from this theorem that no matter what the distribution of a population is, as long as its mean and standard deviation are finite, the sample mean $\overline{X}$ will have approximately a normal distribution $N(\mu_X, \sigma_{\overline{X}} = \sigma_X/\sqrt{n})$ for *sufficiently large* n.

Hence, Theorems 12.1 and 12.2 are valid in large sample settings, where $\sigma_{\overline{X}}$ is estimated by $s_{\overline{X}} = s_X/\sqrt{n}$, and s_X is obtained from a large sample. Now we state the following result.

THEOREM 12.3 *Statistical tests as stated in Theorems 12.1 and 12.2 are valid for the stated hypotheses for large samples from any population with a distribution whose mean and standard deviation are finite.*

Example 12.3 Department stores in a large city claim that their average loss due to shoplifting by teenagers is equal to \$32,000 per year. A sociologist sampled 36 stores and found a sample mean of \$30,000 and a standard deviation of \$6000 per year. Is it likely at the 1% significance level that the population of stores from which the sample was taken has an average loss of \$32,000?

The null hypothesis is stated as H_0: $\mu_X = 32,000$, and for the alternative hypothesis we pick H_A: $\mu_X \neq 32,000$, since the question does not mention the direction of the alternative. Hence the test will be a two-tail Z test. Now,

$$|Z| = \left|\frac{(30,000 - 32,000)}{s/\sqrt{n}}\right| = \left|\frac{-2000}{6000/\sqrt{36}}\right| = 2.0$$

Testing at the 1% level, we see that the value of the test statistic is less than $Z_{0.005} = 2.58$. Hence there is no large discrepancy between the claim and the evidence in the sample.

Example 12.4 (Large sample test for no difference between binomial proportions.) On a television show a panelist stated that the proportion of females saying "yes" to the question "Do you think marriage will be outdated in the year 2000?" is equal to the proportion of males saying "yes." To verify the statement another panelist conducted a telephone poll of 625 "randomly" selected females and 900 "randomly" selected males, and obtained responses in the proportions 0.35 and 0.40, respectively. On the basis of these results, do we accept or reject the panelist's statement?

From Theorem 12.3 it follows that to test H_0: $\mu_X = \mu_Y$ (i.e., $p_X = p_Y$) against H_A: $\mu_X \neq \mu_Y$ (i.e., $p_X \neq p_Y$) we carry out a two-tailed Z test in the following way:

Let $\hat{\sigma}^2_{\overline{X}}$ symbolize "the estimator of $\sigma^2_{\overline{X}}$."

$$\hat{\sigma}^2_{\overline{X}} = s^2_{\overline{X}} = \frac{s^2_X}{n_1} = \frac{(\overline{X})(1 - \overline{X})}{n_1} = \frac{(0.35)(0.65)}{625}$$

$$\hat{\sigma}^2_{\overline{Y}} = s^2_{\overline{Y}} = \frac{s^2_Y}{n_2} = \frac{(\overline{Y})(1 - \overline{Y})}{n_2} = \frac{(0.40)(0.60)}{900}$$

$$\hat{\sigma}_{\overline{X}-\overline{Y}} = \sqrt{\frac{s^2_X}{n_1} + \frac{s^2_Y}{n_2}} = \sqrt{\frac{(0.35)(0.65)}{625} + \frac{(0.40)(0.60)}{900}}$$

$$|Z| = \left| \frac{0.35 - 0.40}{\sqrt{\frac{(0.35)(0.65)}{625} + \frac{(0.40)(0.60)}{900}}} \right| = |-1.99| = 1.99$$

Using a 10% significance level, the critical value in the Z table is equal to $Z_{0.05} = 1.64$. Hence we reject at the 10% level the panelist's statement that the response of both sexes is the same.

A repeated sampling interpretation of the conclusion is that when the same survey is repeated an infinite number of times then the percentage of absolute Z values, which will be greater than the obtained sample value of 1.99, is less than 10%.

Remark. There is an interesting relationship between two-tail hypothesis testing and confidence interval estimation. Rejection (acceptance) of a null hypothesis at the α level means that the corresponding 1 - α confidence interval does not contain (contains) the hypothesized value of the parameter. The reader may verify that in Example 12.3 the hypothesized value of \$32,000 is contained in the 99% confidence interval \$30,000 $\pm$ (2.58)$\left(\frac{2000}{6000/\sqrt{36}}\right)$. In Example 12.4 the hypothesized value is equal to 0, since $\mu_X = \mu_Y$

under the null hypothesis, and it will be seen that 0 is not contained in the 90% confidence interval

$$-0.05 \pm 1.64\sqrt{\frac{(0.35)(0.65)}{625} + \frac{(0.40)(0.60)}{900}}$$

EXERCISES

12.1 State a simple null hypothesis against a simple alternative for a binomial population mean, and sketch the corresponding distributions.

12.2 What sample statistic carries the evidence for or against the null hypothesis in Exercise 12.1?

12.3 State in words the difference between simple and composite hypotheses and illustrate this for a Bernoulli population.

12.4 Suppose that the distribution of reaction time of subjects to a stimulus is normal with a hypothesized mean of 2.30 sec and standard deviation equal to 0.41 sec. A sample of 16 subjects provides an average reaction time of 1.80 sec. Test the stated hypothesis at the 5% level using all three types of alternatives. Provide sketches of the Z distribution, showing the tabled Z values, the calculated Z values, and the significance level.

12.5 Sketch the power for the two-tailed Z test in Exercise 12.4 by calculating the power for several alternative values, and give an explanation of what power is in light of the example.

12.6 In Exercise 12.4 find the calculated tail probabilities corresponding to the Z values of the sample, and draw your conclusions by comparing these with the significance level α.

12.7 The mean weekly expenditure for food per household in a certain province is equal to \$58. A random sample of 64 households in a particular city gave a mean of \$54 and a standard deviation of \$9.80. Is the weekly expenditure for food per household for the city significantly (at $\alpha = 0.05$) lower than the mean for the province? What theorem did you use in answering this question?

12.8 A city executive on the rapid transit board wants to test the (null) hypothesis that 75% of its passengers like to have music piped into the train while they are traveling. What type of error is committed when this hypothesis is erroneously rejected? What type of error is committed when this hypothesis is erroneously accepted?

12.9 Suppose that a sociological testing service is asked to check whether the membership of a particular men's club is "chauvinistic." What type of error is committed if the null hypothesis that they are not chauvinistic is erroneously accepted? What type of error is committed if this hypothesis is erroneously rejected?

12.10 Suppose you work for a brewery in which the advertising department is concerned with the effectiveness of a song it has produced for a television commercial.

(a) What hypothesis are you testing if you commit a Type I error when you conclude erroneously that the song is effective?

(b) What hypothesis are you testing if you commit a Type II error when you conclude erroneously that the song is effective?

12.11 Set up your own data and illustrate Theorem 12.2 using a two-tail Z test. Make a sketch of the Z distribution showing $Z_{\alpha/2}$, α, and your calculated Z value from the two samples.

12.12 Set up a rejection region for testing the following null hypothesis against a one-sided alternative in a large sample setting: The mean age at which females marry is equal to 20 years.

12.13 Two hundred blue and 190 white collar workers were interviewed on the "yes-no" question: "Do you think that a university degree is essential for everyone?" Of the blue collar workers 96 said yes, and 120 yes answers were obtained from the white collar workers. Do the proportions of yes answers in the two underlying populations differ significantly at the 1% level?

12.14 Two groups of preschool children, who were matched for other factors, were given positive and negative "reinforcing" treatments. The scores from a "learning performance" test were:

Negative	Positive
(σ_X = 11.3)	(σ_Y = 13.2)
$\bar{X}$ = 87.4	$\bar{Y}$ = 91.6
n_1 = 49	n_2 = 54

Did the "positive" group score significantly higher (at the 5% level)? Give a repeated sampling interpretation of your conclusion.

12.15 True or false:

(a) The significance level is a test statistic.

(b) The probability of committing a Type II error is equal to 1 - power.

(c) The Z test can be applied to sampling from any population with finite mean and standard deviation as long as the sample size is large.

(d) If the confidence interval for a difference between two population means contains 0 in it, then we reject H_0: $\mu_X = \mu_Y$ against H_A: $\mu_X \neq \mu_Y$.

(e) A composite null hypothesis implies a whole family of underlying distributions.

(f) The null hypothesis specifies the value of the test statistic.

(g) If we reject a true null hypothesis, we make a Type I error.

(h) If the null hypothesis is true, the probability of making a Type II error equals zero.

(i) When we accept a true null hypothesis, we make a Type II error.

(j) If we reject H_0, we cannot make a Type II error.

(k) The probability of drawing extreme values of the sample mean becomes smaller as n increases, in the normal case.

(l) When the obtained test statistic falls within the critical region, we reject H_0.

(m) Employing $\alpha = 0.01$, we reject H_0. If we employed $\alpha = 0.05$, we would also reject H_0.

12.16 In Exercise 12.13 construct a 99% confidence interval for $p_1 - p_2$, and draw the proper conclusion regarding rejection or acceptance of the null hypothesis that $p_1 - p_2 = 0$.

12.17 In Exercise 12.13 test the null hypothesis that the percentage of white collar workers saying yes is equal to 50%. (This is an example of a large sample test for a binomial proportion.)

12.18 Make a true statement (not the definition) about each of the following concepts:

(a) Test statistic
(b) Statistical test
(c) Simple alternative hypothesis
(d) Critical region of a test
(e) Significance level
(f) One-tail Z test
(g) Large sample test
(h) Power of a test
(i) Type II error
(j) Z test for a difference between two means

12.19 A group of 120 freshmen at University A takes a certain standard test and obtains a mean score of 70 with a standard deviation of 14. A group of 80 freshmen at University B takes the same test and obtains a mean score of 75 with a standard deviation of 12. Test the hypothesis that the two groups are random samples from the same population (i.e., that the difference between the mean scores is not significant).

12.20 Suppose two machines, say A and B, are packaging "6-oz" cans of talcum powder, and that 100 cans filled by each machine are emptied and the contents carefully weighed. Suppose the following sample values are found:

Machine	Mean weight (oz)	Standard deviation
A	6.11	0.04
B	6.14	0.05

Are the means significantly different at the 1% level?

12.21 A large corporation wants to choose between two brands of light bulbs on the basis of average life. Brand 1 is slightly less expensive than Brand 2. The company would like to buy Brand 1, unless the average life for Brand 2 is shown to be significantly greater at the 0.05 level of significance. A sample of 100 bulbs from each brand is tested and it is found that $\overline{X}_1 = 985$ hr, $\overline{X}_2 = 1003$ hr, $s_1 = 80$ hr, $s_2 = 60$ hr. What conclusion should be drawn? (*Note*: State the necessary assumptions.)

12.22 In Exercise 12.21 suppose that both brands sell for the same price and that the company has no reason to prefer one over the other, and so show that the difference in sample means is not sufficient to demonstrate that Brand 2 is superior. (*Note*: Lacking any other information or considerations we would no doubt still choose Brand 2.)

12.23 A university investigation, conducted to determine whether car ownership was detrimental to academic achievement, was based upon 2 random samples of 100 students, each drawn from the same student body. The grade point average for the $n_1 = 100$ non-car-owners possessed a mean and variance equal to $\overline{X}_1 = 2.7$ and $s_1^2 = 0.36$, as opposed to an $\overline{X}_2 = 2.54$ and $s_2^2 = 0.40$ for the $n_2 = 100$ car owners. Do the data present sufficient evidence to indicate a difference in the mean achievement between car owners and non-car owners?

12.24 Two groups of mice, one of 50 and the other of 60, are comparable in respect to age, weight, general condition, etc., and both have been given injections of virus. The first group, however, has also been given a certain drug. After 3 days, the number of deaths in the first group was 12 and in the second 19. Is the difference in mortality rates significant?

12.25 Melamed *et al.* (1975) investigated the "personal orientation of religious women" (nuns). Among the many variables under study was X, the degree to which a person accepts aggression. In 1969, a sample of 62 women yielded a mean score of 13.39 with $s_X = 4.0$, and in 1972 a sample of 62 women yielded a mean score of 16.05 with $s_X = 3.4$. Test the null hypothesis that there has been no significant difference, on the average, between the two groups versus the one-sided alternative that in 1972 the scores, on the average, have increased. Also state your assumptions.

12.26 In a referendum submitted to the student body at a college, 850 men and 566 women voted. 530 of the men and 304 of the women voted "yes." Does this indicate a significant difference of opinion on the matter, at the 1% level, between men and women students?

12.27 The records of a hospital show that 52 men in a sample of 1000 men versus 23 women in a sample of 1000 women were admitted because of heart disease. Do these data present sufficient evidence to indicate a higher rate of heart disease among men admitted to the hospital?

12.28 Mr. Davis believes that the proportion p_1 of Conservatives in favor of the death penalty is greater than the proportion p_2 of Liberals in favor of the death penalty. He acquired independent random samples of 200 Conservatives and 200 Liberals, respectively, and found 46 Conservatives and 34 Liberals favoring the death penalty. Does this evidence provide statistical support at the 0.05 level of significance for Mr. Davis' belief?

12.29 Random samples of 200 bolts manufactured by machine A and 200 bolts manufactured by machine B showed 16 and 8 defective bolts, respectively. Do these data present sufficient evidence to suggest a difference in the performance of the machines? Use a 0.05 level of significance.

12.30 This chapter has pointed out the difference between the problem of *estimation* and that of *testing:* in estimation, one has to choose *one* of a large set of numbers as a value for the true parameter; in *testing*, one has to decide between *two* possibilities. Since there are two possibilities there are two types of errors. Moreover, in practice it may be that one of the two errors has much more serious consequences than the other. The problem then reduces to finding a *decision rule* which will ensure that the probability of this error is small. The following sequence of questions constitutes a statistical demonstration of these concepts. You will also discover other interesting characteristics.

1. Suppose a gambler has a stock of biased coins, some with $p = P[H] = 0.4$ and the others $p = 0.6$. He is out with one of these coins to make some money. His game is: the coin is tossed and if a head occurs you pay him \$1 and if a tail occurs he pays you \$1. However, before he plays a game he realizes he doesn't know which coin he has. If he can toss the coin a large number of times he has a small probability of making an error.

   (a) Suppose he tosses the coin 99 times and used the following decision rule: "Choose to play if and only if $X \geq 50$ and refuse to play if and only if $X < 50$, where X is the number of heads observed." Find the probability that he makes the wrong decision (that is, he plays) when in fact $p = 0.4$. Note that if the coin has $p = 0.6$ and he decides that $p = 0.4$ he will not play and no harm is done. But if he decides that $p = 0.6$ when in fact $p = 0.4$ he will play and, of course, he will lose money in the long run.

   (b) Suppose that because of a time problem he tosses the coin a few times and decides on the basis of the number of heads observed whether to play. Suppose he tosses the coin nine times and uses the following decision rule: "Decide to play if $X = 7, 8, 9$."

(i) Find the probability that he makes the wrong decision (and plays) when in fact $p = 0.4$.

(ii) Find the probability that he makes the wrong decision (and does not play) when in fact $p = 0.6$.

(iii) Which error is more serious and why?

2. Suppose that in procedure 1(b) the two parameter values are 0.45 and 0.55 with the same decision rule and $n = 9$.
   (a) Find the probability that he makes the wrong decision (and plays) when in fact $p = 0.45$.
   (b) What influence has the change in parameter values on the probability of his error?
3. Repeat procedure 2(a) and 2(b) for the two values 0.35 and 0.65, respectively.
4. Suppose in procedure 1(b) that the gambler increases the number of tosses to 15 and his decision rule is: "Decide that coin has $p = 0.6$ if $X \geq 10$."
   (a) Find the probability that he makes the error of playing when in fact $p = 0.4$.
   (b) What influence has the change in n on the probability of his error?
5. Suppose we have a situation in which the two possible parameter values are $p_1 = 0.4$ and $p_2 = 0.7$.
   (a) Suppose we decide on one of these values of p after $n = 9$ trials, based on the following decision rule: "Decide on p_2 if 7 or more of the trials are successful." Find the probability that we decide on p_2 when in fact $p = p_1$.
   (b) Repeat part (a) if $p_1 = 0.5$ and $p_2 = 0.7$.
   (c) Suppose $p_1 = 0.5$ and $p_2 = 0.7$. If we decide on one of these possible parameter values after $n = 15$ trials, design a decision rule on choosing p_2 so that the probability of error is less than 0.05.

12.31 Parents were asked to rate their children from 1 (adept at interacting with adults) to 7 (adept at interacting with children). For a sample of 63 children who had "imaginary companions," the mean score was 2.2 with a standard deviation of 1.3, whereas for a sample of 157 children without imaginary companions, the mean score was 2.7 with a standard deviation of 1.4. Do these data suggest that there is a significant difference between the two groups? (This experiment is discussed in detail in Finch *et al.*, 1974.)

12.32 Fagot (1973) reported on a study of sex-related stereotyping in toddlers' behaviors. In the study 45 men and 57 women were interviewed. The proportions of each sex rating the three behaviors as masculine were as follows: for rough house play, men = 0.660, and women = 0.491; for play with transportation toys, men = 0.466 and women = 0.157; and for aggressive behavior, men = 0.800 and women = 0.652. The proportions of each sex rating the three behaviors as feminine were as follows: for play with dolls, men = 0.800 and women = 0.649; for dress up, men = 0.755 and women = 0.631; and for look in the mirror, men = 0.111 and women = 0.035. Do six separate large sample Z tests and state your conclusions.

13

TESTS BASED ON THE t, χ^2, AND F DISTRIBUTIONS

In Chapter 12 we have seen that for a small sample from a normal population or for a large sample from an arbitrary population the only distribution needed to perform a test concerning the mean was the standard normal or Z distribution. If a null hypothesis on the mean of a normal population is to be tested on the basis of a *small* sample, then from what already has been said on confidence interval estimation in Chapter 11 and its relationship to tests of hypotheses in the previous chapter, we know that the t distribution will be important. Similarly it is implied that the chi-square distribution is the key to testing a null hypothesis on the standard deviation or variance of a normal population, while the F distribution is the one used in testing a null hypothesis concerning the ratio of the standard deviations or variances of two independent normal populations.

13.1 ONE-TAIL AND TWO-TAIL t TESTS

We start out the development with the t distribution by considering the following types of hypotheses, which are to be tested on the basis of a *small sample* from a normal population.

(a) H_0: $\mu_X = \mu_0$ (where μ_0 is some specified value of μ)

H_A: $\mu_X > \mu_0$ (i.e., $\mu_X = \mu_1$ such that $\mu_1 > \mu_0$)

(b) H_0: $\mu_X = \mu_0$

H_A: $\mu_X < \mu_0$ (i.e., $\mu_X = \mu_1$ such that $\mu_1 < \mu_0$)

(c) H_0: $\mu_X = \mu_0$

H_A: $\mu_X \neq \mu_0$ (i.e., $\mu_X = \mu_1$ such that $\mu_1 < \mu_0$ or $\mu_1 > \mu_0$)

The test statistic for testing these hypotheses is the t statistic:

$$t = \frac{\overline{X} - \mu_0}{s_{\overline{X}}}$$

which has the t distribution with n - 1 degrees of freedom. The following theorem is the same as Theorem 12.1, except that the Z statistic is replaced by the t statistic and the standard normal distribution is replaced by Student's t distribution with n - 1 degrees of freedom.

THEOREM 13.1 *Statistical tests at the significance level for the hypotheses* (a), (b), *and* (c) *above are given, respectively, by:*

(a) *Reject* H_0 *if* $t > t_{n-1,\alpha}$.
(b) *Reject* H_0 *if* $t < -t_{n-1,\alpha}$.
(c) *Reject* H_0 *if* $|t| > t_{n-1,\alpha/2}$.

The test statistic in each case is

$$t = \frac{\overline{X} - \mu_0}{s_{\overline{X}}} \qquad \text{where } s_{\overline{X}} = \frac{s_X}{\sqrt{n}}$$

and in (c) *we mean* $t > t_{n-1,\alpha/2}$ *or* $t < -t_{n-1,\alpha/2}$. *Also for each case* H_0 *is otherwise accepted.*

The diagrams that correspond to the three tests are given in Figure 13.1.

The interpretation of the three tests in Theorem 13.1 is the usual repeated sampling interpretation. If the sampling is

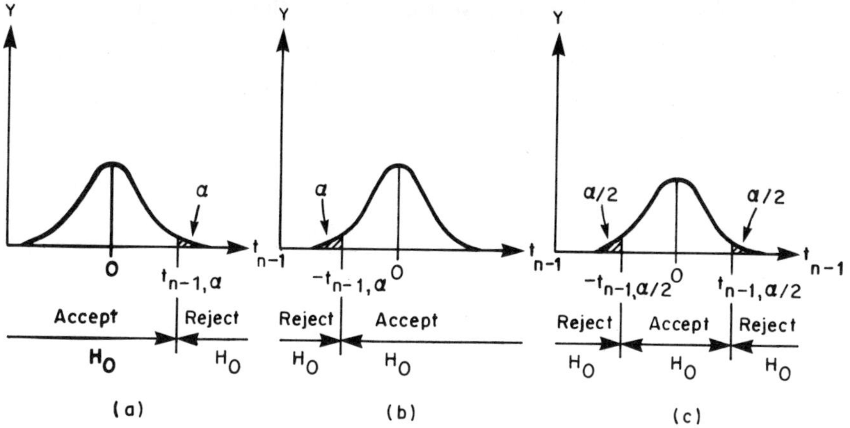

Figure 13.1 t Tests for hypotheses on the mean of a normal population with unknown standard deviation.

repeated an infinite number of times, then we expect to wrongly reject the null hypothesis $\alpha \cdot 100\%$ of the time and accept the null hypothesis wrongly $\beta \cdot 100\%$ of the time.

The tests (a) and (b) are *one-tail t tests*, while (c) is a *two-tail t test* for obvious reasons. The power curves have shapes similar to those given for the Z tests in Figure 12.3. As before, the same test applies when in (a) and (b), $\mu_X = \mu_0$ is replaced by $\mu_X \leq \mu_0$ and $\mu_X \geq \mu_0$, respectively.

Example 13.1 A politician, running for office against an incumbent in an electoral district, accused his opponent of not having done anything regarding prevention of crimes. Specifically, he said that compared to the national rate, the rate for criminal offenses in the district was excessive. The incumbent did a bit of checking by sampling the police files and found for the past year the following rates for criminal offenses (per 1000) for 9 randomly selected communities in the district:

2.4, 4.3, 8.1, 12.5, 4.6, 7.6, 8.9, 3.4, 6.7

Is the politician's claim valid if the national average was equal to 6.2?

To answer this question, we *assume* that the crime rates over all communities in the district follow a normal distribution, which by hypothesis has a mean of 6.2. We are interested in testing:

$$H_0: \quad \mu_X = 6.2 \qquad \text{against} \qquad H_A: \quad \mu_X > 6.2$$

From the sample data we obtain:

$$\bar{X} = 6.50 \qquad s_X = 3.17 \qquad s_{\bar{X}} = \frac{3.17}{\sqrt{9}} = 1.06$$

The value of t calculated from the sample is, by Theorem 13.1, equal to:

$$t = \frac{\bar{X} - \mu_0}{s_{\bar{X}}} = \frac{6.50 - 6.20}{1.06} = 0.28$$

which is far less than $t_{8,0.05} = 1.86$ if we are testing at the 5% significance level. Hence H_0 is accepted at the 5% level, i.e., the crime rate for criminal offenses in the district shows an insignificant difference from the national rate.

13.2 POOLED t TESTS

Let us now concentrate on the two population case, which will lead us to the so-called *pooled t tests*. Consider two independent normal populations which have the same standard deviation, i.e., $\sigma_X = \sigma_Y = \sigma$, say. Therefore, we are interested in the following three types of hypotheses:

(a) $H_0: \quad \mu_X = \mu_Y$

$H_A: \quad \mu_X > \mu_Y$

(b) $H_0: \quad \mu_X = \mu_Y$

$H_A: \quad \mu_X < \mu_Y$

and (c) $H_0: \quad \mu_X = \mu_Y$

$H_A: \quad \mu_X \neq \mu_Y$

Note that if the null hypothesis is true, then we are saying that the two populations are the same, i.e., they have identical distributions. For this, we use the *pooled t test* which utilizes a t variable with $n_1 + n_2 - 2$ degrees of freedom. The test statistic is:

$$t_{n_1+n_2-2} = \frac{\overline{X} - \overline{Y}}{s_{\overline{X}-\overline{Y}}}$$

where

$$s_{\overline{X}-\overline{Y}} = s\sqrt{\frac{1}{n_1} + \frac{1}{n_2}}$$

and s is the pooled sample standard deviation, already used in Chapter 11, which is found by the formula

$$s = \sqrt{\frac{(n_1 - 1)s_X^2 + (n_2 - 1)s_Y^2}{n_1 + n_2 - 2}}$$

$$= \sqrt{\frac{\Sigma(X - \overline{X})^2 + \Sigma(Y - \overline{Y})^2}{n_1 + n_2 - 2}}$$

$$= \sqrt{\frac{[\Sigma X^2 - (1/n_1)(\Sigma X)^2] + [\Sigma Y^2 - (1/n_2)(\Sigma Y)^2]}{n_1 + n_2 - 2}}$$

Hence in analogy with Theorem 12.2 we have the following result.

THEOREM 13.2 *Statistical tests at the significance level* α *for the hypotheses* (a), (b), *and* (c) *are, respectively:*

(a) *Reject* H_0 *if* $t > t_{n_1+n_2-2,\alpha}$.

(b) *Reject* H_0 *if* $t < -t_{n_1+n_2-2,\alpha}$.

(c) *Reject* H_0 *if* $|t| > t_{n_1+n_2-2,\alpha/2}$.

The test statistic is defined by

$$t = \frac{\overline{X} - \overline{Y}}{s_{\overline{X}-\overline{Y}}}$$

In each case H_0 *is otherwise accepted.*

Example 13.2 In a behavioral experiment with 20 mice, which were matched for other factors, 2 treatments were randomly allocated to 10 mice each. The following data represent scores from a passive-avoidance conditioning test:

Treatment 1	Treatment 2
16	13
13	15
14	14
15	15
15	16
16	13
16	14
15	12
16	13
14	15
$\bar{X} = 15$	$\bar{Y} = 14$
$\Sigma(X - \bar{X})^2 = 10$	$\Sigma(Y - \bar{Y})^2 = 14$

Under the assumption that these are samples from normal populations, is there a difference between the treatment means?

We have the following testing problem:

$$H_0: \quad \mu_X = \mu_Y \qquad \text{against} \qquad H_A: \quad \mu_X \neq \mu_Y$$

and the test statistic by Theorem 13.2 requires the calculation of $s_{\bar{X}-\bar{Y}}$. This is found by utilizing the formula for the pooled standard deviation, i.e.,

$$s = \sqrt{\frac{10 + 14}{18}} = 1.155$$

so that

$$s_{\bar{X}-\bar{Y}} = 1.155\sqrt{\frac{1}{10} + \frac{1}{10}} = 0.516$$

Hence the calculated t value is equal to

$$t = \frac{15 - 14}{0.516} = 1.94$$

which at the 5% level is smaller than $t_{18,0.025} = 2.101$. Therefore there is no significant difference at the 5% level between the two treatments.

There is another celebrated t test known as the *paired t test.* This is discussed in Chapter 15, "Design and Analysis of Experiments."

13.3 CHI-SQUARE TESTS ON THE STANDARD DEVIATION OR VARIANCE

Except for a brief mention of a null hypothesis on the standard deviation of heights of females of a given ethnic group in a certain country, at the beginning of Chapter 12, we have not touched upon *tests of hypotheses concerning the standard deviation of a population.*

Consider a normal population $N(\mu_X, \sigma_X)$; hypotheses similar to those for the mean can be made about the standard deviation σ_X. These are:

(a) H_0: $\sigma_X = \sigma_0$

H_A: $\sigma_X > \sigma_0$

(b) H_0: $\sigma_X = \sigma_0$

H_A: $\sigma_X < \sigma_0$

(c) H_0: $\sigma_X = \sigma_0$

H_A: $\sigma_X \neq \sigma_0$

An obvious test statistic is the sample standard deviation s_X, and if this shows a large discrepancy relative to the null hypothesis, then we should reject it. It is easier to work with the statistic $(n - 1)s_X^2/\sigma_0^2 = \Sigma(X - \overline{X})^2/\sigma_0^2$, because we know already that it has a chi-square distribution with (n - 1) degrees of freedom. Besides, going from s_X to χ^2 requires a simple squaring of s_X, multiplying by (n - 1), and dividing through by σ_0^2. We are already familiar with this from what was done in Chapter 11 on confidence interval estimation of σ_X.

We have the following theorem:

THEOREM 13.3 *Statistical tests at the significance level* α *for the hypotheses* (a), (b), *and* (c) *are given by:*

(a) *Reject* H_0 *if* $\chi^2 > \chi^2_{n-1,\alpha}$.

(b) *Reject* H_0 *if* $\chi^2 < \chi^2_{n-1,1-\alpha}$.

(c) *Reject* H_0 *if* $\chi^2 > \chi^2_{n-1,\alpha/2}$ *or* $\chi^2 < \chi^2_{n-1,1-\alpha/2}$.

In each case the test statistic is

$$\chi^2 = \frac{(n - 1)s_X^2}{\sigma_0^2}$$

Also in each case H_0 *is otherwise accepted.*

Note that these tests are also valid if we make null hypotheses about the variance σ_X^2 rather than σ_X. The first two tests are called *one-tail* χ^2 *tests*, while the last one is known as the *two-tail* χ^2 *test*. Their sketches are presented in Figure 13.2.

Example 13.3 It is conjectured that the standard deviation of the ages of delinquent teenagers in a province is equal to 2.1 years.

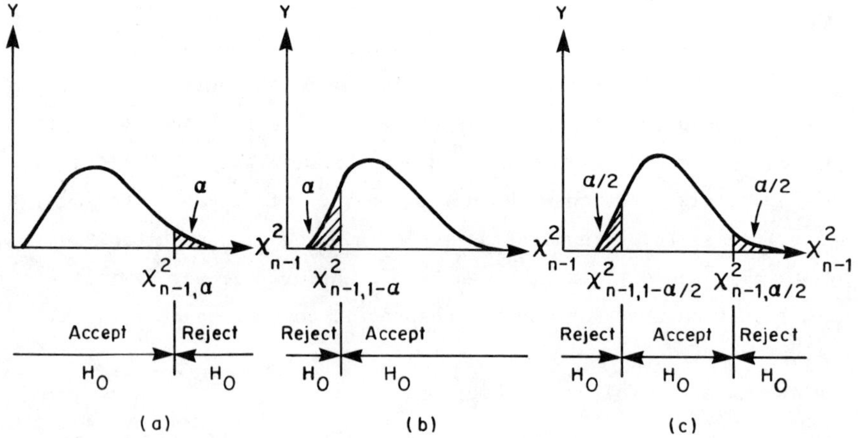

Figure 13.2 χ^2 tests for hypotheses on the standard deviation of a normal population.

A random sample of 15 records in a city gave the following ages in years:

13.3, 16.9, 14.2, 19.1, 14.8, 17.2, 13.9, 15.9,
16.3, 17.6, 19.6, 14.3, 16.7, 14.6, 13.8

Is there a significant departure between the province-wide variability in ages and the city variability as exhibited in the sample?

Assuming that ages of delinquent teenagers are normally distributed, we have the following to test:

$$H_0: \quad \sigma_X = 2.1 \qquad (\text{i.e., } \sigma_X^2 = 4.41)$$

$$H_A: \quad \sigma_X \neq 2.1 \qquad (\text{i.e., } \sigma_X^2 \neq 4.41)$$

The calculated χ^2 value is equal to:

$$\frac{(n - 1)s_X^2}{\sigma_0^2} = \frac{(14)(3.84)}{4.41} = 12.19$$

which is between $\chi^2_{14,0.975} = 5.63$ and $\chi^2_{14,0.025} = 26.12$, and hence according to Theorem 13.3 there is no significant departure (at the 5% level) in variability of ages between the hypothesized value of 2.1 and the evidence in the sample.

13.4 CHI-SQUARE GOODNESS-OF-FIT TEST

We now present two popular applications of the χ^2 distribution. The first one is called the *chi-square goodness-of-fit test* and the second one is known as the *chi-square test of independence*. We first give the general setting for the goodness-of-fit test.

Consider a variable, be it qualitative or quantitative, which has h classes attached to it. In the qualitative cases the classes are simply the *levels* which may appear singly as a class or combined into classes. In the quantitative case the classes are determined in the usual way by class limits. A sample of n units is selected from the infinite population for which a distribution is hypothesized. The number of units falling in the classes are called *observed frequencies* and the frequencies calculated by multiplying n with the probabilities of the classes, obtained from the hypothesized distribution, are called *expected frequencies*.

There will be *discrepancies* between the observed and expected frequencies, which are measured by the absolute differences between them. It is much more convenient, as will be seen shortly, to measure the discrepancies by the squares of the absolute differences or, for that matter, by the squares of the differences divided through by their expected frequencies. If we sum these last quantities, then in essence we are proposing a test statistic for judging *goodness-of-fit* of the hypothesized distribution. This measure has the obvious characteristic that it will be "large" if the fit is poor. Schematically, the general situation is as depicted in Table 13.1. (In the table, h is simply the number of classes.)

Table 13.1 Observed and Expected Frequencies for a Chi-Square Goodness-of-Fit Test

Class	Observed frequency = O	Hypothesized probability = p	Expected frequency = E	$(O - E)^2/E$
1	O_1	p_1	E_1	$(O_1 - E_1)^2/E_1$
2	O_2	p_2	E_2	$(O_2 - E_2)^2/E_2$
3	O_3	p_3	E_3	$(O_3 - E_3)^2/E_3$
⋮	⋮	⋮	⋮	⋮
h	O_h	p_h	E_h	$(O_h - E_h)^2/E_h$
Totals	n	1	n	$\sum_{i=1}^{h} \frac{(O_i - E_i)^2}{E_i}$

The following theorem is true in repeated sampling, and it can be proved mathematically.

THEOREM 13.4 *In repeated sampling, the test statistic*

$$\chi^2_{h-1-m} = \sum_{i=1}^{h} \frac{(O_i - E_i)^2}{E_i}$$

has approximately the chi-square distribution with $h - 1 - m$ *degrees of freedom if* n *is sufficiently large, where* m *is the number of independent parameters estimated from the data and used in the probability distribution to provide the probabilities, and hence, the expected frequencies.*

If no parameters are estimated from the data, then we have $h - 1$ degrees of freedom. As a working rule, a sample size which will give a good approximation to the χ^2 distribution is one which results in an expected frequency of 5 or more for each class.

The null hypothesis and the alternative to it for testing goodness-of-fit are stated as:

H_0: $E_i = np_i$ $\quad i = 1, 2, \ldots, h$

H_A: At least one E_i is not equal to np_i

Since the χ^2 statistic in Theorem 13.4 is a test statistic, because we know its distribution (approximately), we are at once led to the following theorem.

THEOREM 13.5 *The chi-square goodness-of-fit test at significance level* α *for testing* H_0: $E_i = np_i$ *against* H_A: *at least one* E_i *is not equal to* np_i, *is to reject* H_0 *if the sample*

$$\chi^2 = \sum \frac{(O_i - E_i)^2}{E_i} > \chi^2_{h-1-m,\alpha}$$

and to accept H_0 *otherwise.*

We now give several applications of Theorem 13.5.

Example 13.4 Five hundred randomly selected married women expecting their first child were interviewed on their preference regarding the sex of their first child. Of these, 220 preferred "baby boys," and the rest of them "baby girls." Does this finding clash with the notion that women do not have a preference as to the sex of their first child?

The null hypothesis of no preference implies a Bernoulli probability distribution with $p_1 = 1/2$ and $p_2 = 1 - p_1 = 1/2$. Stated in its general form, we have to test:

$$H_0: \quad E_1 = (500)\left(\frac{1}{2}\right) = 250 \qquad \text{and hence } E_2 = 250$$

$$H_A: \quad \text{At least one } E_i \text{ is not equal to 250} \qquad (\text{i.e., } E_1 \neq 250)$$

This leads us to the following setup for performing the test:

Classes	O	p	E	(O - E)	$(O - E)^2$	$(O - E)^2/E$
Baby boy	220	1/2	250	-30	900	3.60
Baby girl	280	1/2	250	30	900	3.60
						$\chi^2 = 7.20$

No parameters were estimated from the data, so that we have $2 - 1 = 1$ degrees of freedom. Setting the significance level at $\alpha = 5\%$ we see that the value of $\chi^2_{1,0.05} = 3.84$, which is exceeded by 7.20. Hence by Theorem 13.5 we reject H_0 at the 5% level, that women have no preference regarding the sex of their first child; i.e., the fit is poor for a Bernoulli distribution with $p = 1/2$.

Since the sample size in the example is large, we can confirm our conclusions by using a two-tail Z test. The reader is invited to complete this analysis.

Example 13.5 In Section 7.2 we discussed the fitting of a normal distribution to a binomial distribution, and to an observed frequency distribution. The latter example related to IQ scores obtained from 50 sixth graders of the so-called "faster" or "A" group. The observed and expected frequencies were already calculated in Table 7.1, and we reproduce them below for the purpose of applying the chi-square goodness-of-fit test.

Classes	O	p	E	(O - E)	$(O - E)^2$	$(O - E)^2/E$
102.5-107.5	8	0.1276	6.380	1.620	2.624	0.411
107.5-112.5	10	0.2703	13.515	-3.515	12.355	0.914
112.5-117.5	20	0.3317	16.585	3.415	11.622	0.701
117.5-122.5	8	0.2028	10.140	-2.140	4.580	0.452
122.5-127.5	4	0.0681	3.405	0.595	0.354	0.104
Totals	50	1.0	50.0	0.0	--	2.582

In this example, two independent parameters, namely, the mean and the standard deviation, were estimated from the data. Hence we have 5 - 1 - 2 = 2 degrees of freedom for χ^2. Taking the significance level α to be 5%, we see from the chi-square table that $\chi^2_{2,0.05} = 5.99$. Since $2.582 < 5.99$, we conclude that the null hypothesis:

H_0: X is normally distributed

against

H_A: X is not normally distributed

is accepted at the 5% significance level.

Note that the last class has an expected frequency of 3.405, which is less than 5. If this class is combined with the fourth one, then we have 4 - 1 - 2 = 1 degrees of freedom. The reader may verify that the same conclusion is reached with four classes. Also the probabilities for the first and last class are calculated as $P[X < 107.5]$ and $P[X > 122.5]$ to make the total 1.0.

Example 13.6 One hundred randomly selected police officers in a large metropolitan city were interviewed regarding their attitude toward jail sentences for offenders caught smoking marijuana. The results were:

Type of sentence	Number of police officers
1. Should get 5 or more years	32
2. Should get 1 to 5 years	22
3. Should get 1 day to a year	26
4. Should not be prosecuted at all	20
Total	100

Test the hypothesis that the probabilities are the same for each class in the population of police officers in the city.

The null hypothesis and its alternative are:

H_0: $p_1 = p_2 = p_3 = p_4 = \frac{1}{4}$ (i.e., $E_i = (n)\left(\frac{1}{4}\right) = 25$, for each i)

H_A: At least one p_i is not equal to $\frac{1}{4}$ (i.e., at least one E_i is not equal to 25)

The required calculations necessary to execute the test are as follows:

Class	O	E	(O - E)	$(O - E)^2$	$(O - E)^2/E$
1	32	25	7	49	1.96
2	22	25	-3	9	0.36
3	26	25	1	1	0.04
4	20	25	-5	25	1.00
Total	100	100	0	--	3.36

From the χ^2 table, we obtain at the 5% significance level, that $\chi^2_{3,0.05} = 7.82$. Since $3.36 < 7.82$, we accept H_0 at the 5% significance level.

Remark. When the number of classes, i.e., h, is relatively large, then there frequently occurs the situation that the expected frequencies are less than 5 for some classes. The remedy is to combine classes such that the resulting expected frequencies for the classes are greater than or equal to 5.

13.5 CHI-SQUARE TEST OF INDEPENDENCE

As mentioned earlier, there is another important chi-square test, namely, the *chi-square test of independence*. Consider a two-variable (also called *bivariate*) case, where each variable has classes associated with them. A sample of n units is selected from the infinite underlying population for which, under the hypothesis that the classifications are *independent*, a distribution is specified. If there are r classes for the first variable and c classes for the second variable, then there will be (r) • (c) cells, for each of which a probability is found from the specified distribution under the hypothesis of independence. If these probabilities are multiplied by n, then we obtain, as before, the expected cell frequencies. The sample of n observed units, which are classified into the cells, gives rise to the observed cell frequencies. These are then compared with the expected cell frequencies to discover whether the discrepancies lead to rejection or acceptance of the null hypothesis of independence.

From our Chapter 5 on probability we know that the probability of the intersection of two events is equal to the product of the probabilities of the events, if the events are independent. This means that if the two classifications are independent and we provide the marginal probabilities of each classification, then the cell probabilities (which are probabilities associated with intersection of two events) can be filled by the products of the corresponding marginal probabilities. This is done in Table 13.2, where, for example, the probability p_1q_1 in the cell corresponding to class 1 of X and class 1 of Y is equal to the product of the marginal probabilities p_1 and q_1.

Table 13.2 Cell Probabilities Obtained from the Marginals under the Hypothesis of Independence

		Classes for Y 1	2	• • •	c	Marginal probabilities
Classes for X	1	p_1q_1	p_1q_2	• • •	p_1q_c	$p_{1.}$
	2	p_2q_1	p_2q_2	• • •	p_2q_c	$p_{2.}$
	⋮	⋮	⋮		⋮	⋮
	r	p_rq_1	p_rq_2	• • •	p_rq_c	$p_{r.}$
Marginal probabilities		q_1	q_2	• • •	q_c	1

It is the hypothesis of independence which allows us to fill the cells with the cell probabilities, which in turn provide the expected frequencies when they are multiplied by n. But where do the marginal probabilities come from? In most practical settings they are not specified in advance, but rather are *estimated* from the observed marginal frequencies. Denote the observed frequencies as given in Table 13.3.

If in this table a unit is picked randomly from among the n units, then its chances of belonging to class 1 of X are equal to $O_{1.}/n$, and its chances of belonging to class 1 of Y are equal to $O_{.1}/n$. Hence the estimated probability that it belongs to both class 1 of X and class 1 of Y is equal to the product

$$\frac{O_{1.}}{n} \cdot \frac{O_{.1}}{n}$$

under the hypothesis of independence. Therefore the estimated expected cell frequency for the first cell is equal to:

$$\hat{E}_{11} = \frac{O_{1.}}{n} \cdot \frac{O_{.1}}{n} \cdot n = \frac{(O_{1.})(O_{.1})}{n}$$

Table 13.3 Observed Frequencies for the r × c Cells and Their Marginal Totals

		Classes for Y 1	2	• • •	c	Marginal totals
Classes for X	1	O_{11}	O_{12}	• • •	O_{1c}	$O_{1.}$
	2	O_{21}	O_{22}	• • •	O_{2c}	$O_{2.}$
	⋮	⋮	⋮		⋮	⋮
	r	O_{r1}	O_{r2}	• • •	O_{rc}	$O_{r.}$
Marginal totals		$O_{.1}$	$O_{.2}$	• • •	$O_{.c}$	$O_{..} = n$

In other words, the estimated expected frequency of the first cell is equal to the observed first row total times the observed first column total divided by n.

This result can be generalized to any arbitrary cell; indeed,

$$\hat{E}_{ij} = \frac{(O_{i.})(O_{.j})}{n} \qquad \begin{array}{l} i = 1, 2, \ldots, r \\ j = 1, 2, \ldots, c \end{array}$$

Note that in the process we have estimated (r - 1) + (c - 1) independent probabilities, and hence (r - 1) + (c - 1) independent parameters of the distribution.

The test statistic which measures the magnitude of the discrepancies between the observed frequencies O_{ij} and the estimated expected frequencies $\hat{E}_{ij}$ is the variable:

$$\sum_{i=1}^{r} \sum_{j=1}^{c} \frac{(O_{ij} - \hat{E}_{ij})^2}{\hat{E}_{ij}}$$

The following theorem applies when repeated sampling of size n is performed.

THEOREM 13.6 *In repeated sampling, the test statistic*

$$\chi^2_{(r-1)\cdot(c-1)} = \sum_{i=1}^{r} \sum_{j=1}^{c} \frac{(O_{ij} - \hat{E}_{ij})^2}{\hat{E}_{ij}}$$

has approximately a chi-square distribution with (r - 1) • (c - 1) *degrees of freedom.*

The degrees of freedom, d.f. = (r - 1) • (c - 1), are obtained by observing that there are (r) • (c) cells, which can be filled with (r) • (c) - 1 independent probabilities, because they must add up to 1. However, we have estimated (r - 1) + (c - 1) independent ones. Therefore:

$$\begin{aligned} \text{d.f.} &= rc - 1 - [(r - 1) + (c - 1)] \\ &= rc - 1 - r + 1 - c + 1 \\ &= rc - r - c + 1 \\ &= (r - 1) \cdot (c - 1) \end{aligned}$$

The null hypothesis and its alternative to test independence of the X and Y classifications in terms of the expected frequencies is stated as:

H_0: $E_{ij} = np_i q_j$ $i = 1, 2, \ldots, r;\ j = 1, 2, \ldots, c$

H_A: At least one E_{ij} is not equal to $np_i q_j$

The null hypothesis states that the two classifications are independent, while the alternative states that they are dependent.

The statistical test of independence of the two classifications is given in the following theorem:

THEOREM 13.7 *The chi-square test of independence of two classifications, i.e., for testing*

H_0: $E_{ij} = np_i q_j$

against

H_A: At least one E_{ij} is not equal to $np_i q_j$

at the significance level α *is to reject* H_0 *if*

$$\chi^2 = \sum \sum \frac{(O_{ij} - \hat{E}_{ij})^2}{\hat{E}_{ij}} > \chi^2_{(r-1)(c-1),\alpha}$$

and to accept H_0 *otherwise.*

As before, each of the estimated expected cell frequencies should be greater than or equal to 5 for the χ^2 approximation to be functional.

The test of independence for two classifications in Theorem 13.7 is called by its more proper name as the *test of independence for two cross-classifications*, because we are dealing with a two-way cross-classification table. It is also known as the *test of independence for a two-way contingency table*. It can be generalized to three and higher cross-classifications.

Example 13.7 A study was conducted to test the dependency of marriage adjustment relative to income by classifying 400 couples according to their income levels and marriage adjustment ratings. The observed frequencies were:

		Marriage adjustment				Total
		Low	Medium	High	Very high	
Income ($)	> 30,000	18	29	70	115	232
	12,000-30,000	17	28	30	41	116
	< 12,000	11	10	11	20	52
		46	67	111	176	400

We must test:

$$H_0: \quad E_{ij} = (400)p_i q_j \qquad i = 1, 2, 3; \; j = 1, 2, 3, 4$$

against

$$H_A: \quad \text{At least one } E_{ij} \text{ is not equal to } (400)p_i q_j$$

where p_i is the marginal probability of the ith income class and q_j is the marginal probability of the jth rating of marriage adjustment. The calculation of the estimated expected frequency $\hat{E}_{11}$ is obtained as

$$\hat{E}_{11} = \frac{(232)(46)}{400} = 26.68$$

The other estimated expected frequencies are similarly obtained and are given here:

		Marriage adjustment				Total
		Low	Medium	High	Very high	
Income ($)	> 30,000	26.68	38.86	64.38	102.08	232.0
	12,000-30,000	13.34	19.43	32.19	51.04	116.0
	< 12,000	5.98	8.71	14.43	22.88	52.0
		46.0	67.0	111.0	176.0	400.0

The calculated value of χ^2 is then obtained as:

$$\chi^2 = \sum_{i=1}^{3} \sum_{j=1}^{4} \frac{(O_{ij} - \hat{E}_{ij})^2}{\hat{E}_{ij}} = \frac{(18 - 26.68)^2}{26.68} + \frac{(29 - 38.86)^2}{38.86}$$

$$+ \cdots + \frac{(20 - 22.88)^2}{22.88} = 19.35$$

There are (3 - 1)(4 - 1) = 6 degrees of freedom, so that the tabled value of χ^2 at the 5% significance level is equal to $\chi^2_{6,0.05}$ = 12.59. Since 19.35 > 12.59, we reject H_0; i.e., income and marriage adjustment are dependent. Therefore, the data support the fact that as income increases, marriage adjustment rates significantly increase. This is the nature of the dependency.

13.6 F TEST ON EQUALITY OF TWO STANDARD DEVIATIONS OR VARIANCES

The chi-square distribution provided us with tests of hypotheses on the standard deviation or variance of one normal population. The F distribution permits us to test hypotheses on the ratio of the standard deviations or variances of two independent normal populations. Let $N(\mu_X, \sigma_X)$ and $N(\mu_Y, \sigma_Y)$ be two independent normal populations; then we are interested in the following practical test:

(a) H_0: $\sigma_X = \sigma_Y$ (i.e., $\sigma_X^2 = \sigma_Y^2$, or $\dfrac{\sigma_X^2}{\sigma_Y^2} = 1$)

H_A: $\sigma_X > \sigma_Y$ (i.e., $\sigma_X^2 > \sigma_Y^2$, or $\dfrac{\sigma_X^2}{\sigma_Y^2} > 1$)

The evidence for rejecting or accepting H_0 is in the ratio s_X^2/s_Y^2, because when this is "much larger" than 1, then we should reject H_0, and when it is "close" to 1 then H_0 should be accepted. If the null hypothesis is true then $\sigma_X^2 = \sigma_Y^2$, so that from Chapter 9 it follows that the test statistic

$$F_{n_1-1,n_2-1} = \frac{s_X^2/\sigma_Y^2}{s_Y^2/\sigma_Y^2} = \frac{s_X^2}{s_Y^2}$$

has the F distribution with $n_1 - 1$ and $n_2 - 1$ degrees of freedom. Hence we have the following theorem:

THEOREM 13.8 *A statistical test at the significance level* α *for testing*

$$H_0: \quad \sigma_X = \sigma_Y \qquad (\text{i.e., } \sigma_X^2 = \sigma_Y^2)$$

against

$$H_A: \quad \sigma_X > \sigma_Y \qquad (\text{i.e., } \sigma_X^2 > \sigma_Y^2)$$

is to reject H_0 *if*

$$F = \frac{s_X^2}{s_Y^2} > F_{n_1-1,n_2-1,\alpha}$$

and H_0 *is accepted otherwise.*

Recall that the F table in the Appendix is constructed in such a way that $\nu_1 = n_1 - 1$ corresponds to the numerator sample variance s_X^2, and $\nu_2 = n_2 - 1$ corresponds to the denominator sample variance s_Y^2. The F table gives values for 5% and 1% significance levels, which are right-hand tail probabilities.

The reader is invited to extend Theorem 13.8 for testing the following two types of hypotheses (which were purposely left out above):

(b) $H_0: \quad \sigma_X^2 = \sigma_Y^2, \qquad H_A: \quad \sigma_X^2 < \sigma_Y^2$

(c) $H_0: \quad \sigma_X^2 = \sigma_Y^2, \qquad H_A: \quad \sigma_X^2 \neq \sigma_Y^2$

The tests for (a) and (b) are *one-tail* F *tests*, while that for (c) is a *two-tail* F *test*.

Finally, the F tests can be generalized to test any ratio between two variances, e.g., H_0: $\sigma_X^2/\sigma_Y^2 = c$ against H_A: $\sigma_X^2/\sigma_Y^2 \neq c$, where c is a specified positive constant.

Example 13.8 Two groups of 20 grade 9 students were randomly selected, one from public high schools and one from private high schools. They were administered a mathematics aptitude test and the results were:

X: 15, 26, 30, 24, 18, 34, 43, 20, 18, 8,
22, 25, 31, 26, 30, 63, 26, 18, 24, 17

Y: 30, 37, 35, 28, 17, 45, 33, 18, 24, 18,
30, 30, 32, 33, 38, 42, 40, 20, 23, 25

Do public school grade 9 students exhibit a larger variability in mathematics aptitude test scores than similar students from private schools?

From the data we find:

$$s_X^2 = 134.62 \qquad s_Y^2 = 68.41$$

$$F = \frac{134.62}{68.41} = 1.97$$

The F in the table at the 5% significance level is $F_{19,19,0.05} = 2.17$. Since $1.97 < 2.17$, we accept H_0: $\sigma_X^2 = \sigma_Y^2$ against H_A: $\sigma_X^2 > \sigma_Y^2$ at the 5% significance level. So, we can conclude that the two populations are homoscedastic.

Remark. It should have become evident by now that some answers to certain examples are highly subject to the value of α selected. Technically speaking, according to the Neyman-Pearson theory of testing hypotheses (which is in fact the theory adopted in these chapters), the steps of the analysis should be as follows:

1. State H_0 and H_A. (This is essentially a statement of the purpose of the experiment.)
2. Choose α.
3. State the decision rule.

4. Obtain the data. (This requires a design or plan which adheres to the assumptions embedded in the test statistic contained in the decision rule. It may also require secondary tests on testable assumptions!)
5. Calculate the value of the test statistic and find the so-called critical value of the test (or values if it is a two-tail test).
6. State conclusions and recommendations.

One important feature of this analysis is that the selection of α occurs *before* the data are obtained, not after the value of the test statistic is found.

EXERCISES

13.1 True or false:

(a) t tests can be applied to test hypotheses concerning the mean from any population whatsoever.

(b) A sample sometimes gives a negative value of χ^2.

(c) If the alternative hypothesis is stated as $H_A: \theta \neq \theta_0$, then a two-tail test of $H_0: \theta = \theta_0$ is called for.

(d) When the confidence interval for the difference between two means contains zero, then we reject the hypothesis $H_0: \mu_X = \mu_Y$ against the alternative $H_A: \mu_X \neq \mu_Y$.

(e) The test statistic for $H_0: \sigma_X^2 = \sigma_Y^2$ against $H_A: \sigma_X^2 \neq \sigma_Y^2$ in the case of two independent normal populations depends on σ_X^2/σ_Y^2.

13.2 The weight printed on the label of a bottle containing an expensive drug reads 50 g. A sample of 6 bottles is examined and the weights found are equal to: 48.2, 49.6, 50.4, 49.8, 49.4, 50.2. Is the manufacturer cheating his customers? (Use $\alpha = 5\%$. In your answer you must state all necessary assumptions.)

13.3 Twenty lines, each exactly 6 in. long, were drawn. A student estimated by eye the center of each line. The distance in inches of each point, so estimated, from the left-hand of the line was measured, with the following results:

2.97	3.23	3.00	2.98
3.11	2.98	3.06	3.07
2.97	3.02	2.98	2.07
3.18	2.92	3.00	3.03
3.13	3.13	2.94	3.20

Is there any reason to believe that the student was making a systematic error? (*Hint*: $\overline{X} = 3.0485$, $s_X = 0.0894$.)

13.4 Sixteen faculty members at a university were interviewed, and it was found that they had an average of 3.6 children and a standard deviation of 1.10 children. Do faculty members at this university have a significantly (at the 1% level) higher mean than the national average of 2.4? (In your answer you must state all necessary assumptions.)

13.5 A soft drink vending machine in the University Center is regulated so that it discharges an average of 7 oz per cup. Assume the amount of drink per cup is normally distributed. A random sample of 17 drinks from this machine yielded an average content of $\overline{X} = 6.7$ oz with a standard deviation of $s_X = 0.48$ oz. Test the null hypothesis that $\mu_X = 7$ oz against the alternative hypothesis that $\mu_X < 7$ oz at the 0.05 level of significance. What possible conclusions can be made about this soft drink vending machine?

13.6 Two groups of trainees were rated on their achievements. The results were:

Group A: 83, 44, 58, 76, 63, 71

Group B: 59, 83, 62, 47, 35

Is there a significant difference between the two groups? (In your answer you must state all necessary assumptions and test any if possible!)

13.7 In Exercise 13.4, test the null hypothesis H_0: $\sigma_X = 2$ against H_A: $\sigma_X < 2$.

13.8 A random sample of 100 undergraduate students were drawn from a B.A. Honors program in order to test whether the males and females were equally represented. It was found that there were 58 females and 42 males. What is your conclusion regarding the sex ratio? (Use $\alpha = 5\%$.)

13.9 A nationwide sample of 2900 grade 12 students was asked to agree or disagree with the statement: "People should not be allowed to vote unless they are paying property taxes." The responses of the two sexes were tabulated separately:

	Agree	Disagree	No opinion
Male	640	360	310
Female	710	290	590

Is the response dependent upon the sex of the students? (Use $\alpha = 1\%$.)

13.10 In Exercise 13.6 test the null hypothesis H_0: $\sigma_B^2 = \sigma_A^2$ (at the 5% level), against H_A: $\sigma_B^2 > \sigma_A^2$.

13.11 Ten housewives in each of 1024 cities were interviewed regarding their "yes-no" opinion concerning sex education in elementary schools. The results were:

No. of yes answers	No. of cities
0	2
1	11
2	50
3	115
4	196
5	258
6	200
7	105
8	63
9	16
10	8

Do the data fit the binomial distribution with $p = 1/2$ at the significance level $\alpha = 0.01$?

13.12 Make an *incorrect* statement concerning each of the following concepts:

(a) Use of the t distribution in testing H_0: $\mu_X = \mu_0$ against H_A: $\mu_X \neq \mu_0$.

(b) Use of the χ^2 distribution in testing $H_0: \sigma_X = \sigma_0$ against $H_A: \sigma_X > \sigma_0$.

(c) Use of the χ^2 distribution in testing goodness-of-fit.

(d) Use of the χ^2 distribution in testing independence of two classifications.

(e) Use of the F distribution in testing $H_0: \sigma_X^2 = \sigma_Y^2$ against $H_A: \sigma_X^2 < \sigma_Y^2$.

13.13 Set up your own data and fit a normal distribution to it using the chi-square goodness-of-fit test procedure at the 5% level.

13.14 From the tables of the F and χ^2 distributions, discover a relationship which makes the χ^2 distribution a special case of the F distribution.

13.15 (A very challenging problem!) Discover the relationship between the F and t variables by studying their tables or their definitions.

13.16 Sketch the distribution of the test statistic used in Exercise 13.6 and show the value found from the sample, the tabled value, the significance level, and the rejection and acceptance regions.

13.17 Do as in 13.16 for Exercises 13.7 and 13.10.

13.18 Show that the χ^2 test statistic for a 2 × 2 contingency table reduces to:

$$\frac{(O_{11}O_{22} - O_{12}O_{21})^2 n}{(O_{11} + O_{12})(O_{21} + O_{22})(O_{11} + O_{21})(O_{12} + O_{22})}$$

13.19 In the literature on χ^2 tests for independence, there is described "Yates' correction for continuity." When does this apply? Illustrate your answer.

13.20 Describe concisely what is meant by one-tail F tests in contrast to two-tail F tests.

13.21 Answer Exercises 11.12 to 11.17 by using hypothesis testing techniques. Do the results concur with the confidence interval estimation approach?

13.22 Two small samples of herring were measured for length (in millimeters), with the following results:

1. 192, 179, 181, 193, 215, 181, 178
2. 173, 194, 194, 187, 168, 186, 176, 191, 191, 178, 185, 160

Do the populations differ significantly in average length at the 5% level? (A complete analysis is expected!)

13.23 The following data refer to a survey of physicians (Cornfield, 1956). Data were obtained on a group of patients with lung cancer, and a group without lung cancer. The question is: "Do you detect any connection between smoking and lung cancer among physicians?" Analyze these data.

	Controls	Lung cancer patients	Totals
Smokers	32	60	92
Nonsmokers	11	3	14
Total	43	63	106

13.24 The question "Should tax money be spent on nursery schools?" was asked of 1500 persons, and the answers were grouped according to age of person. The data were as follows:

Response	20-34	34-54	Over 54	Total
Favorable	160	189	70	419
No opinion	36	52	28	116
Unfavorable	380	420	165	965
Total	576	661	263	1500

Does this show any association between age group and response?

13.25 Analyze the following (classical) data on the frequency of criminality among twin brothers or sisters of criminals.

Twins	Not convicted	Convicted
Identical	3	10
Nonidentical	15	2

13.26 A thousand individuals are classified according to sex and whether they are color-blind:

	Male	Female
Color-blind	34	8
Not color-blind	446	512

Analyze the data.

13.27 A group of 90 men and 40 women is asked to express a preference for the aromas of two pipe tobaccos. The results are shown:

	Brand A	Brand B
Women	14	26
Men	13	77

Can you conclude from these data that preference for Brand A or B is related to sex? (A valid statistical procedure should support your answer!)

13.28 In a recent survey this question was posed: "Do you think that French Canada should be permitted a greater degree of autonomy?" When the responses were tabulated by region, the following frequencies were observed:

	Agree	Undecided	Disagree
Ontario	89	79	297
Prairies	118	130	350
British Columbia	241	140	248
Maritimes	37	59	197

(a) Compute the expected frequencies, assuming independence.

(b) Do the responses of Ontario persons differ significantly from those of Maritimes persons on this question?

(c) Compare the responses of Prairies people with those of British Columbia.

13.29 A geneticist has carried out a crossing experiment between two F_1 hybrids and obtains an F_2 progeny of 90 offspring, 80 of which appear to be wild type and 10 mutants. Are these data consistent with a 3:1 hypothesis (ratio of phenotypes expected)?

13.30 The following data (from MacMillan, 1931) show the outcome of a dihybrid cross in tomato genetics, in which the expected ratio of phenotypes is 9:3:3:1. Do a χ^2 goodness-of-fit test.

Phenotypes	O	P(X)
Tall, cut leaf	926	9/16
Tall, potato leaf	288	3/16
Dwarf, cut leaf	293	3/16
Dwarf, potato leaf	104	1/16
Total	1611	

13.31 In a genetic experiment involving a cross between two varieties of the bean *Phaseolus vulgaris*, Smith (1939) obtained the following results:

Phenotype	O	E	Ratio
Purple, buff	63	67.8	9
Purple, testaceous	31	22.6	3
Red, buff	28	22.6	3
Red, testaceous	12	7.5	1
Purple	39	45.2	6
Ox-blood red	16	15.1	2
Buff	40	45.2	6
Testaceous	12	15.1	2
Total	241	241.1	

Do these observations follow Smith's law?

13.32 A recent survey of one of the 15 Ontario universities regarding the country of birth of university professors yielded the percentage indicated in column 1 of the following table. The corresponding percentages of the actual origin (country of birth) for the whole Canadian population, according to the 1961 census, are given in column 2. Analyze the data and interpret your conclusions. (*Hint*: Pool some countries.)

Country of birth	1(%)	2(%)
Canada	34	84.4
England	36	3.5
U.S.	16	1.5
Italy	0	1.4
Scotland	2	1.3
Germany	4	1.0
U.S.S.R.	0	1.0
Other	8	5.9

13.33 Lazarsfeld (1955) mentions that "there does exist a relation between education and age," and the following data on 2300 people support this belief:

	Age	
Education	(< 40 years)	(≥ 40 years)
High	600	400
Low	400	900

The education break is between those who completed high school and those who did not. Do these data in fact support the relationship at the 5% level of significance?

13.34 Wilkinson (1968) has described one form of the World Series Pool:

> A system frequently used because it is simple and because it fits the needs of small and medium-sized organizations is to sell chances on the digits from zero to nine. The price of each digit is the same, and the numbers are generally "drawn from a hat" to insure "fairness." At the conclusion of each game, the scores are added together and the pot is awarded to the person holding the end digit of the sum. For example, a person holding the digit 2 would be the winner if the score were 2-0, 7-5, 8-4, 9-3, 10-2, 11-1, 12-0, 12-10, etc. Ordinarily there will be a new draw for each game because most participants intuitively recognize that not all numbers have an equal chance of winning.

The data below are taken from 300 World Series games.

Winner	0	1	2	3	4	5	6	7	8	9
Observed	15	27	23	49	31	41	27	38	23	26

Test the null hypothesis of equal probability. What recommendations can you suggest?

13.35 Wynne and Hartnagel (1975) investigated the relationship between race and plea negotiation. Do their data below suggest that the two variables are independent?

Negotiation	White	Native
Yes	494	28
No	1113	198

13.36 You have learned six testing procedures in this chapter utilizing three sampling distributions. For each test, state precisely the general formula for calculating the degrees of freedom of the test statistic.

14

REGRESSION AND CORRELATION

In the development of the subject of statistical inference, we have so far dealt with point estimation, confidence interval estimation, and tests of hypotheses concerning parameters of infinite populations, either in the single variable or independent bivariate setting. No mention yet has been made regarding *relationships* in the bivariate case. A very important topic in statistics, known as *regression theory*, is concerned with problems associated with *regression*, i.e., a situation where particular values of a variable are used in making *predictions* regarding the means of another variable at those values. The word "regression" comes from the fact that an earlier researcher tried to predict the mean heights of sons by going back to the heights of fathers, i.e., by regressing mean heights of sons on heights of fathers. This was done by making an (X, Y) plot of the heights of fathers, X, and the heights of sons, Y, which resulted in a *scatter diagram*. The concept has been broadened considerably in the present-day literature, and regression theory now falls under the branch of statistics called *linear and nonlinear models*.

The simplest possible relationship is the *linear relationship*; i.e., the relationship in the population is represented by a *straight line*. This line is estimated by selecting n pairs of sample values and then plotting a line through the scatter diagram. Clearly, many lines can be drawn through the points, but that is not the issue. The issue is to come up with one line, which is the best in some sense. Such a line is the *least-squares* line, and in this chapter we will only be concerned with this type of line.

Correlation measures the strength of the joint variability among the two random variables relative to their marginal variabilities. It has an interesting interpretation in the case of linear regression. *Linear correlation* measures the strength of a linear relationship, and our development will be limited to this case only, because of its popularity and usefulness.

Let us now go into a formal development of the topics by beginning with linear regression.

14.1 LINEAR REGRESSION MODELS

In studying the relationship between variables, it is imperative to be clear about the setting and structure of the two variables. We consider the following two cases:

Case A. The variable X is a *controllable quantitative variable*, i.e., a variable whose range of values is under complete control of the investigator, and hence is not a random variable. A controllable variable is called a *fixed factor*, and its values are referred to as *levels*. For each level of the fixed factor X, we observe a random variable Y, which has some underlying distribution. This means that there is an infinite population of values of Y for each given X. If an investigation is conducted using a set of n levels of X, call them $X_1, X_2, \ldots, X_n$, then at each X_i we have a distribution of Y with a certain mean and standard deviation. Indicate the mean of Y at the arbitrary value X by the symbol $\mu_{Y|X}$ and the standard deviation by $\sigma_{Y|X}$. Assume that the distributions have a common standard deviation (are homoscedastic); i.e., $\sigma_{Y|X_1} = \sigma_{Y|X_2} = \cdots = \sigma_{Y|X_n} = \sigma_{Y|X}$, say. Now assume further that the means $\mu_{Y|X_1}, \mu_{Y|X_2}, \ldots, \mu_{Y|X_n}$ all lie on a straight line with the equation $\mu_{Y|X} = \alpha + \beta X$, where α is the intercept and β the slope.

This development has led us to our first straight-line regression model, which is sketched in Figure 14.1 under the added assumption (for illustration purposes only!) that the distribution of Y for each X_i is normal. In this setting the factor X is always the *independent* variable and Y is the *dependent* variable.

Case B. If both X and Y are *random variables*, then they form a bivariate population with a joint probability distribution. For any given value of X, the random variable Y has a conditional distribution, with mean $\mu_{Y|X}$. In the normal case there are an infinite number of X values and hence an infinite number of *conditional* distributions for Y. Assume that the conditional distributions have the same standard deviation and call it $\sigma_{Y|X}$. Finally, assume that the means of the conditional distributions, i.e., the $\mu_{Y|X}$'s, all lie on a straight line with the equation $\mu_{Y|X} = \alpha + \beta X$. Hence we have obtained a second linear regression model.

Note that the only difference between Cases A and B is the fact that in B both X and Y are random variables with a joint distribution, while in A the only random variable is Y. Also note that in B the roles of X and Y can be interchanged, but in A the interchange

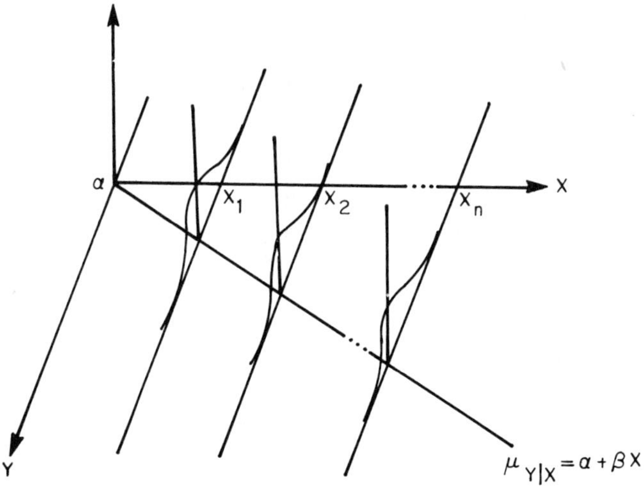

Figure 14.1 Straight-line model for X being a factor and Y a random variable.

will not make any sense whatsoever. A remarkable fact, the proof of it being beyond the scope of this book, is that when X and Y have a joint normal probability distribution (which, by the way, looks like a Mexican sombrero), then there is automatically a line connecting the means of the conditional distributions; i.e., one need not assume that they lie on a line. Also there is no need to assume that the conditional distributions have the same standard deviation, since this will be the case. When the line in this normal setup is drawn, a picture similar to Figure 14.1 will occur, and the only distinction that results is that we have an infinite number of conditional distributions rather than n.

DEFINITION 14.1 *A linear regression model is given by the equation* $\mu_{Y|X} = \alpha + \beta X$, *where* (i) X *is a factor and* Y *a random variable, whose distribution for any given value of* X, *say* X*, *has a mean equal to* $\mu_{Y|X^*} = \alpha + \beta X^*$ *and a standard deviation equal to* $\sigma_{Y|X^*}$; *or* (ii) *both* X *and* Y *are random variables with a joint distribution, such that for a given value of* X, *say* X*, *the random variable* Y *has a conditional distribution with mean equal to* $\mu_{Y|X^*} = \alpha + \beta X^*$ *and a standard deviation* $\sigma_{Y|X^*}$.

A linear regression model is also called a *simple linear regression model* of Y on X or *straight-line model* of Y on X. The line $\mu_{Y|X} = \alpha + \beta X$ is often referred to as the *population regression line* or *true regression line* of Y on X. The parameters α and β are called the *intercept* and *slope* of the true regression line.

Let us now illustrate Definition 14.1 with some examples.

Example 14.1 In an experiment on the effects of the drug cocaine on the behavior of rats, Dougherty and Pickens (1973) used three concentrations X_1, X_2, and X_3 and made measurements with respect to a particular *response* variable Y. It is assumed that the regression of Y on X is linear.

This example falls under the setting A, since X is a fixed factor whose levels are determined by the researcher, i.e., they *do not* arise randomly, and Y is a random variable, since each level of X repeated on a large number of rats will produce a population of responses. This is so because not all extraneous factors can be controlled. Hence we will have a distribution of Y values for each level of X. The assumption that the true regression line of Y on X is linear means that the line $\mu_{Y|X} = \alpha + \beta X$ goes through the means of the distributions.

Example 14.2 Lawton and Seim (1973) studied the relationship between IQ and reading comprehension in grade 6 students. They assumed the regression of reading comprehension scores, Y, on IQ scores, X, was linear. Does their analysis fall in setting A or B?

This is a clear cut example of setting B, since both X and Y are random variables, and hence they have a joint distribution. For each given value of X there is a population of Y values, namely, the conditional distribution of Y for that given value of X. The assumption of linear regression implies that there is a line, $\mu_{Y|X} = \alpha + \beta X$, passing through the means of the conditional distributions of Y.

Since both X and Y are random variables, it follows that we can interchange the roles of X and Y, i.e., take the linear regression of X on Y.

Remark. The parameters α and β of the population regression line $\mu_{Y|X} = \alpha + \beta X$ have the following interpretations. The intercept α is the value of the mean $\mu_{Y|X}$ when X assumes the value 0, while the slope β is the change in the mean $\mu_{Y|X}$ associated with one unit increase in X. Another more general interpretation of the slope is that β is the ratio of a change in $\mu_{Y|X}$ values over the corresponding change in X values. We will come back to these interpretations after estimation of α and β.

14.2 LEAST-SQUARES ESTIMATION OF THE SLOPE AND INTERCEPT OF A REGRESSION LINE

The problem of estimation in simple linear regression under both settings A and B is to find the "best" estimators of the parameters α, β, and $\sigma^2_{Y|X}$ on the basis of a sample which gives rise to n pairs of values of X and Y, i.e.,

	X	Y
Pair 1	X_1	Y_1
Pair 2	X_2	Y_2
Pair 3	X_3	Y_3
⋮	⋮	⋮
Pair n	X_n	Y_n

A plot of these n pairs of values results in the *scatter diagram* in Figure 14.2.

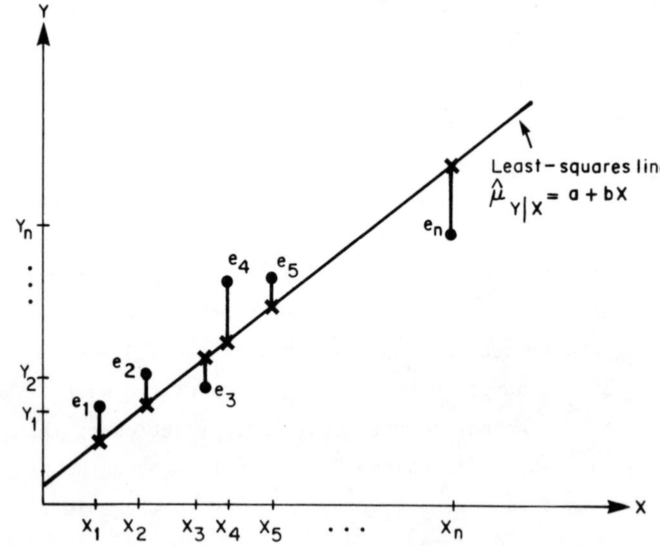

Figure 14.2 Scatter diagram of (X, Y) values and the least-squares line $\hat{\mu}_{Y|X} = a + bX$: •, observed values; ×, predicted values.

We have drawn a line through the scatter diagram, which is called the *least-squares line*. For each value of X_i it gives a predicted or estimated value $\hat{\mu}_{Y|X_i} = a + bX_i$. When an observed value Y_i is compared with a predicted value $\hat{\mu}_{Y|X_i}$, the result is the *deviation* or *error* $e_i = Y_i - \hat{\mu}_{Y|X_i}$. For the n pairs we have the following tabulation of errors:

Observed value	Predicted value	Error
Y_1	$\hat{\mu}_{Y\|X_1} = a + bX_1$	$e_1 = Y_1 - (a + bX_1)$
Y_2	$\hat{\mu}_{Y\|X_2} = a + bX_2$	$e_2 = Y_2 - (a + bX_2)$
Y_3	$\hat{\mu}_{Y\|X_3} = a + bX_3$	$e_3 = Y_3 - (a + bX_3)$
Y_4	$\hat{\mu}_{Y\|X_4} = a + bX_4$	$e_4 = Y_4 - (a + bX_4)$
⋮	⋮	⋮
Y_n	$\hat{\mu}_{Y\|X_n} = a + bX_n$	$e_n = Y_n - (a + bX_n)$

The least-squares line has the property that no other line will have a smaller sum of squared errors, i.e., $\sum_{i=1}^{n} e_i^2 = \sum_{i=1}^{n} [Y_i - (a + bX_i)]^2$ is a minimum, and hence its name least-squares line. [The method of least squares was first used by K. F. Gauss (1777-1855).] But what about a and b? These are given in the following theorem:

THEOREM 14.1 *The least-squares estimator of the population regression line* $\mu_{Y|X} = \alpha + \beta X$ *is given by* $\hat{\mu}_{Y|X} = a + bX$ *where:*

$$b = \frac{\sum_{i=1}^{n} (X_i - \overline{X})(Y_i - \overline{Y})}{\sum_{i=1}^{n} (X_i - \overline{X})^2}$$

and

$$a = \overline{Y} - b\overline{X}$$

The formula for b given in Theorem 14.1 is the defining formula. For calculation purposes one uses the equivalent computational formula:

$$b = \frac{\sum_{i=1}^{n} X_i Y_i - (1/n)\left(\sum_{i=1}^{n} X_i\right)\left(\sum_{i=1}^{n} Y_i\right)}{\sum_{i=1}^{n} X_i^2 - (1/n)\left(\sum_{i=1}^{n} X_i\right)^2}$$

For the sake of simplicity we introduce the following notation:

$$SS(X) = \Sigma(X_i - \overline{X})^2 = \Sigma X_i^2 - \frac{1}{n}(\Sigma X_i)^2$$

$$SS(Y) = \Sigma(Y_i - \overline{Y})^2 = \Sigma Y_i^2 - \frac{1}{n}(\Sigma Y_i)^2$$

$$SS(XY) = \Sigma(X_i - \overline{X})(Y_i - \overline{Y}) = \Sigma X_i Y_i - \frac{1}{n}(\Sigma X_i)(\Sigma Y_i)$$

Therefore we can write

$$b = \frac{SS(XY)}{SS(X)}$$

After a and b are calculated from the sample data, the line $\hat{\mu}_{Y|X} = a + bX$ can be drawn through the scatter. Since a and b are estimators of the population intercept α and the population slope β, respectively, it follows that they have the same interpretations as given already for α and β, with the only difference being that they are sample quantities. Thus a is the estimated value of the mean $\mu_{Y|X}$ when X = 0, and when X is increased with one unit, the corresponding estimated change in the mean $\mu_{Y|X}$ is b units. Figure 14.3 provides illustrations for various values of a and b.

The following theorem defines an estimator of the remaining parameter $\sigma^2_{Y|X}$.

THEOREM 14.2 *An unbiased estimator of the variance* $\sigma^2_{Y|X}$ *is given by the formula:*

$$s^2_{Y|X} = \frac{1}{n-2}\sum_{i=1}^{n} e_i^2 = \frac{1}{n-2}\sum_{i=1}^{n} (Y_i - a - bX_i)^2$$

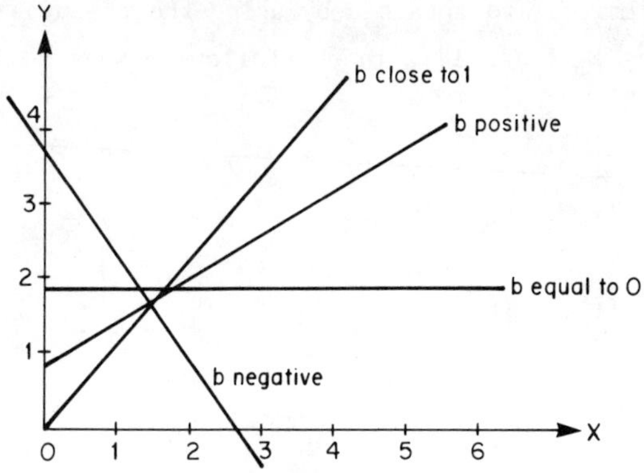

Figure 14.3 Regression line with various intercepts and slopes.

For ease of calculation the following formula is recommended:

$$s^2_{Y|X} = \frac{1}{n - 2} [SS(Y) - b \cdot SS(XY)]$$

The standard deviation $s_{Y|X}$ is found by taking the square root of $s^2_{Y|X}$. It can be shown that the estimators a and b are unbiased estimators of α and β.

Example 14.3 The following experiment is artificial, and the intent is to illustrate technique. However, such experiments are common. For analogous drug experiments the interested students are referred to MacInnes and Uphouse (1973), Dougherty and Pickens (1973), or Bolster and Schuster (1973).

An experiment was conducted with six concentrations of the drug Xyolin, and the response times in seconds of the experimental subjects were as follows:

Concentration (%)	0	1	2	3	4	5
Response time (sec)	2	4	6	7	11	14

Assuming a linear regression model of type A, let us find point estimates of the true intercept α, the true slope β, and the variance $\sigma^2_{Y|X}$.

The least-squares estimates are obtained by using the computational formulas of Theorems 14.1 and 14.2 in the following way:

X	Y	X^2	Y^2	XY
0	2	0	4	0
1	4	1	16	4
2	6	4	36	12
3	7	9	49	21
4	11	16	121	44
5	14	25	196	70
15	44	55	422	151

$n = 6$

$\Sigma X = 15 \qquad \Sigma X^2 = 55$

$\Sigma Y = 44 \qquad \Sigma Y^2 = 422$

$\Sigma XY = 151$

Therefore:

$$b = \frac{SS(XY)}{SS(X)}$$

$$= \frac{\Sigma XY - (\Sigma X)(\Sigma Y)/n}{\Sigma X^2 - (\Sigma X)^2/n}$$

$$= \frac{151 - (15)(44)/6}{55 - (15)^2/6}$$

$$= \frac{41}{17.50} = 2.34$$

and

$$a = \overline{Y} - b\overline{X}$$

$$= \frac{\Sigma Y - b \cdot \Sigma X}{n}$$

$$= \frac{44 - (2.34)(15)}{6}$$

$$= 1.48$$

The scatter diagram with the regression line $\hat{\mu}_{Y|X} = 1.48 + 2.34X$ drawn through it is presented in Figure 14.4.

Hence when the drug is absent the estimated mean response time is equal to 1.48 sec, and when the concentration is increased to 1%, then the resulting increase in the mean response time is estimated to be 2.34 sec.

An unbiased estimate of $\sigma^2_{Y|X}$ is:

$$s^2_{Y|X} = \frac{1}{n-2}\,[SS(Y) - b \cdot SS(XY)]$$

$$= \frac{1}{4}\left[\left(422 - \frac{(44)^2}{6}\right) - 2.34\left(151 - \frac{(15)(44)}{6}\right)\right]$$

$$= \frac{1}{4}\,[99.33 - 95.94] = 0.85\ (\text{sec})^2$$

Hence the estimated standard deviation is:

$$s_{Y|X} = \sqrt{0.85} = 0.92 \text{ sec}$$

DEFINITION 14.2 *A least-squares estimator of the mean of* Y *for the given value* X* *of* X *is given by* $\hat{\mu}_{Y|X^*} = a + bX^*$.

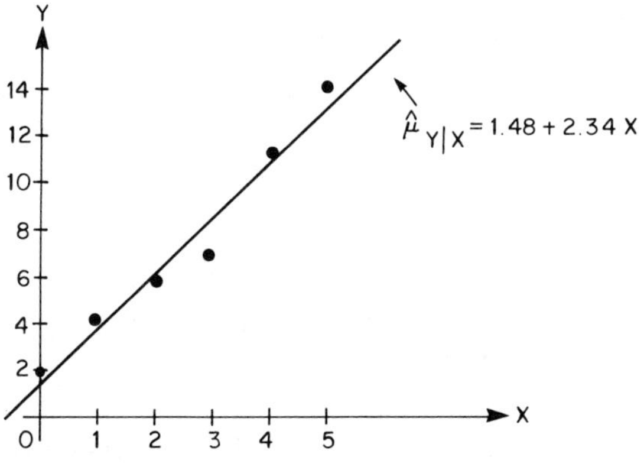

Figure 14.4 Linear regression of estimated mean response time on the drug Xyolin.

In Example 14.3, if we pick $X^* = 2.5$, then the estimated mean response time for this value is equal to $\hat{\mu}_{Y|2.5} = 1.48 + (2.34)(2.5) = 7.33$ sec. Note that estimates were used to draw the regression line, and that the intercept is an estimate for $X = 0$.

We know that in repeated sampling we will get different values of a, b, and the estimator $\hat{\mu}_{Y|X^*}$; i.e., these estimators are random variables. This means that we should provide their estimated standard errors. These are given by the following expressions:

$$s_a = s_{Y|X}\sqrt{\frac{\Sigma X^2/n}{SS(X)}}$$

$$s_b = s_{Y|X}/\sqrt{SS(X)}$$

$$s_{\hat{\mu}_{Y|X^*}} = s_{Y|X}\sqrt{\frac{1}{n} + \frac{(X^* - \overline{X})^2}{SS(X)}}$$

For Example 14.3 we obtain the following results:

$$s_a = 0.92\sqrt{\frac{55/6}{17.50}} = 0.67 \text{ sec}$$

$$s_b = 0.92\sqrt{\frac{1}{17.50}} = 0.22 \text{ sec}$$

and

$$s_{\hat{\mu}_{Y|3.0}} = 0.92\sqrt{\frac{1}{6} + \frac{(3.0 - 2.5)^2}{17.50}} = 0.39 \text{ sec}$$

for $X^* = 3.0$.

14.3 CONFIDENCE INTERVAL ESTIMATION AND TESTS OF HYPOTHESES IN LINEAR REGRESSION

So far in the development we have not assumed that the population of Y values for each given value of X is normal. The following theorem is concerned with the distribution of a, b, and $\hat{\mu}_{Y|X^*}$ when normality is imposed in both settings A and B of linear regression.

THEOREM 14.3 *The estimators* a, b, *and* $\hat{\mu}_{Y|X*}$ *in repeated sampling are, respectively,*

$$N\left(\mu_a = \alpha,\ \sigma_a^2 = \sigma_{Y|X}^2\left(\frac{\Sigma X_i^2/n}{SS(X)}\right)\right)$$

$$N\left(\mu_b = \beta,\ \sigma_b^2 = \frac{\sigma_{Y|X}^2}{SS(X)}\right)$$

and

$$N\left(\mu_{\hat{\mu}_{Y|X*}} = \mu_{Y|X*},\ \sigma_{\hat{\mu}_{Y|X*}}^2 = \sigma_{Y|X}^2\left(\frac{1}{n} + \frac{(X* - \overline{X})^2}{SS(X)}\right)\right)$$

This theorem then leads us to the following crucial theorem, which is needed for small sample confidence interval estimation and small sample tests of hypotheses.

THEOREM 14.4 *Each of the statistics* $(a - \alpha)/s_a$, $(b - \beta)/s_b$, *and* $(\hat{\mu}_{Y|X*} - \mu_{Y|X*})/s_{\hat{\mu}_{Y|X*}}$ *has the t distribution with* n - 2 *degrees of freedom for given* α, β, *and* $\mu_{Y|X*}$.

By the procedure outlined in Chapters 11 and 13 we obtain the following pair of theorems (notice that we have used the symbol $\alpha*$ for the significance level, in order not to confuse it with the intercept α):

THEOREM 14.5 *A* $\gamma = 1 - \alpha*$ *confidence interval for* α, β, *and* $\mu_{Y|X*}$ *is given, respectively, by:*

$$(L,\ U) = (a - t_{n-2,\alpha*/2} \cdot s_a,\ a + t_{n-2,\alpha*/2} \cdot s_a)$$

$$(L,\ U) = (b - t_{n-2,\alpha*/2} \cdot s_b,\ b + t_{n-2,\alpha*/2} \cdot s_b)$$

$$(L,\ U) = (\hat{\mu}_{Y|X*} - t_{n-2,\alpha*/2} \cdot s_{\hat{\mu}_{Y|X*}},\ \hat{\mu}_{Y|X*} + t_{n-2,\alpha*/2} \cdot s_{\hat{\mu}_{Y|X*}})$$

THEOREM 14.6 *Tests for the hypotheses on* α, β, *and* $\mu_{Y|X^*}$ *are as follows:*

Case	H_0	H_A	Decision rule: Reject H_0 if	Test statistic
1	$\alpha = \alpha_0$	$\alpha > \alpha_0$	$t > t_{n-2,\alpha^*}$	$t = (a - \alpha_0)/s_a$
2	$\alpha = \alpha_0$	$\alpha < \alpha_0$	$t < -t_{n-2,\alpha^*}$	$t = (a - \alpha_0)/s_a$
3	$\alpha = \alpha_0$	$\alpha \neq \alpha_0$	$\|t\| > t_{n-2,\alpha^*/2}$	$t = (a - \alpha_0)/s_a$
4	$\beta = \beta_0$	$\beta > \beta_0$	$t > t_{n-2,\alpha^*}$	$t = (b - \beta_0)/s_b$
5	$\beta = \beta_0$	$\beta < \beta_0$	$t < -t_{n-2,\alpha^*}$	$t = (b - \beta_0)/s_b$
6	$\beta = \beta_0$	$\beta \neq \beta_0$	$\|t\| > t_{n-2,\alpha^*/2}$	$t = (b - \beta_0)/s_b$
7	$\mu_{Y\|X^*} = \mu_0$	$\mu_{Y\|X^*} > \mu_0$	$t > t_{n-2,\alpha^*}$	$t = \dfrac{\hat{\mu}_{Y\|X^*} - \mu_0}{s_{\hat{\mu}_{Y\|X^*}}}$
8	$\mu_{Y\|X^*} = \mu_0$	$\mu_{Y\|X^*} < \mu_0$	$t < -t_{n-2,\alpha^*}$	$t = \dfrac{\hat{\mu}_{Y\|X^*} - \mu_0}{s_{\hat{\mu}_{Y\|X^*}}}$
9	$\mu_{Y\|X^*} = \mu_0$	$\mu_{Y\|X^*} \neq \mu_0$	$\|t\| > t_{n-2,\alpha^*/2}$	$t = \dfrac{\hat{\mu}_{Y\|X^*} - \mu_0}{s_{\hat{\mu}_{Y\|X^*}}}$

All null hypotheses are accepted otherwise.

The interesting null hypotheses are H_0: $\alpha = 0$ and H_0: $\beta = 0$. In the former case we are testing whether the true regression line goes through the origin, and in the latter case we are testing whether the true regression line is parallel to the X axis (i.e., no variation in $\mu_{Y|X}$ values due to X, or equivalently, no linear regression).

For Example 14.3, let us construct 95% confidence intervals for the true slope β and the true mean $\mu_{Y|X^*}$ when $X^* = 3.0$. From Theorem 14.5 we have immediately the following results:

For

$$\beta: \quad (L, U) = (2.34 - (2.776)(0.22),\ 2.34 + (2.776)(0.22)) = (1.73, 2.95)$$

and for

$$\mu_{Y|3.0}: \quad (L, U) = (8.50 - (2.776)(0.39),\ 8.50 + (2.776)(0.39)) = (7.42, 9.58)$$

Using the same example, to test H_0: $\beta = 0$ against H_A: $\beta \neq 0$ we see from Theorem 14.6 that with $\alpha^* = 0.05$

$$|t| = \left|\frac{2.34 - 0}{0.22}\right| = 10.64$$

which is greater than 2.776, and hence H_0 is rejected. This conclusion agrees with the confidence interval for β, which does not contain 0. These results imply that there is a significant regression.

If confidence intrevals are calculated for the true means for various X^*'s, then these can be plotted above and below the $\hat{\mu}_{Y|X^*}$'s, and by connecting the points we obtain what is known in statistics as *confidence bands for the true line*. In an exercise at the end of this chapter we give the reader an opportunity to sketch such bands.

14.4 THE PEARSON PRODUCT-MOMENT CORRELATION COEFFICIENT

The concept of *correlation* makes sense only in case B of linear regression, i.e., both variables are random variables. The Pearson product-moment correlation coefficient, or simply correlation coefficient, measures the strength or degree of the linear regression. The symbol ρ (the Greek letter rho) is reserved for the population correlation coefficient. The following theorem gives an expression of ρ and its range:

THEOREM 14.7 *For model* B *of linear regression* $\rho = \beta\sigma_X/\sigma_Y$, *where* β *is the population slope of the regression line* $\mu_{Y|X} = \alpha + \beta X$. *Also,* $-1 \leq \rho \leq 1$.

From $\rho = \beta\sigma_X/\sigma_Y$ it follows that ρ and β have the same sign; e.g., when ρ is negative, then β is negative, and vice versa. It is beyond the scope of this book to show that ρ is invariant under interchange of the random variables in the regression model; i.e., the correlation between X and Y is the same as between Y and X. When ρ is either +1 or -1 there is *perfect linear regression*, and the joint distribution of X and Y is then degenerate. Interesting cases are in between. When $\rho = 0$ then $\beta = 0$, so that the population regression line is parallel, and hence there is no linear regression.

The problem, of course, is to find an estimator for ρ based on a sample of n pairs. This is given in the next theorem:

THEOREM 14.8 *The sample product-moment correlation coefficient* $r = bs_X/s_Y$ *is a biased estimator of* ρ.

The formula of $r = bs_X/s_Y$ can be rewritten in the following ways:

$$r = \frac{\Sigma(X_i - \overline{X})(Y_i - \overline{Y})}{\sqrt{\Sigma(X_i - \overline{X})^2\Sigma(Y_i - \overline{Y})^2}} = \frac{SS(XY)}{\sqrt{SS(X) \cdot SS(Y)}}$$

$$= \frac{\Sigma X_i Y_i - (1/n)(\Sigma X_i)(\Sigma Y_i)}{\sqrt{[\Sigma X_i^2 - (1/n)(\Sigma X_i)^2][\Sigma Y_i^2 - (1/n)(\Sigma Y_i)^2]}}$$

It can be shown that $-1 \le r \le 1$, and, as before, r and b have the same sign; when $r = 0$, then b is also 0 and vice versa. The four diagrams in Figure 14.5 illustrate how r measures the strength of the sample linear regression.

Example 14.4 Suppose 10 father-son pairs of mature men were chosen at random and their heights measured. If X and Y refer, respectively, to the fathers' and sons' heights in inches, then suppose the results were:

Pair	1	2	3	4	5	6	7	8	9	10
X	69	70	69	68	70	73	69	67	69	64
Y	68	69	72	67	70	71	72	66	71	65

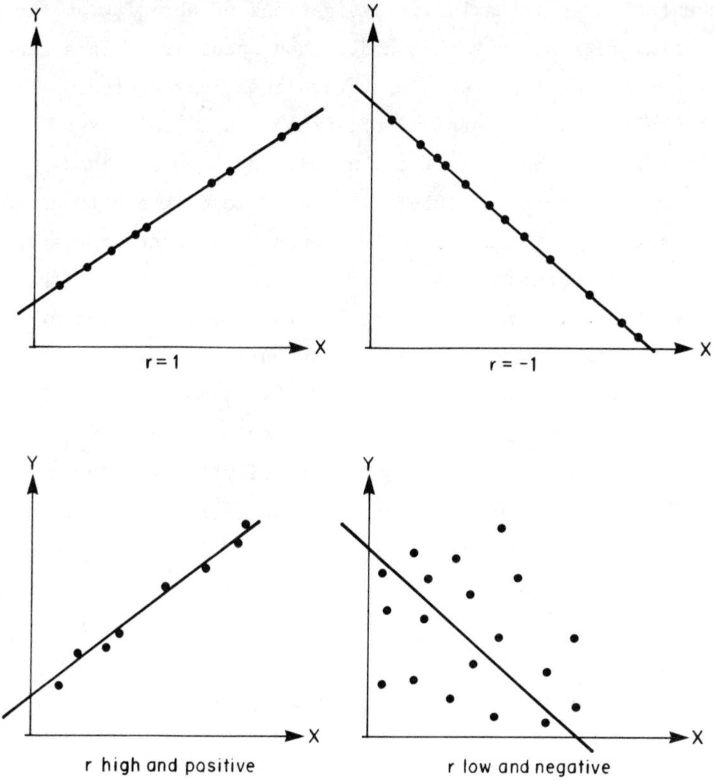

Figure 14.5 Illustrations of the meaning of r.

The value of r is

$$r = \frac{SS(XY)}{\sqrt{SS(X) \cdot SS(Y)}} = \frac{37.2}{\sqrt{(47.6)(56.9)}} = 0.715$$

This indicates a fairly strong, positive, linear relationship between Y and X. The estimated strength of the regression line $\hat{\mu}_{Y|X} = 15.3 + 0.78X$ is equal to 0.715.

The historical incident (which is well documented) that led to the word *regression* is itself of value in your understanding of it. The English biometrician Sir Francis Galton (1822-1911), one of the first men ever to study individual differences in a systematic fashion, invented it while making investigations of heredity on the "law of universal regression." He said, "Each peculiarity in a man is shared by his kinsmen, but on the average in a less degree." He collected the heights of parents and offspring and actually submitted them to the kind of analysis that has already been outlined. His friend, Karl Pearson (1857-1936), collected more than a thousand records of heights of members of family groups. The scatter diagram used Y = son's height against X = father's height. He noticed that tall fathers do tend to have tall sons, but the average height of sons of a group of tall fathers is less than their father's height. For n = 1078 sample pairs he actually found $\hat{\mu}_{Y|X} = 33.73 + 0.516X$. There is a regression, or going back, of the son's height toward the average of all men. Galton called this trend of the offspring "falling back" toward the general mean the "law of filial regression." Today we speak of "regression toward the mean." Galton used the slope of the line, b, as an index of this regression. Today it is called the "regression coefficient." Pearson developed r as we know it and called it the "correlation index" and used it as a measure of the degree of regression. So historically the regression line came first.

14.5 TESTS OF HYPOTHESES AND CONFIDENCE INTERVAL ESTIMATION FOR ρ

Testing the null hypothesis H_0: $\rho = 0$ is equivalent to testing H_0: $\beta = 0$, because of the relationship $\rho = \beta\sigma_X/\sigma_Y$. Under the assumption of normality a test for H_0: $\beta = 0$ automatically provides a test for H_0: $\rho = 0$. Hence it is actually not necessary to introduce a new test for H_0: $\rho = 0$. However, for completeness' sake we state the following theorem, which is a consequence of Theorem 14.6:

THEOREM 14.9 *Under the assumption of bivariate normality, the statistic* $r/\sqrt{(1 - r^2)/(n - 2)}$ *has in repeated sampling a t distribution with* n - 2 *degrees of freedom when* $\rho = 0$, *so that for testing the hypotheses* (a) H_0: $\rho = 0$, H_A: $\rho > 0$; (b) H_0: $\rho = 0$, H_A: $\rho < 0$; *and* (c) H_0: $\rho = 0$, H_A: $\rho \neq 0$ *we have, respectively, the tests*

(a) *reject* H_0 *if* $t > t_{n-2,\alpha}$; (b) $t < -t_{n-2,\alpha}$; *and* (c) $|t| > t_{n-2,\alpha/2}$, *where* $t = r/s_r$ *and* $s_r = \sqrt{(1 - r^2)/(n - 2)}$.

Example 14.5 To illustrate Theorem 14.9, consider Example 14.4. To test H_0: $\rho = 0$ versus H_A: $\rho \neq 0$, the value of the test statistic is

$$t = \frac{0.715}{\sqrt{[1 - (0.715)^2]/8}} = \frac{0.715}{0.247} = 2.9$$

which is larger (in absolute value) than $t_{8,0.025} = 2.306$. Hence the null hypothesis, that the strength of the linear relationship is zero, is rejected at the 5% level of significance.

Now, the reader may query, why did we not immediately tackle testing a null hypothesis of the type H_0: $\rho = \rho_0$. After all, H_0: $\rho = 0$ is just a special case of this. There is trouble, because the statistic $(r - \rho_0)/\sqrt{(1 - r^2)/(n - 2)}$ does not have the t distribution when $\rho_0 \neq 0$.

The following theorems are due to R. A. Fisher. Theorem 14.10 implicitly defines the so-called *Fisher z transformation*.

THEOREM 14.10 *Under the assumption of normal linear regression, the variable* $z = (1/2)\ln[(1 + r)/(1 - r)]$ *is approximately*

$$N\left(\mu_z = \frac{1}{2}\ln\frac{1 + \rho}{1 - \rho},\ \sigma_z = (n - 3)^{-1/2}\right)$$

THEOREM 14.11 *Approximate statistical tests for*

(a) H_0: $\rho = \rho_0$, H_A: $\rho > \rho_0$
(b) H_0: $\rho = \rho_0$, H_A: $\rho < \rho_0$
(c) H_0: $\rho = \rho_0$, H_A: $\rho \neq \rho_0$

are given, respectively, by

(a) *Reject* H_0 *if* $Z > Z_\alpha$.
(b) *Reject* H_0 *if* $Z < -Z_\alpha$.
(c) *Reject* H_0 *if* $|Z| > Z_{\alpha/2}$.

The test statistic is $Z = (z - \mu_0)/(n - 3)^{-1/2}$ *and* $\mu_0 = (1/2)\ln[(1 + \rho_0)/(1 - \rho_0)]$. *The null hypotheses are accepted otherwise.*

Table A.7 gives transformed z values for given r values, so that there is no problem of messy calculations.

Example 14.6 Kaufman and Kaufman (1972) found that for a random sample of first grade children, the Stanford Achievement Test had a correlation of $r = 0.64$ with the Gesell School Readiness Test. Suppose $n = 50$ and someone claimed that $\rho = 0.50$. Do the data support this claim?

Under the assumption that the grades have a bivariate normal distribution, it follows from Theorem 14.11 and Table A.7 that:

$$|Z| = \left|\frac{z - \mu_0}{\sqrt{1/(n - 3)}}\right| = \left|\left(\frac{1}{2}\ln\frac{1 + 0.64}{1 - 0.64} - \frac{1}{2}\ln\frac{1 + 0.50}{1 - 0.50}\right)\cdot\sqrt{47}\right|$$

$$= |(0.758 - 0.549)\sqrt{47}| = |1.43| = 1.43$$

which is less than $Z_{0.025} = 1.96$. Hence we do not reject H_0: $\rho = 0.5$, and therefore the data lack evidence to reject the claim.

Fisher's z transformation can also be used to put confidence intervals on ρ. The trick is to construct via Theorem 14.10 a confidence interval for $\mu_z = (1/2)\ln[(1 + \rho)/(1 - \rho)]$ and then use Table A.7 in an inverse manner. We thus have the following theorem:

THEOREM 14.12 *An approximate* $\gamma = 1 - \alpha$ *confidence interval for* $\mu_z = (1/2)\ln[(1 + \rho)/(1 - \rho)]$ *is given by* $(L, U) = (z - Z_{\alpha/2}\cdot\sqrt{1/(n - 3)},\ z + Z_{\alpha/2}\cdot\sqrt{1/(n - 3)})$, *so that an approximate* $\gamma = 1 - \alpha$ *confidence interval for* ρ *is given by the inverse z transformation of* L *and* U.

Example 14.7 To illustrate Theorem 14.12, let $r = 0.2$ and $n = 25$. Using Table A.7, we obtain the following 95% confidence interval $(L, U) = (0.203 - (1.96)(1/\sqrt{22}),\ 0.203 + (1.96)(1/\sqrt{22})) = (-0.214, 0.620)$, and looking at the body of Table A.7 we have immediately the 95% confidence interval $(L, U) = (-0.21, 0.55)$ for ρ.

EXERCISES

14.1 Give two examples, from literature of your own choice, of both cases of linear regressions discussed in this chapter.

14.2 True or false:

(a) The correlation coefficient is sometimes greater than 1.

(b) The slope b can be greater than β.

(c) If X is increased by one unit, then $\hat{\mu}_{Y|X}$ is increased by r units.

(d) There exist exact tests for H_0: $\beta = 0$ and H_0: $\rho = 0$ against H_A: $\beta > 0$ and H_A: $\rho \neq 0$, respectively.

(e) The slope b is positive if and only if r is positive.

(f) The regression line may be defined as a straight line which makes the squared deviations around it maximal.

(g) The statistic r is the appropriate measure of correlation when the variables are related in a nonlinear fashion.

(h) The strength of the linear relationship between two variables is indicated by the coefficient of correlation.

(i) The regression coefficient b can take on values only from -1.00 to +1.00.

(j) The slope of the regression line depends on the units of X and Y.

(k) The vertical distance from a point to the regression line represents the deviation of a point from the predicted value of μ_Y in Case A.

(l) There are other methods for obtaining a regression line relating Y to X, besides the least-squares approach.

14.3 Consider the following sample of 15 pairs of measurements (in grams) on 15 randomly selected experimental subjects:

X	12	20	17	11	8	8	4	12	9	12	16	11	10	13	15
Y	74	147	147	75	46	59	20	90	74	77	144	110	99	109	109

(a) Make a scatter diagram of the data.

(b) Find the least-squares regression line and superimpose it on the scatter diagram.

(c) Give an interpretation of the calculated slope and intercept.

(d) Construct a 99% confidence interval for β and give a repeated sampling interpretation.

(e) Is the slope β significantly different from zero? (Use $\alpha = 1\%$.)

(f) Estimate $\mu_{Y|X*}$ for some X* and find a confidence interval for it.

14.4 For example 14.3 draw the 95% confidence bands by calculating several confidence intervals.

14.5 For the data in Exercise 14.3:

(a) Find r and give an interpretation.

(b) Is there a significant correlation? (Use $\alpha = 5\%$.)

(c) Test the null hypothesis H_0: $\rho = 0.4$ against H_A: $\rho \neq 0.4$.

(d) Find a 95% confidence interval for ρ.

14.6 From another statistical text find the least-squares line for a regression through the origin; i.e., $\mu_{Y|X} = \beta X$. Illustrate with your own data. Calculate r and interpret it and the least-squares estimate b.

14.7 *Spearman's rank correlation* is calculated in the following way:

Values		Ranks		$D = \lvert R_X - R_Y \rvert$	D^2
X	Y	R_X	R_Y		
X_1	Y_1	R_{X_1}	R_{Y_1}	$\lvert R_{X_1} - R_{Y_1} \rvert$	D_1^2
X_2	Y_2	R_{X_2}	R_{Y_2}	$\lvert R_{X_2} - R_{Y_2} \rvert$	D_2^2
⋮	⋮	⋮	⋮	⋮	⋮
X_n	Y_n	R_{X_n}	R_{Y_n}	$\lvert R_{X_n} - R_{Y_n} \rvert$	D_n^2

[Charles Spearman (1863-1945) was an English psychologist.] The ranks are assigned in such a way that 1 corresponds to the highest value and n to the lowest value. Tied ranks are assigned the average. Then:

$$r = 1 - \frac{6\Sigma D_i^2}{n(n^2 - 1)}$$

Find r for the data in Exercise 14.3 and explain what you have found.

14.8 Does the regression line $\hat{\mu}_{Y|X} = a + bX$ go through the points $(\overline{X}, \overline{Y})$? Illustrate your answer for the data in Exercise 14.3.

14.9 From other texts, find the reason why r^2 is called the *coefficient of determination*, and why $(1 - r^2)$ is called the *coefficient of alienation*.

14.10 Explain the difference between regression and correlation.

14.11 The concepts of correlation are applicable to quantitative data, but developments have taken place with respect to *measures of association for attribute data*. List three such coefficients from other texts and illustrate them with data of your own choice.

14.12 Multiple choice:

1. Which of the following values of r will result in the highest degree of linear relationship?

   (a) 0.00
   (b) 0.18
   (c) 0.58
   (d) -0.34
   (e) -0.71

2. The regression line for predicting $\mu_{Y|X}$ from known values of X when r equals 0.00:

   (a) Forms a 45° angle with the X axis
   (b) Forms a 45° angle with the regression line for predicting μ_X from values of Y
   (c) Is $\overline{X}$
   (d) Is $\overline{Y}$
   (e) Passes through all the paired scores

3. As n increases, r

   (a) Increases
   (b) Decreases
   (c) Increases to a point and then decreases
   (d) Is independent of n in the sense of increase or decrease
   (e) None of the above

4. If we obtain a negative r, this means that:

   (a) Individuals scoring high on one variable tend to score low on a second variable .
   (b) Individuals scoring high on one variable tend to score high on a second variable.
   (c) There is no relationship between the two variables.
   (d) The relationship is in the opposite direction to the one predicted.
   (e) We have made an error.

5. Which of the following might explain a low correlation?

   (a) Little or no relationship between the two variables.
   (b) The variables are related in a nonlinear fashion.
   (c) The range of values of one of the variables is restricted.
   (d) All of the above.
   (e) None of the above.

6. If the correlation between body weight and annual income was high and positive, we could conclude that

   (a) High incomes cause people to eat more food.
   (b) Low incomes cause people to eat less food.
   (c) High-income people spend a greater proportion of their income on food.
   (d) All of the above.
   (e) None of the above.

7. When the scores on one variable increase as the scores on the second variable decrease, there is a __________ relationship.

   (a) Positive
   (b) Negative
   (c) Zero
   (d) Infinite
   (e) Skewed

8. A correlation between college entrance exam grades and scholastic achievement was found to be -1.08. On the basis of this you would tell the university that

   (a) The entrance exam is a good predictor of success.

(b) They should hire a new statistician.
(c) The exam is a poor predictor of success.
(d) Students who do best on this exam will make the worst students.
(e) Students at this school are underachieving.

14.13 A student union president ranked 20 liberal arts professors on their neatness and their teaching proficiency during last spring semester. He performed a correlation analysis on his results.

(a) What correlation coefficient (i.e., Pearson or Spearman) should he use?

(b) If he obtained a correlation coefficient of -0.39, what can he conclude?

14.14 Five mature healthy females were selected in a study of Ellis and Cohn (1975) and the following data were obtained. Analyze these data.

Subject	1	2	3	4	5
Weight (kg)	89.5	78.6	61.8	59.1	52.7
Height (m)	1.587	1.720	1.614	1.549	1.564

14.15 The following data provide the actual speed records attained in the Indianapolis Memorial Day automobile races (1911-1941) in miles per hour. (*Note*: X = year - 1911, Y = speed - 70 mph.)

X	Y
0	4.6
1	8.7
2	5.9
3	12.5
4	19.8
5	13.3
8	18.1
9	18.6
10	19.6
11	24.5
12	21.0
13	28.2
14	31.1
15	25.9
16	27.5

(Table continued)

X	Y
17	29.5
18	27.6
19	30.4
20	26.6
21	34.1
22	34.2
23	34.9
24	36.2
25	39.1
26	43.6
27	47.2
28	45.0
29	44.3
30	45.1

$\Sigma X = 452$ $\quad\quad \overline{X} = 15.586$

$\Sigma Y = 797.1$ $\quad\quad \overline{Y} = 27.486$

$\Sigma X^2 = 9370$ $\quad\quad SS(X) = 2325.034483$

$\Sigma Y^2 = 25{,}949.07$ $\quad\quad SS(Y) = 4039.814483$

$\Sigma XY = 15{,}395.0$ $\quad\quad SS(XY) = 2971.234483$

(Note that in 1917 and 1918, no race was run, so that n = 29.)

(a) What is the average speed in miles per hour over this time span?

(b) Is this Case A or Case B?

(c) Find the regression line and plot it on the scatter diagram.

(d) Using this line, predict the speed in 1973, 1974, and 1975. (This is called *extrapolation*, and it has some obvious disadvantages.) Also find the 95% confidence interval for each of these years. Do your estimates come close to the actual recorded speed records for these years? Plot the point and interval estimates and the actual values on the scatter diagram. Any conclusions?

14.16 (a) If r is close to zero, would knowledge of X be of any help in estimating $\mu_{Y|X}$?

(b) If r is close to 1, answer part (a).

(c) Why would one be interested in testing whether the regression coefficient is zero?

(d) What information about the causal relationship between X and Y does ρ provide?

(e) What is the essential difference between $r = +1$ and $r = -1$?

(f) The Spearman's rank correlation coefficient is sometimes used, where the regular correlation coefficient could be used. Why?

14.17 Write a true statement about the following:

(a) A line drawn freehand through the scatter diagram.

(b) Y intercept of a line.

(c) You computed a 90% confidence interval for β to be $0.26 \le \beta \le 0.80$.

(d) s_a and s_b.

14.18 We are given data for nine regions to obtain information on per capita income and regional expenditure per student for education. Our interest is in determining the strength of the relationship between the two variables.

(a) What regression model would you suggest?

(b) What are the variables in the problem?

(c) Discuss the procedure you would use in solving the above problem.

14.19 The data given in Tables 1 and 2 were taken from a study by Jenkins (1963). This study involved selected socioeconomic characteristics of many neighborhoods in New York City in 1960. n refers to specific neighborhoods in Brooklyn (Table 1) and Manhattan (Table 2). X refers to the median family income and Y refers to the juvenile delinquency rate per 1000 youths aged 7 to 20. The variable X is often used as a criterion of need for community subsidy for services. Y is a measure of social disorganization reflecting special problems. (The actual Y rates here concern youngsters who were arrested or referred to the police or children's court in 1961.)

TABLE 1

n	X	Y
1	5767	48
2	4545	67
3	5469	54
4	5170	85
5	4550	101
6	5776	57
7	4595	80
8	5532	34
9	4719	58
10	5743	51
11	5836	39
12	6998	23
13	6308	20
14	6370	18
15	6235	19
16	5352	49
17	6929	21
18	6286	25
19	7351	16
20	6870	26

TABLE 2

n	X	Y
1	4799	36
2	7910	46
3	4119	63
4	5149	67
5	8853	56
6	8352	33
7	9746	24
8	5770	66
9	7180	49
10	5064	55
11	4949	63
12	3999	101
13	3712	87
14	6428	25
15	5662	40

For the data of Tables 1 and 2, answer the following:

(1) Is there any relationship between juvenile delinquency and family income?

(a) Does the problem fall under setting A or setting B?

(b) What are the assumptions?

(c) Compute $\hat{\mu}_{Y|X} = a + bX$ and draw it through a scatter diagram.

(2) Obtain estimates of the population correlation coefficients of the data of both tables. Compare them.

(3) What conclusions can be reached from this analysis?

14.20 The following data are from a more extensive study, due to Jenkins (1963). We have taken 20 Brooklyn neighborhoods and listed the values of X, the juvenile delinquncy rate (see Exercise 14.19), and Y, the percentage of total population which is Puerto Rican, Black, and other nonwhite groups in 1960.

	1	2	3	4	5	6	7	8	9	10
X	48	67	54	85	101	57	80	34	58	51
Y	7	48	13	40	82	35	46	12	34	16

	11	12	13	14	15	16	17	18	19	20
X	39	23	20	18	19	49	21	25	16	26
Y	9	1	1	1	1	8	3	7	1	2

Is there any statistical evidence of a relationship between the juvenile delinquency rate and the proportion of nonwhite peoples in neighborhoods of Brooklyn in 1960?

14.21 Ellis and Cohn (1975) obtained the following data on 9 randomly chosen adult male subjects:

Subject	Age (years)	Weight (kg)	Height (m)	TBK (g)	TBCa (g)
1	43	100.0	1.830	161.3	1348
2	36	90.0	1.855	180.6	1366
3	35	86.4	1.690	148.2	1081
4	33	83.2	1.778	140.7	1169
5	51	76.4	1.793	142.0	1161
6	39	72.3	1.598	118.9	1036
7	49	70.5	1.750	119.8	1042
8	37	64.1	1.730	115.5	907
9	30	61.8	1.640	100.4	876
Mean	39.2	78.3	1.740	136.4	1110
SD	7.1	12.6	0.085	25.2	171.5

(a) TBK and TBCa refer to total body potassium and calcium content, respectively. They obtained the linear relationship $\hat{\mu}_{TBK} = -17.1 + 0.1383\text{TBCa}$ with $r = 0.941$. Verify these results.

(b) Are there any significant linear regressions or correlations in these data?

14.22 For a sample of 234 young children Masters and Mokros (1973) found a correlation of $r = 0.25$ in a controlled choice preference experiment. Test the null hypothesis that $\rho = 0$ using a 1% level of significance.

14.23 Smith and Hanna (1975) have observed the following data on 14 randomly chosen subjects; where W = body weight (kg) and Y = body surface area (square meters) and X = age (years).

	W	Y	X
	79.9	1.88	22
	59.1	1.67	20
	84.5	1.98	19
	62.4	1.70	24
	53.4	1.56	20
	61.4	1.66	18
	73.5	1.81	20
	84.5	2.05	20
	50.4	1.53	18
	69.0	1.85	27
	59.9	1.79	18
	74.8	1.97	22
	66.6	1.84	22
	97.3	2.20	20
Mean	69.76	1.82	20.7

(a) Obtain scatter diagrams of W versus X, W versus Y, and Y versus X.

(b) Obtain the regression lines.

(c) Test if the true regression coefficients are significantly different from zero.

(d) For those regression coefficients which are significantly different from zero, find the sample correlation coefficient and then test if $\rho = 0$.

14.24 Gallup (1975) obtained the following data from a large survey sample on the question, "Do you favor registration of all firearms?"

X = Income ($)	Y = % in favor
3,000-less than 5,000	66
5,000-less than 7,000	70
7,000-less than 10,000	66
10,000-less than 15,000	71
15,000-less than 20,000	66

X = Age (years)	Y = % in Favor
18-24	76
25-29	67
30-49	60
50-59	64

For these two data sets:

(a) Plot Y versus X = midpoint of interval.

(b) Is there evidence to suggest a significant correlation?

14.25 The following article actually appeared in the *Toronto Telegram* on May 25, 1971:

SALARY INCREASES AS APPLICANTS' HEIGHT GOES UP

A job applicant's starting salary depends more on how far above the floor his brain is than on what the brain contains, a University of Pittsburgh administrator says.

"Men over six feet make about 10 percent more than men under six feet on the first job," says Leland Deck, director of labor relations in the university's personnel department.

Mr. Deck said a 1967 survey of Pitt business school graduates found a 4 percent higher starting salary for men over six feet and a survey of 1970 graduates found the bonus upped 10 percent.

In fact the tallest graduate surveyed last year had the lowest grade point average, but got the highest starting salary.

"The sad thing about this all is that we are not talking about laborers...I am talking about guys who are being recruited by corporations because they are presumably bright, have got a brain and the brain has been trained.

"Yet they are not examining the brain, they are just hiring on the length of the spine."

Suggest a possible scatter diagram which would reflect this report, and comment on this article.

15

DESIGN AND ANALYSIS OF EXPERIMENTS

As pointed out in Chapter 1, the discipline of statistics is concerned with the collection, analysis, interpretation, and presentation of data. The first aspect, namely, the collection of data, clearly requires a plan or *design* on how to obtain the relevant data. There are often situations where no elaborate plan is needed, because the data can be readily obtained from existing files of governmental, commercial, research, and other institutions. If no data are readily available, then the projected investigations will call for either a *survey* or an *experiment*. There is a distinction between a survey and an experiment.

In a survey the phenomenon under investigation is observed or measured under existing conditions, i.e., without exercising any control on factors that may influence it. If the objective of the investigation is to observe responses by influencing factors, then we have a controlled experiment. For example, we may conduct a survey of sociological conditions of a particular population, or we can carry out an experiment to measure the effect of different doses of a drug on reaction time of humans.

The whole subject of design and analysis of experiments was put on a rigorous statistical basis by R. A. Fisher (1935) in his well-known book, *The Design of Experiments*. Since then, numerous authors have made significant contributions in this area. A good source of the development up to 1968 can be found in Federer and Balaam (1972).

Before defining some of the notions of experimental design, consider a psychological experiment to compare the change in pupil size of eight homosexual versus eight heterosexual males when viewing pictures of a female. For an original experiment of this sort the reader is referred to E. H. Hess *et al.* (1965). Suppose that for illustration purposes the percentage changes in pupil sizes were as follows:

Heterosexual		Homosexual	
Subject	Change (%)	Subject	Change (%)
1	+6.2	1	-30.6
2	+14.1	2	+11.4
3	-1.6	3	+19.3
4	+40.4	4	-8.2
5	-20.1	5	+22.1
6	+10.8	6	+16.9
7	+4.8	7	-3.7
8	+16.5	8	+12.8

The factor under study in this experiment is "sexual orientation of the male" with the two levels or treatments being "heterosexual" and "homosexual." The subjects form the experimental units and the response or variable of interest is percentage change of pupil size. In order to analyze the data statistically, certain assumptions are necessary. If the data can be viewed as random samples from two independent normal populations having the same variance, then from Chapter 13, we know that a pooled t test will provide us with the answer on whether to reject or accept the null hypothesis of no difference between the mean percentage changes in the pupil sizes of the two populations.

Suppose now that the experimenter enlarged the experiment by including in the study eight heterosexual and eight homosexual females. In this setup there are four treatments, seen to be arising from two factors; namely, "sexual orientation of the male" and "sexual orientation of the female," each at two levels. The null hypothesis of interest is that the mean percentage changes in the pupil sizes of the four populations are the same. Making the assumption that the data are random samples from four independent normal populations with the same variance will not lead to the use of the pooled t test, because this test is only valid for two populations. This chapter develops the *analysis of variance* technique to analyze data not only of the type illustrated above, but also those coming from experiments where the subjects or units are grouped into blocks.

From the above we note that in an experiment we typically begin with one or more factors, whose levels or combinations of levels give rise to a set of treatments. For each treatment there is a set of experimental units which is assumed to be a random sample from a population of units. On each unit the effects of factors are noted by making measurements on certain variables. If the assumption of randomness of the units is valid, then for each treatment the measurements themselves may be viewed as a random sample from a population of measurements. A further typical assumption made in the analysis of variance technique is the assumption of normality of the measurements. The purpose of the analysis of the data from an experiment is to reach a conclusion regarding the relative merit of the treatments by analyzing the differences between the treatments. To do this in an unbiased manner, the assumption of random selection of units for a treatment or the random allocation of the treatments to the units is crucial. The random selection or random allocation process is known as *randomization*.

Let us now lay the groundwork by providing some necessary definitions.

15.1 PRELIMINARY DEFINITIONS

DEFINITION 15.1 *A factor is a variable such that the effects of its levels are of interest to the experimenter in an experiment.*

A factor can be *fixed* or *random*. In the first case, the experimenter chooses the levels to be studied without having to resort to a random choice, while in the second case the levels are selected randomly from a population of levels. A factor can be *qualitative* or *quantitative*. For example, if an experimenter wishes to study the effects of the two methods "with audio visual" and "without audio visual" of teaching mathematics to grade 7 students, then we have a fixed qualitative factor under study. On the other hand, if an experimenter selects randomly five concentrations from a population of possible concentrations of a drug to study its effects on behavior of mice, then we are dealing with a random quantitative factor.

In any given experiment several factors may be studied simultaneously, by using combinations of their levels. This leads us to the concept of a *treatment*.

DEFINITION 15.2 *A treatment is a combination of levels of factors.*

As an example, suppose that 4 concentrations of drug A are to be studied in combination with 5 concentrations of drug B in an experiment. This will give rise to 4 × 5 = 20 treatments, and the experimenter may run all 20 or use a subset, depending on the aims of the experiment.

DEFINITION 15.3 *A comparative experiment is an experiment with more than one treatment under study, while an absolute experiment is an experiment with a single treatment under study.*

Note that in both the comparative and the absolute experiment we may have several factors appearing, because a treatment is in general a combination of levels of factors.

DEFINITION 15.4 *A treatment design is a plan describing the complete structure of the treatments.*

In the example of 4 concentrations of drug A and 5 concentrations of drug B, when all 20 combinations are used the treatment design is called a *complete* 4 × 5 *factorial* and the plan is given by listing all 20 combinations. If a subset is used, then the treatment design is known as an *incomplete or fractional* 4 × 5 *factorial* and the plan is given by listing the subset.

DEFINITION 15.5 *An experimental unit is a unit to which a treatment is allocated with the objective of measuring its effect relative to a variable of interest.*

DEFINITION 15.6 *A sampling unit is the smallest unit on which measurements are made.*

It follows that experimental units and sampling units can be the same or different units. An example is given below.

If treatments are allocated randomly to a given set of similar experimental units, then it is clear that this will eliminate *allocation biases*. This means that differences between treatments can be assessed on a fair and square basis, because no treatment has been systematically favored over the other. Random allocation is called *randomization* in the theory of experimental design. When treatments are allotted randomly to all the experimental units, then one refers to this as *unrestricted randomization*. As can be observed, this situation corresponds to drawing random samples for each treatment. Random allocation of treatments to experimental units which are grouped into homogeneous blocks on some basis is called *restricted randomization*. We will be discussing these concepts shortly.

DEFINITION 15.7 *An experimental design is a plan describing the allocation of the treatments to the experimental units.*

For example, if six litters of three mice are available to study three treatments, t_1, t_2, and t_3, we can have, among others, the following two designs:

A. Allocate randomly the three treatments to the litters of mice, such that each treatment appears on two litters.
B. Allocate randomly the three treatments to the mice in each litter.

Schematically, these two designs may look like:

Litter no.

	1	2	3	4	5	6
	t_2	t_1	t_2	t_3	t_3	t_1
A	t_2	t_1	t_2	t_3	t_3	t_1
	t_2	t_1	t_2	t_3	t_3	t_1

Litter no.

	1	2	3	4	5	6
	t_3	t_2	t_2	t_3	t_1	t_1
B	t_1	t_1	t_3	t_1	t_2	t_3
	t_2	t_3	t_1	t_2	t_3	t_2

Under the assumption that the smallest units of measurement are the mice, we see that in design A the experimental units are the litters and the sampling units are the mice. Design A is an unrestricted randomized design. In design B both the experimental and sampling units are the same, and the design is a restricted randomized design. Which design is to be preferred?

Clearly, design A is inferior because differences in response due to the treatments cannot be separated from the differences in response due to the litters. One says in such a case that treatment differences are *confounded* with litter differences. In design B, treatment differences in responses can be assessed unencumbered by differences in response due to litters. This is so, because litter effects are eliminated when treatment differences in response are analyzed in each litter. Hence treatment differences are *unconfounded* with litter differences.

DEFINITION 15.8 *A replicate in an experimental design is a subset of homogeneous units to which all the treatments are allocated, such that each treatment appears at least once.*

The concept of replication should not be confused with *repetition* of a treatment. By this we mean the number of times a treatment appears in a design. In design A there is no replication, however, each treatment is repeated six times. In design B there are, by Definition 15.8, six replications, namely, the six litters, and each treatment is repeated six times. The reader is invited to construct a design such that the number of replications is different from the numbers of repetitions of the treatments.

Experimental units in nature are generally not homogeneous, because it is not possible to control all influencing factors before the treatments are applied to the units. Homogeneity is an idealistic term, because it is practically never achieved. If one treatment is applied to units, conceived to be homogeneous but in reality not, the response will differ from unit to unit, although they have been treated the same. This repetition-to-repetition variability of the response is part of what is called, in design jargon, *experimental error*. We define this concept formally as:

DEFINITION 15.9 *Experimental error in a design is variability in the response caused by all uncontrolled sources, which by their occurrence mask the true effects of the treatments.*

Notice that experimental error not only includes variability in response of units receiving the same treatment, which is due to the inherent variability among the units, but also variability caused by factors beyond the control of the experimenter. There are some obvious ways to reduce the magnitude of experimental error, namely:

1. Use of experimental material which is as homogeneous as possible.
2. Use of blocks or groups whose units are as homogeneous as possible.
3. Use of variables which are highly correlated with the one under study, so that adjustments for these can be made.
4. If experimental material is plentiful, then as much replication should be made as possible.

5. Use of that design which will lead to the lowest experimental error.
6. Removal of outlying data after valid testing and explanation.

Replication and repetition provide the basis for obtaining an estimate of experimental error. They also affect the size of the experimental error and hence the precision with which differences between treatments can be estimated. Hence the number of replicates and repetitions depends not only on the variability among experimental units, but also on the degree of precision desired. We have met this when we dealt with sample size calculations in one of the earlier chapters. The size and shape of experimental units also affect replication and repetition.

In practice it is often impossible to guarantee equal repetition for each treatment. *Unequal repetition* has the effect that differences between treatment means are estimated with unequal precision, because it affects the denominators of the standard errors. As will be seen shortly, the analysis of such designs is more complicated than designs with equal repetition.

We now study two basic randomized designs and their analyses.

15.2 THE COMPLETELY RANDOMIZED DESIGN

DEFINITION 15.10 *A completely randomized design with equal repetition* r *is a plan in which* v *treatments are randomly allocated to* $v \cdot r$ *experimental units, such that each treatment appears on* r *units.*

The construction of a completely randomized design (crd) with equal repetition can be done as follows. Identify the v treatments with the letters A, B, C, ..., V and write each letter on r chips or pieces of paper. This means that there will be $r \cdot v$ chips altogether. Place these chips in an urn and mix well. Draw the first chip and assign the letter on it to the first unit in the list of $r \cdot v$ experimental units. Do not replace the drawn chip. Proceed in this manner until the last chip has been drawn and the letter on it has been allocated to the last experimental unit.

Example 15.1 To illustrate the procedure, suppose that 3 concentrations of a drug are to be tested on 12 experimental mice, who have been matched for other factors. If each concentration is to be given to 4 mice, then the procedure might result in the completely randomized design shown in Figure 15.1.

In this design the first chip drawn had the third concentration C on it, and this treatment was given to mouse number 1. The rest of the design is explained in the same way.

Another procedure of construction, which is more closely tied to drawing random samples, and which utilizes the table of random numbers, is carried out by numbering the v • r experimental units and by assigning r units corresponding to the first r random numbers drawn between 1 and v • r to treatment A, the next r random numbers drawn among the remaining ones are assigned to B, etc., until the last remaining r numbers are assigned to treatment V. (The use of a random numbers table is explained in Chapter 16, "Sample Surveys.") Undoubtedly, the reader can provide other procedures for coming up with a completely randomized design.

A completely randomized design may be viewed as drawing a random sample of r units from each of v infinite populations. Each

1 C	2 B	3 C
4 A	5 C	6 B
7 B	8 A	9 A
10 A	11 B	12 C

Figure 15.1 A crd with v = 3 treatments and r = 4 repetitions.

infinite population comprises all the possible units receiving a particular treatment. Since the population values or the sample values of a variable can be classified according to treatments, it may be viewed as a *one-way classification design*. A much better name for the crd is *zero-way elimination of heterogeneity design*, because the units were not grouped at all to remove a source of heterogeneity. This name has the fundamental assumption of homogeneity built into it, i.e., from the sample data the differences between treatment means can be estimated free of interfering factors.

15.3 ANALYSIS OF A COMPLETELY RANDOMIZED DESIGN

The crd gives rise to drawing a sample of size r from v infinite populations. The sample data then can be organized in a one-way table, as given in Table 15.1. In the table X_{ij} refers to the jth sample measurement of the variable X for the ith treatment, i = 1, 2, ..., v and j = 1, 2, ..., r. Thus X_{11} is the first sample measurement for treatment 1, etc. The rest of the symbols have the following explanations:

Table 15.1 Data from a Completely Randomized Design with Equal Repetition

	Treatments				
	1	2	...	V	
	X_{11}	X_{21}	...	X_{v1}	
	X_{12}	X_{22}	...	X_{v2}	
	$\vdots$	$\vdots$		$\vdots$	
	X_{1r}	X_{2r}	...	X_{vr}	
Total	$X_{1.}$	$X_{2.}$	...	$X_{v.}$	$X_{..}$
Mean	$\overline{X}_{1.}$	$\overline{X}_{2.}$	...	$\overline{X}_{v.}$	$\overline{X}_{..}$

$$X_{i.} = \sum_{j=1}^{r} X_{ij} = \text{total of r measurements for the ith treatment}$$

$$\overline{X}_{i.} = \frac{1}{r} \sum_{j=1}^{r} X_{ij} = \text{mean of r measurements for the ith treatment}$$

$$X_{..} = \sum_{i=1}^{v} \sum_{j=1}^{r} X_{ij} = \text{grand total of all } r \cdot v \text{ measurements}$$

$$\overline{X}_{..} = \frac{X_{..}}{r \cdot v} = \text{grand mean of all measurements}$$

We now make the assumption that the treatments arise from *fixed factors* and that the sample measurements came from v independent normal populations. Explicitly we have:

$X_{11}, X_{12}, \ldots, X_{1r}$ are from $N(\mu_1, \sigma^2)$

$X_{21}, X_{22}, \ldots, X_{2r}$ are from $N(\mu_2, \sigma^2)$

. . . .

$X_{i1}, X_{i2}, \ldots, X_{ir}$ are from $N(\mu_i, \sigma^2)$

. . . .

$X_{v1}, X_{v2}, \ldots, X_{vr}$ are from $N(\mu_v, \sigma^2)$

From Definition 15.9 it follows that σ^2 is a measure of experimental error in each of the v populations. It is often referred to as *experimental error variance*. The objective of performing a crd is twofold, namely, tests of hypotheses concerning the treatment means $\mu_1, \mu_2, \ldots, \mu_v$, and confidence interval estimation of a difference between two means.

R. A. Fisher developed a tool known as *analysis of variance* (ANOVA), which consists of decomposing the total sum of squares of the measurements (i.e., total variability) into component sum of squares (component variabilities) according to the sources in the design. The ANOVA for the completely randomized design is given in Table 15.2.

Table 15.2 ANOVA Table for a CRD with Equal Repetition

Source of variation	Degrees of freedom	Sum of squares	Mean squares
Treatments	$v - 1$	$SS(T) = \sum_{i=1}^{v} X_{i.}^2/r - X_{..}^2/vr$	$MS(T) = SS(T)/(v - 1)$
Error	$v(r - 1)$	SS(E) (By subtraction)	$MS(E) = SS(E)/v(r - 1)$
Total	$rv - 1$	$TSS = \sum_{i=1}^{v} \sum_{j=1}^{r} X_{ij}^2 - X_{..}^2/vr$	--

It is traditional to indicate sum of squares by SS and mean squares by MS. Thus, TSS, SS(T), SS(E), MS(T), and MS(E) are self-explanatory. The degrees of freedom for experimental error (or simply error) come from the fact that there are $(r - 1)$ degrees of freedom among the measurements for each treatment, so that altogether there are $v(r - 1)$ degrees of freedom. It can also be found by subtracting the degrees of freedom for treatments from the degrees of freedom for the total. For the sum of squares we have given the computational form, rather than the definitional form.

The following theorems are based on the ANOVA table:

THEOREM 15.1 *In repeated sampling from* v *independent normal populations the statistic* MS(T)/MS(E) *has the* F *distribution with* $\nu_1 = v - 1$ *and* $\nu_2 = v(r - 1)$ *degrees of freedom if there are no differences between the treatment means* $\mu_1, \mu_2, \ldots, \mu_v$.

THEOREM 15.2 *A statistical test at the significance level* α *for testing the null hypothesis* H_0: *there are no differences between the treatment means* $\mu_1, \mu_2, \ldots, \mu_v$ *against* H_A: *there are differences between two or more of the treatment means* $\mu_1, \mu_2, \ldots, \mu_v$ *is to reject* H_0 *if*

$$F = \frac{MS(T)}{MS(E)} > F_{v-1,v(r-1),\alpha}$$

and to accept H_0 *otherwise.*

THEOREM 15.3 *In repeated sampling from* v *independent normal populations the statistic*

$$\frac{(\overline{X}_{i.} - \overline{X}_{i'.}) - (\mu_i - \mu_{i'})}{s_{\overline{X}_{i.} - \overline{X}_{i'.}}}$$

where $i \neq i'$ *and*

$$s_{\overline{X}_{i.} - \overline{X}_{i'.}} = \sqrt{MS(E)\left(\frac{1}{r} + \frac{1}{r}\right)}$$

has the t *distribution with* $v(r - 1)$ *degrees of freedom.*

THEOREM 15.4 *A* $\gamma = 1 - \alpha$ *confidence interval for the difference* $\mu_i - \mu_{i'}$ *is given by*

$$(L, U) = ((\overline{X}_{i.} - \overline{X}_{i'.}) - t_{v(r-1),\alpha/2} \cdot s_{\overline{X}_{i.} - \overline{X}_{i'.}},$$

$$(\overline{X}_{i.} - \overline{X}_{i'.}) + t_{v(r-1),\alpha/2} \cdot s_{\overline{X}_{i.} - \overline{X}_{i'.}})$$

We now illustrate the results of Theorems 15.2 and 15.4 with an example.

Example 15.1 (continued). Suppose in the design of Example 15.1 the variable which was measured was the time (in seconds) after injection to reach drowsiness with the following results:

	A	B	C	
	42	40	44	
	45	40	40	
	46	38	40	
	49	44	34	
$X_{i.}$	182	162	158	$502.0 = X_{..}$
$\overline{X}_{i.}$	45.5	40.5	39.5	$41.83 = \overline{X}_{..}$

From the formulas in ANOVA Table 15.2, we obtain:

$$\frac{(X^2_{..})}{12} = 21{,}000.33$$

$$TSS = (42^2 + 45^2 + \cdots + 34^2) - 21{,}000.33 = 177.67$$

$$SS(T) = \frac{1}{4}(182^2 + 162^2 + 158^2) - 21{,}000.33 = 82.67$$

$$SS(E) = TSS - SS(T) = 177.67 - 82.67 = 95.0$$

Therefore we have

Source	d.f.	SS	MS	F
Treatments	2	82.67	41.34	3.91
Error	9	95.00	10.56	--
Total	11	177.67	--	--

Since $F = 3.91 < F_{2,9,0.01} = 8.02$, we conclude by Theorem 15.2 that there are no significant differences between the three means. Note that this means that the 99% confidence intervals on differences between two means will contain 0.

Since $s_{\overline{X}_{i.}-\overline{X}_{i'.}} = \sqrt{10.56(1/4 + 1/4)} = 2.30$, then a 99% confidence interval for the difference $\mu_1 - \mu_3$ (see Theorem 15.4) is:

$$\begin{aligned}(L, U) &= (6.0 - (3.250)(2.30), 6.0 + (3.250)(2.30)) \\ &= (-1.48, 13.48)\end{aligned}$$

Remark. The F test is a broad test. It does not pinpoint which differences are significant in case the null hypothesis is rejected. Of course, this can be done by constructing confidencee intervals and by noting whether 0 is contained in them or not. Moreover, zero will not be in a confidence interval if the difference $\overline{X}_{i.} - \overline{X}_{i'.}$ is greater than $t_{v(r-1),\alpha/2} \cdot s_{\overline{X}_{i.}-\overline{X}_{i'.}}$. This quantity is called the *least significant difference* or, abbreviated, LSD. Therefore a difference between two means will be significant at the α level if it is greater than the LSD. In the crd with equal repetition there is only one LSD, and all confidence intervals are of the form $(L, U) = ((\overline{X}_{i.} - \overline{X}_{i'.}) - LSD, (\overline{X}_{i.} - \overline{X}_{i'.}) + LSD)$.

15.4 COMPLETELY RANDOMIZED DESIGN WITH UNEQUAL REPETITIONS

In this case treatment 1 is repeated on r_1 units, treatment 2 on r_2 units, ..., treatment v is repeated on r_v units. This leads us to the summary table and ANOVA shown in Tables 15.3 and 15.4.

We now generalize Theorems 15.1, 15.2, 15.3, and 15.4.

THEOREM 15.5 *In repeated sampling the statistic* MS(T)/MS(E) *has the* F *distribution with* $\nu_1 = v - 1$ *and* $\nu_2 = \sum_{i=1}^{v} r_i - v$ *degrees of freedom if there are no differences between the treatment means* $\mu_1, \mu_2, \ldots, \mu_v$.

THEOREM 15.6 *A statistical test at the significance level* α *for testing the null hypothesis* H_0: *there are no differences between the treatment means* $\mu_1, \mu_2, \ldots, \mu_v$ *against* H_A: *there are differences between two or more of the treatment means* $\mu_1, \mu_2, \ldots, \mu_v$ *is to reject* H_0 *if*

$$F = \frac{MS(T)}{MS(E)} > F_{v-1,\Sigma r_i - v,\alpha}$$

Table 15.3 Data from a CRD with Unequal Repetitions

	Treatments				
	1	2	...	v	
	X_{11}	X_{21}	...	X_{v1}	
	X_{12}	X_{22}	...	X_{v2}	
	⋮	⋮		⋮	
	X_{1r_1}	X_{2r_2}	...	X_{vr_v}	
Total	$X_{1.}$	$X_{2.}$	...	$X_{v.}$	$X_{..}$
Mean	$\bar{X}_{1.}$	$\bar{X}_{2.}$	...	$\bar{X}_{v.}$	$\bar{X}_{..}$

Table 15.4 ANOVA for a CRD with Unequal Repetitions

Source of variation	Degrees of freedom	Sum of squares	Mean squares
Treatments	$v - 1$	$\sum_{i=1}^{v} X_{i.}^2/r_i - \left(X_{..}^2/\sum_{i=1}^{v} r_i\right)$	$SS(T)/(v - 1)$
Error	$\Sigma r_i - v$	By subtraction	$SS(E)/(\Sigma r_i - v)$
Total	$\Sigma r_i - 1$	$\sum_{i=1}^{v} \sum_{j=1}^{r_i} X_{ij}^2 - \left(X_{..}^2 / \sum_{i=1}^{v} r_i\right)$	--

THEOREM 15.7 *In repeated sampling the statistic*

$$\frac{(\overline{X}_{i.} - \overline{X}_{i'.}) - (\mu_i - \mu_{i'})}{s_{\overline{X}_{i.}-\overline{X}_{i'.}}}$$

where $i \neq i'$ *and*

$$s_{\overline{X}_{i.}-\overline{X}_{i'.}} = \sqrt{MS(E)\left(\frac{1}{r_i} + \frac{1}{r_{i'}}\right)}$$

has the t *distribution with* $\Sigma r_i - v$ *degrees of freedom.*

THEOREM 15.8 *A* $\gamma = 1 - \alpha$ *confidence interval for the difference* $\mu_i - \mu_{i'}$ *is given by*

$$(L, U) = ((\overline{X}_{i.} - \overline{X}_{i'.}) - t_{\Sigma r_i - v, \alpha/2} \cdot s_{\overline{X}_{i.}-\overline{X}_{i'.}},$$
$$(\overline{X}_{i.} - \overline{X}_{i'.}) + t_{\Sigma r_i - v, \alpha/2} \cdot s_{\overline{X}_{i.}-\overline{X}_{i'.}})$$

The LSD for declaring a difference between two means significant at α level is given by:

$$LSD = (t_{\Sigma r_i - v, \alpha/2})\sqrt{MS(E)\left(\frac{1}{r_i} + \frac{1}{r_{i'}}\right)}$$

Note now that this quantity is dependent upon the repetitions of the treatments under consideration.

Example 15.2 Aptitude test scores in a subject are given below. They were obtained by administering a test to randomly selected individuals from four age groups, who were matched for other factors:

Age groups

1	2	3	4
45	35	34	41
46	33	34	41
49	68	35	44
44		34	43
184		33	41
		170	42
			44
			41
			41
			378

Calculations

$v = 4$	
$r_1 = 4$	$X_{1.} = 184$
$r_2 = 2$	$X_{2.} = 68$
$r_3 = 5$	$X_{3.} = 170$
$r_4 = 9$	$X_{4.} = 378$
$\Sigma r_i = 20$	$X_{..} = 800$

$$\text{TSS} = 45^2 + 46^2 + \cdots + 41^2 - \frac{800^2}{20} = 464$$

$$\text{SS(T)} = \frac{184^2}{4} + \frac{68^2}{2} + \frac{170^2}{5} + \frac{378^2}{9} - \frac{800^2}{20} = 432$$

$$\text{SS(E)} = 464 - 432 = 32$$

ANOVA Table

Source	d.f.	SS	MS	F
Treatments	3	432	144	72
Error	16	32	2	
Total	19	464	--	--

Since $F_{3,16,0.01} = 5.29$, there are significant differences between the treatment means. The reader is invited to calculate the LSDs and detect the significant differences.

15.5 THE RANDOMIZED COMPLETE BLOCK DESIGN

There are certain disadvantages associated with a completely randomized design. Two of these are:

1. The design can be effectively used only if the experimental units are homogeneous. This means that the units should be matched for all other interfering factors, which is seldom the case.
2. When the number of treatments is large the crd requires a large number of experimental units, so that securing homogeneous material may become an arduous task.

To alleviate these disadvantages, the idea of forming groups of homogeneous experimental units comes immediately to our minds; and this is precisely what we are going to discuss next.

When there is a relatively large variability in the experimental material, it is always possible to select a design which will be more efficient than the completely randomized design. To achieve this we group or stratify the material on the basis of certain variables such that the units in groups are relatively homogeneous. When it is possible to form groups or blocks of units and the treatments are randomly allocated in each block, then we have created a new design, which is known as a *randomized complete block design* or rcbd. Its definition is given next.

DEFINITION 15.11 *A randomized complete block design with equal repetition* b *for each of* v *treatments is a plan in which* v • b *experimental units are grouped in* b *blocks of* v *units and the* v *treatments are randomly allocated to the units in each block.*

Notice that a *block* satisfies Definition 15.8 of a replicate, and in the literature it is often referred to by this name. rcbd's originated in agriculture, where plots of land for experimental purposes tend to exhibit wide variability across a field with respect to fertility and other soil factors. Blocks are best determined by carrying out a *uniformity trial* with the experimental units; i.e., by treating all the units alike one may observe the inherent variability among them and group them accordingly. Blocks in experimentation can be groups of plots of land, groups of days, groups of locations, groups of animals of the same makeup, schools of a certain type, blue collar workers making the same income, undergraduate students taking the same academic program, etc.

Although rcbd's fall in the class of *two-way classification designs*, because the data can be classified in two ways, it does not imply that we are interested in differences between block means. It belongs more properly in the class of *one-way elimination of heterogeneity designs*, because the idea behind formation of blocks is to eliminate heterogeneity caused by an extraneous factor.

Some of the advantages of a rcbd are:

1. When the variability between block means is large, then the gain in precision over the crd is large. The reason for this is that in the ANOVA of the rcbd the sum of squares for blocks is extracted from the sum of squares of error, and hence will lead to a substantially smaller error mean square.
2. There are no restrictions on the number of treatments (although for a large number of treatments, blocks of homogeneous units might become a problem). It should be clear that the number of blocks must be greater than or equal to 2.
3. The analysis is easy, when there are no missing data in a rcbd.

Example 15.3 Earlier we talked about the experiment performed to study the effects of three treatments, using six litters of three mice each, by allocating the three treatments randomly to the mice in each block. The litters are the blocks, because we expect them (the mice) to be homogeneous within a litter. The result of a particular randomization might lead to the earlier indicated plan:

Block number

1	2	3	4	5	6
C	B	B	C	A	A
A	A	C	A	B	C
B	C	A	B	C	B

15.6 ANALYSIS OF A RANDOMIZED COMPLETE BLOCK DESIGN

Data from an experiment in a rcbd with v treatments and b blocks lead to the two-way Table 15.5. In this table, X_{ij} is the treatment in the jth block for the ith treatment, $i = 1, 2, \ldots, v$, $j = 1, 2, \ldots, b$, and

$$X_{i.} = \sum_{j=1}^{b} X_{ij} = \text{total for the ith treatment}$$

$$\overline{X}_{i.} = \frac{X_{i.}}{b} = \text{mean for the ith treatment}$$

$$X_{.j} = \sum_{i=1}^{v} X_{ij} = \text{total for the jth block}$$

$$X_{..} = \sum_{i=1}^{v} \sum_{j=1}^{b} X_{ij} = \text{grand total of all measurements}$$

We will need the ANOVA for development of tests and confidence intervals. This is given in Table 15.6.

Table 15.5 Data from a RCBD with v Treatments and b Blocks

		1	2	...	b	Total	Mean
Treatments	1	X_{11}	X_{12}	...	X_{1b}	$X_{1.}$	$\bar{X}_{1.}$
	2	X_{21}	X_{22}	...	X_{2b}	$X_{2.}$	$\bar{X}_{2.}$
	⋮	⋮	⋮		⋮	⋮	⋮
	v	X_{v1}	X_{v2}	...	X_{vb}	$X_{v.}$	$\bar{X}_{v.}$
		$X_{.1}$	$X_{.2}$	...	$X_{.b}$	$X_{..}$	$\bar{X}_{..}$

Table 15.6 The ANOVA Table for a RCBD with v Treatments in b Blocks

Source of variation	Degrees of freedom	SS	MS
Blocks	b - 1	$\Sigma X_{.j}^2/v - X_{..}^2/bv$	SS(B)/(b - 1)
Treatments	v - 1	$\Sigma X_{i.}^2/b - X_{..}^2/bv$	SS(T)/(v - 1)
Error	(b - 1)(v - 1)	By subtraction	SS(E)/(b - 1)(v - 1)
Total	bv - 1	$\sum_{i=1}^{v} \sum_{j=1}^{b} X_{ij}^2 - X_{..}^2/bv$	--

Assume now that the treatments are derived from a fixed factor, and note that in a rcbd we have a sample of size 1 from each of the v • b infinite populations, which we assume to have different means but the same variance σ^2. We further assume that they are normal populations. The sampling structure is thus equal to:

X_{11} from $N(\mu_{11}, \sigma)$

X_{21} from $N(\mu_{21}, \sigma)$

$\vdots$

X_{vb} from $N(\mu_{vb}, \sigma)$

The means of treatments over the blocks are defined as:

$$\bar{\mu}_{1.} = \frac{1}{b} \sum_{j=1}^{b} \mu_{1j}$$

$$\bar{\mu}_{2.} = \frac{1}{b} \sum_{j=1}^{b} \mu_{2j}$$

$$\vdots$$

$$\bar{\mu}_{v.} = \frac{1}{b} \sum_{j=1}^{b} \mu_{vj}$$

and we are interested in testing the null hypothesis that there are no differences between these means and putting confidence intervals on the differences between them.

The following theorems solve our problem:

THEOREM 15.9 *In repeated sampling the statistic* MS(T)/MS(E) *has the* F *distribution with* $\nu_1 = v - 1$ *and* $\nu_2 = (b - 1)(v - 1)$ *degrees of freedom if there are no differences between the treatment means* $\bar{\mu}_{1.}, \bar{\mu}_{2.}, \ldots, \bar{\mu}_{v.}$.

THEOREM 15.10 *A statistical test at the significance level* α *for testing the null hypothesis* H_0: *there are no differences between the treatment means* $\bar{\mu}_{1.}, \bar{\mu}_{2.}, \ldots, \bar{\mu}_{v.}$ *against* H_A: *there are differences between two or more of the treatment means* $\bar{\mu}_{1.}, \bar{\mu}_{2.}, \ldots, \bar{\mu}_{v.}$ *is to reject* H_0 *if* $F = MS(T)/MS(E) > F_{v-1,(b-1)(v-1),\alpha}$.

THEOREM 15.11 *In repeated sampling the statistic*

$$\frac{(\bar{X}_{i.} - \bar{X}_{i'.}) - (\bar{\mu}_{i.} - \bar{\mu}_{i'.})}{s_{\bar{X}_{i.} - \bar{X}_{i'.}}}$$

where $i \neq i'$ *and*

$$\bullet \quad s_{\bar{X}_{i.} - \bar{X}_{i'.}} = \sqrt{MS(E)\left(\frac{1}{b} + \frac{1}{b}\right)}$$

has the t *distribution with* $(b - 1)(v - 1)$ *degrees of freedom.*

THEOREM 15.12 *A* $\gamma = 1 - \alpha$ *confidence interval for the difference* $\bar{\mu}_{i.} - \bar{\mu}_{i'.}$ *is given by*

$$(L, U) = ((\bar{X}_{i.} - \bar{X}_{i'.}) - t_{(b-1)(v-1),\alpha/2} \cdot s_{\bar{X}_{i.} - \bar{X}_{i'.}},$$

$$(\bar{X}_{i.} - \bar{X}_{i'.}) + t_{(b-1)(v-1),\alpha/2} \cdot s_{\bar{X}_{i.} - \bar{X}_{i'.}})$$

Finally, the LSD for testing significance of a difference between two means is given by:

$$LSD = (t_{(b-1)(v-1),\alpha/2})\sqrt{MS(E)\left(\frac{1}{b} + \frac{1}{b}\right)}$$

Remark. The case of a rcbd for $v = 2$ treatments is known as a *paired experiment* with b pairs. There is no need to develop this separately. The traditional *paired t test* is simply obtained by calculating the LSD for $v = 2$ and seeing whether the difference between the two means exceeds it.

Example 15.4 Five schizophrenic patients were observed during 1-hr periods for 6 days in order to observe their behavior relative to others when they were brought together in a common area. The scores given by an attending psychiatrist were:

		Period						Total	Mean
		1	2	3	4	5	6		
Patients	1	8	12	11	31	22	9	93	15.50
	2	23	26	21	15	24	23	132	22.00
	3	41	22	21	46	15	22	167	27.83
	4	28	48	16	44	38	43	217	36.17
	5	67	33	59	49	57	36	301	50.17
Total		167	141	128	185	156	133	910	27.00

Assume that the 30 underlying populations are normal; then we need the following calculations to perform the test and construct confidence intervals for treatment differences:

$$\text{TSS} = 8^2 + 13^2 + \cdots + 36^2 - \frac{910^2}{30} = 34{,}840 - 27{,}603.33$$

$$= 7236.67$$

$$\text{SS(B)} = \frac{167^2 + 141^2 + \cdots + 133^2}{5} - \frac{910^2}{30}$$

$$= 28{,}080.8 - 27{,}603.33 = 477.47$$

$$\text{SS(T)} = \frac{93^2 + 132^2 + \cdots + 301^2}{6} - \frac{910^2}{30}$$

$$= 31{,}942 - 27{,}603.33 = 4338.67$$

$$\text{SS(E)} = 7236.67 - 477.47 - 4338.67 = 2420.53$$

$$\text{MS(T)} = \frac{4338.67}{4} = 1084.67$$

$$\text{MS(E)} = \frac{2420.53}{20} = 121.03$$

$$F = \frac{1084.67}{121.03} = 8.96$$

$$\text{LSD at 5\%} = (2.086)\sqrt{(121.03)\left(\frac{2}{6}\right)} = 13.25$$

ANOVA Table

Source	d.f.	SS	MS	F
Blocks	5	477.47		
Treatments	4	4338.67	1084.67	8.96
Error	20	2420.53	121.03	
Total	29	7236.67		

Since $F = 8.96 > F_{4,20,0.05} = 2.87$, we conclude that there are significant differences between the patients relative to their social behavior. There are significant differences between P_1 and P_2, P_1 and P_5, P_2 and P_4, P_2 and P_5, P_3 and P_5, and P_4 and P_5.

For those interested in other schizophrenic experiments, see Wyatt *et al.* (1972), Hanley *et al.* (1972), or Bull and Venables (1974).

Remark. There are many other designs, besides the two discussed in this chapter. The reader is encouraged to read more about Latin Square designs, hierarchical designs, factorial treatment designs, etc.

EXERCISES

15.1 What is a nonrandomized experimental design? Give an example. What is the major disadvantage of such a design?

15.2 Give an illustration of your own choice of a treatment design.

15.3 In the literature find an experimental design and describe its treatment design.

15.4 Give an example of an experiment with a qualitative factor.

15.5 Design an experiment where the differences between treatments are confounded with differences of an extraneous factor.

15.6 What is meant by replication? Give your own example.

15.7 What is experimental error in an absolute experiment?

15.8 What purposes are served by randomization?

15.9 A completely randomized design to test the influence of 5 diets on gain of weight of 20 experimental subjects resulted in the following data:

A	B	C	D	E
42	40	44	41	34
45	40	40	43	35
46	38	40	40	34
49	44	34	40	33

(a) State how the randomization was done.

(b) Sketch a possible layout resulting from the randomization.

(c) Complete the ANOVA table and carry out the F test at the 5% level.

(d) Calculate the LSD and conclude which differences between the treatment means are significant if the result in (c) is significant.

(e) Obtain a confidence interval for $\mu_A - \mu_B$.

15.10 Carry out your own crd with two treatments with unequal repetitions. Perform the F test at the 1% level. What relationship is there between the F test and the pooled t test of Chapter 13?

15.11 The following data were obtained from a rcbd:

		Blocks 1	2	3	4
Treatments	1	8	10	9	30
	2	20	24	20	20
	3	39	19	18	41
	4	26	45	15	38
	5	65	31	59	42

(a) Select treatments and blocks from your area of interest for the above data.

(b) Sketch a possible layout of the rcbd and describe how it was obtained.

(c) Carry out the ANOVA and do the F test at the 5% level.

(d) Construct a 95% confidence interval for the difference between the means of treatments 3 and 4.

(e) Give a repeated sampling interpretation to (d).

15.12 Formulate the null hypothesis and the alternatives to it in terms of equality of the means (using the symbols for the means) for the crd and rcbd.

15.13 Describe under which conditions a crd and a rcbd are proper designs for performing a comparative experiment.

15.14 For treatments 1 and 2 analyze Exercise 15.11 using a paired t test.

15.15 What is the effect of unequal repetitions on the standard error of a difference between two means?

15.16 True or false:

(a) A design is a plan on how to obtain data.

(b) There is no significant difference between a survey and an experiment.

(c) Often more than one treatment is allocated to an experimental unit.

(d) Randomization is a process of allocating treatments to experimental units in order to eliminate bias.

(e) A random factor is a special case of a fixed factor.

(f) The number of replications equals the number of repetitions of a treatment.

(g) Experimental error cannot be minimized in a given experiment.

(h) In an ANOVA table, a mean square is always variance divided by degrees of freedom.

(i) $\overline{X}_{..} = X_{..} \cdot v \cdot \sum_{i=1}^{v} r_i$ in a crd.

(j) TSS = SS(T) + SS(E) in a crd.

(k) In an ANOVA table, if the value of the test statistic F is less than 1, then there really is no need to look up the critical F value.

(l) In a crd, if H_0 is accepted or rejected, your analysis is complete.

(m) In the analysis of experiments LSD is a type of drug treatment.

(n) $\frac{1}{r} \sum_{i=1}^{v} X_{i.}^2 = r \sum_{i=1}^{v} (\overline{X}_{i.})^2$, for a crd.

(o) If v = 2 in a rcbd, then $F = t^2$, where t is the value of the test statistic in the paired t test.

15.17 Three methods of group-encounter techniques were to be compared with respect to the mean level of group interaction. A total of 21 group leaders participated in the study, which was done at the same training camp. Six were assigned to method 1, 7 to method 2, and 8 to method 3. After 1 session (half a day), leaders were scored on their ability to achieve meaningful group interaction. The following table presents the data. Test the hypothesis that the three methods of instruction achieved the same mean level of group interaction.

Method 1	Method 2	Method 3
71	81	92
29	82	93
78	80	84
74	79	90
75	84	87
36	83	98
	80	92
		90

15.18 A firm is introducing a new after shave lotion called "Ape." It wishes to study the effect on sales of using two different counter displays, A and B, and two different prices, C and D, in promoting the product. For each combination of display and price, three stores are selected at random to sell the product. The number of sales in each of the stores is recorded for a trial period. Results are as follows:

Combination of display and price of "Ape"

Store	A, C	A, D	B, C	B, D
1	65	73	60	90
2	85	71	55	85
3	55	74	58	84

Complete the two-way analysis of variance for these data.

15.19 Tsushima and Hogan (1975) investigated the differences in verbal ability and school achievement of bilingual and monolingual children in grades 3, 4, and 5. American children attending a Department of Defence School at Camp Zama, Japan were classified as bilingual or monolingual based on information provided by parents. Verbal ability was determined by using the Lorge-Thorndike Intelligence Test.

		Monolingual	Bilingual
	3	40.8	40.3
Grade	4	48.1	47.8
	5	45.8	45.9

The values in this table represent the mean scores for the group of children in that group. Analyze the data.

15.20 The following experiment can be found in Williams and Goulet (1975). They studied the performance of nursery school children in two controlled experiments involving *cued* and *constrained free recall*. Under cued instructions, subjects could recall items in any order they wished; whereas under constrained instructions, subjects were required to recall items by category. The mean number of items recalled over the seven training trials is presented in the following table:

	Trials						
Treatments	1	2	3	4	5	6	7
Constrained 1-7	6.33	7.67	8.17	9.33	9.33	9.25	9.42
Constrained 1-4	7.17	9.00	9.92	10.00	6.83	7.41	7.83
Cued 1-7	4.92	5.67	6.42	6.42	6.25	7.50	7.58
Cued 1-4	4.92	6.33	6.50	6.25	6.91	6.83	6.75
Control	5.33	5.75	6.50	5.75	6.08	6.67	6.58

What conclusions can be reached from these data?

15.21 Ansari (1975) studied three groups of impotent males and obtained the following ANOVA table regarding their ages:

Source	d.f.	SS	MS	F
Groups	2	364.74	182.37	20.2
Error	62	559.41	9.02	
Total	64	924.15		

Do these data support the null hypothesis that there is no significant difference in the mean ages of the three groups?

15.22 Is there a significant difference between the heterosexual and homosexual males from the data at the beginning of this chapter?

16

SAMPLE SURVEYS

All the developments with respect to estimation and tests of hypotheses relied heavily on the assumption that sampling was done from an infinite population, which in most settings was assumed to be normal. In this chapter we consider problems of estimation of parameters of a *finite* population, using some random sampling techniques. It must be understood that no underlying probability distribution is assumed for a finite population, because a probability distribution implies an infinite population.

In Chapter 15, it was pointed out that when a *survey* of a population is conducted, then the variables of interest are measured without exercising any control on factors that may influence it, while in an *experiment* changes of the influencing factors are introduced to study changes in the variables of interest. Thus, there is a definite distinction between a "controlled" experiment and a survey. It was also pointed out in Chapter 1 that a finite population consisting of N units can be either completely or partially enumerated relative to one or more variables. In the first case one usually speaks of a *census*, while in the second case the commonly used term is *sample survey*. In most cases a complete enumeration is not ecomomical or not feasible, so that a sample survey is called for. There are many practical problems when carrying out a sample survey, e.g., how to obtain a complete list of individuals, how to implement a sampling design, construction and debugging of questionnaires, and editing for errors after the survey. We will not go into these, but rather expose the reader to some basic sampling techniques and the estimation of population parameters associated with them.

16.1 PROBABILITY AND NONPROBABILITY SAMPLING

When a sample is drawn from a finite population such that the probability of including any particular unit in the sample is known, then we speak of *probability sampling*. Such a sample provides a probabilistic basis for deriving sample estimates and estimates of their accuracy. Probability sampling is characterized by the following:

1. The sample is drawn by a method of random selection.
2. Every unit of the population has a known probability of being selected in the sample.
3. The estimates based on the sample take account of the probabilities of selecting the units in the sample.

Note that in probability sampling we do not require that the probability of selection be equal for all units of the population. All we need is that these probabilities be known.

From the above it follows that when a sample is drawn from a population such that the probability for including any particular unit is not known, then the sample will be a *nonprobability sample*. In this case there is no probabilistic basis, and the results of probability theory cannot be applied.

Let us first describe some types of nonprobability sampling. Many of these types are often used in practice, and clearly are defective in some manner.

The first one is the *purposely biased sample*. In this case the sample is drawn in such a way that the units in the sample provide the desired response. This method of sampling is often used by politicians, organizations promoting certain issues or products, unions, etc.

The second type is known as the *convenience sample.* This is defined to be a sample which is convenient to take without regard to representativeness of the population from which the sample is drawn. For example, an interviewer in a household survey might decide to interview households closest to his house or the place where he is staying.

The third type is called the *judgment sample.* Here the decision of which units to include in the sample is left to the judgment of the organizer of the survey, because the claim is that he "knows" the units in the population so he is "best" qualified to determine which units whould be in the sample to "best" represent the population.

The fourth type is the *quota sample.* In this case it is left to the discretion of the enumerator to select units from the population so that he can meet his assigned quota. The actual selection is usually done using judgment or convenience sampling.

We have given four types of nonprobability sampling methods. Undoubtedly, the reader can find many more in statistical practice. One common property underlying these methods is that they are all subjective in nature and hence are *biased methods.* The accuracy of sample estimates cannot be assessed, because there is no probabilistic basis for estimating the sampling variability of the estimates from sample to sample. We will, for obvious reasons, not deal further with nonprobability sampling in this chapter.

We now proceed with the discussion of some probability sampling techniques.

16.2 SIMPLE RANDOM SAMPLING

If we wish to obtain a random sample of n units from a list of N units (such a list is called a *sampling frame*), then it is necessary to clarify what is meant by *simple random sampling* (SRS), because this is the simplest procedure encountered in practice.

DEFINITION 16.1 *By simple random sampling is meant the drawing of a sample from a sampling frame in which* (a) *the units in the sample have been drawn independently from each other and* (b) *each unit in the sample has been drawn with the same probability.*

From Chapter 5, on probability, we know that a random sample can be drawn in two ways, namely, with and without replacement. This leads us to two types of simple random sampling, namely, *simple random sampling with replacement* and *simple random sampling without replacement*. In the first case a sample of n units is drawn such that at every stage any unit is drawn independently in the sample and has equal probability of being selected irrespective of whether it was selected earlier or not. In simple random sampling without replacement the n units are drawn such that at each stage of selection any unit is drawn independently in the sample with equal probability from units not selected earlier.

In Chapter 5 we found that the size of the sample space in the case of sampling n units from among N units with replacement was equal to N^n, and in the case without replacement was equal to $C_n^N = N!/n!(N - n)!$. Hence we may also define the two methods of simple random sampling as follows:

DEFINITION 16.2 *Simple random sampling with replacement is a technique of drawing a simple random sample of* n *units from among* N *such that each of the possible* N^n *samples has the same chance* $1/N^n$ *of being selected as the sample.*

It can be shown that the probability that a particular unit is selected in the sample is equal to $1 - [(N - 1)/N]^n$ in the case of sampling with replacement.

As an illustration, consider the case where the population has N = 3 units, and a sample of size n = 2 is selected with replacement. The set of all possible samples is equal to:

$$S = \{(1, 1), (2, 2), (3, 3), (1, 2), (2, 1), (1, 3), (3, 1), (2, 3), (3, 2)\}$$

and the set contains $N^n = 3^2 = 9$ samples. The probability of any one of the samples being selected is equal to $1/N^n = 1/3^2 = 1/9$.

The probability that any particular unit, say unit 1, will be included in the sample is equal to:

$$P\{(1, 1), (1, 2), (2, 1), (1, 3), (3, 1)\} = \frac{5}{9}$$

which agrees with the result $1 - [(N - 1)/N]^n = 1 - (2/3)^2 = 1 - 4/9 = 5/9$.

DEFINITION 16.3 *Simple random sampling without replacement is a technique of drawing a simple random sample of* n *units from among* N *units such that each of the* C_n^N *possible samples has the same chance* $1/C_n^N$ *of being selected as the sample.*

It can be shown that the probability that a particular unit will be selected in the sample is equal to n/N. This ratio, by the way, is known as the *sampling fraction*.

As an example, consider the previous setting with N = 3 and n = 2. The sample space is equal to:

$$S = \{(1, 2), (1, 3), (2, 3)\}$$

and it has exactly $C_2^3 = 3!/2!1! = 3$ samples. The probability that any one of the samples is selected is $1/C_2^3 = 1/3$, and the probability that any particular unit, say unit 1, is selected in the sample is

$$P\{(1, 2), (1, 3)\} = \frac{2}{3}$$

which agrees with the general result n/N = 2/3.

In many textbooks the term "simple random sampling" is reserved for the without replacement case. Since the without replacement case does not produce duplicate measurements, it is more efficient than the with replacement case. We will limit our discussion to the without replacement case.

16.3 USE OF RANDOM NUMBERS IN DRAWING A SIMPLE RANDOM SAMPLE

To facilitate the process of drawing a simple random sample, either with or without replacement, from a sampling frame, statisticians have constructed *tables of random numbers*. Table A.8 presents 1750 randomly assorted digits. The digits in the rows were obtained by filling each position independently with a digit from among 0, 1, 2, ..., 9.

To illustrate the use of the table, suppose that an enumeration district contains N = 480 households and a 10% simple random sample without replacement is desired. This means that 48 households should be drawn from the sampling frame in such a way that once a household is drawn it is struck from the list before the next drawing is made. Using tables of random numbers, this process of drawing can be translated in terms of drawing 48 random numbers. The first step in using the tables of random numbers is to number the units in the sampling frame with three digits, i.e., 000 to 479. Select arbitrarily a page in a book containing tables of random numbers. Suppose that the selected page is that of Table A.8. Now select arbitrarily a column of the page, say column 6, and select arbitrarily a row, say 11. Reading the number corresponding to the selected row and column, we find number 7. Select the first three digits and start reading the three digit numbers below these. The number 694 does not represent a sampling unit, since it is greater than 480. The next one, 334, is a sampling unit, 522 is not, but 139 is, etc. Continue until you have obtained 48 units. Since we are sampling without replacement, duplications should be ignored when encountered.

If sampling is done with replacement, then repetitions are allowed, so when duplications are encountered they should be included.

16.4 POPULATION PARAMETERS FOR SIMPLE RANDOM SAMPLING WITHOUT REPLACEMENT

Before discussing sample estimation of population parameters it is necessary to state, with the aid of Chapters 4 and 10, the definitions of the population mean, the population total, the population variance, the population standard deviation, the population relative variance, and the population coefficient of variation.

Let:

N = size of the population, i.e., the number of units in the population

X_i = the value of the variable X for the ith unit, i = 1, 2, ..., N

DEFINITION 16.4 *The population total* τ_X *is the sum of all the values* $X_1, X_2, \ldots, X_N$*; i.e.,* $\tau_X = \sum_{i=1}^N X_i$*, and the population mean* μ_X *is the population total divided by* N*; i.e.,* $\mu_X = (1/N) \sum_{i=1}^N X_i = \tau_X/N$.

DEFINITION 16.5 *The population variance is equal to* $\sigma_X^2 = (1/N) \sum_{i=1}^N (X_i - \mu_X)^2$*, and the population standard deviation is given by* $\sigma_X = \sqrt{(1/N) \sum_{i=1}^N (X_i - \mu_X)^2}$.

The expression for σ_X^2, and hence σ_X, is the defining formula, which is equal to the computing formula

$$\sigma_X^2 = \frac{1}{N}[\Sigma X_i^2 - \frac{1}{N}(\Sigma X_i)^2] = \frac{1}{N}[\Sigma X_i^2 - \frac{1}{N}\tau_X^2]$$

Sometimes the divisor N - 1 is used for the population variance, but we will refrain from doing this.

DEFINITION 16.6 *The relative population variance is equal to* σ_X^2/μ_X^2 *, and the population coefficient of variation is equal to* $\sigma_X/\mu_X \cdot 100\%$.

Note that the parameters in Definitions 16.4, 16.5, and 16.6 can be obtained if a complete enumeration of the population is done. Since complete enumerations are expensive and most often impractical, the objective is to estimate them via a sample survey.

16.5 ESTIMATION OF PARAMETERS IN SIMPLE RANDOM SAMPLING WITHOUT REPLACEMENT

The following statistics based on a sample $X_1, X_2, \ldots, X_n$ are commonly used in simple random sampling without replacement:

$$\overline{X} = \frac{1}{n} \sum_{i=1}^{n} X_i = \text{sample mean of } X$$

$$N\overline{X} = \frac{N}{n} \sum_{i=1}^{n} X_i = N \text{ times the sample mean of } X$$

$$s_X^2 = \frac{1}{n-1} \sum_{i=1}^{n} (X_i - \overline{X})^2 = \frac{1}{n-1}\left[\sum_{i=1}^{n} X_i^2 - \frac{1}{n}\left(\sum_{i=1}^{n} X_i\right)^2\right]$$

$$= \text{sample variance of } X$$

$$s_X = \sqrt{s_X^2} = \text{sample standard deviation of } X$$

We have met the concept of *expected value of an estimator* in Chapter 10 in the infinite population setting. In a finite population setting this concept is easier to grasp, and we define it formally in the following way.

DEFINITION 16.7 *The expected value of an estimator* $\hat{\theta}$ *of a parameter* θ *is equal to the mean of the values of* $\hat{\theta}$ *over all possible random samples.*

In the simple random sampling without replacement case the mean of the estimator $\hat{\theta}$ is taken over the sample space S containing precisely C_n^N samples of size n. If we use the notation $\hat{\theta}_s$ to indicate the value of $\hat{\theta}$ for the sample s, then Definition 16.7 can be restated as:

$$\text{Expected value of } \hat{\theta} = \frac{1}{C_n^N} \sum \hat{\theta}_s$$

Using the concept of *unbiasedness* of an estimator, which was also developed in Chapter 10, we obtain the following important theorem.

THEOREM 16.1 *The estimators* $\overline{X}$, $N\overline{X}$, *and* $[(N - 1)/N]s_X^2$ *are unbiased estimators of* μ_X, τ_X, *and* σ_X^2, *respectively.*

Let us demonstrate the unbiasedness of $\overline{X}$ for the case $N = 3$ and $n = 2$ and $X_1 = 1$, $X_2 = 3$, $X_3 = 6$, say. The sample space $S = \{(X_1, X_2), (X_1, X_3), (X_2, X_3)\}$, so that the following means are obtained:

$$\text{Sample 1:} \quad \overline{X}_1 = \frac{X_1 + X_2}{2} = \frac{1 + 3}{2} = 2$$

$$\text{Sample 2:} \quad \overline{X}_2 = \frac{X_1 + X_3}{2} = \frac{1 + 6}{2} = 3.5$$

$$\text{Sample 3:} \quad \overline{X}_3 = \frac{X_2 + X_3}{2} = \frac{3 + 6}{2} = 4.5$$

Hence the expected value of $\overline{X}$ is equal to

$$\frac{1}{C_2^3} \sum \overline{X}_s = \frac{1}{3}(2 + 3.5 + 4.5) = \frac{10}{3}$$

which is precisely equal to the population mean $\mu_X = (X_1 + X_2 + X_3)/3 = (1 + 3 + 6)/3 = 10/3$. We thus have shown the truth of Theorem 16.1 for $\overline{X}$ in this example. The reader is invited to do the same for $N\overline{X}$ and $[(N - 1)/N]s_X^2$.

The *sampling error* of an unbiased estimator $\hat{\theta}$ of θ for a particular sample is given by $\hat{\theta} - \theta$. Since the average of these errors over all possible samples is equal to 0, we cannot utilize it as a measure of sampling error. Statisticians have adopted the squared deviations averaged over all possible samples as a measure of sampling error, and this they have termed the *sampling variance* of $\hat{\theta}$, i.e., the sampling variance of $\hat{\theta}$ is the expected value of the squared sampling error of $\hat{\theta}$. We thus have:

DEFINITION 16.8 *The sampling variance of* $\hat{\theta}$ *is equal to* $\sigma^2_{\hat{\theta}} = (1/C^N_n)\Sigma(\hat{\theta}_s - \theta)^2$ *in simple random sampling without replacement.*

This definition leads us to the following theorem:

THEOREM 16.2 *The sampling variances of* $\overline{X}$ *and* $N\overline{X}$ *are, respectively, equal to* $\sigma^2_{\overline{X}} = [(N - n)/n(N - 1)]\sigma^2_X$ *and* $\sigma^2_{N\overline{X}} = N^2\sigma^2_{\overline{X}} = [N^2(N - n)/n(N - 1)]\sigma^2_X$.

The *standard errors* of $\overline{X}$ and $N\overline{X}$ are therefore:

$$\sigma_{\overline{X}} = \sqrt{\frac{(N - n)}{n(N - 1)}}\,\sigma_X$$

and

$$\sigma_{N\overline{X}} = \sqrt{\frac{N^2(N - n)}{n(N - 1)}}\,\sigma_X$$

Since an unbiased estimator of σ^2_X is given by $[(N - 1)/N]s^2_X$ (see Theorem 16.1), then

$$s^2_{\overline{X}} = \frac{(N - n)}{n(N - 1)}\,\frac{(N - 1)}{N}\,s^2_X$$

$$= \frac{N - n}{nN}\,s^2_X$$

is an unbiased estimator of $\sigma^2_{\overline{X}}$. We are thus led to the following theorem.

THEOREM 16.3 *Unbiased estimators of the sampling variances* $\sigma^2_{\overline{X}}$ *and* $\sigma^2_{N\overline{X}}$ *are, respectively, given by* $s^2_{\overline{X}} = [(N - n)/nN]s^2_X$ *and* $[N^2(N - n)/nN]s^2_X$.

Hence *estimators of the standard errors* of $\overline{X}$ and $N\overline{X}$ are:

$$s_{\overline{X}} = \sqrt{\frac{N - n}{nN}}\,s_X$$

and

$$s_{N\overline{X}} = \sqrt{\frac{N^2(N - n)}{nN}}\,s_X = N\sqrt{\frac{N - n}{nN}}\,s_X$$

By Theorems 16.1 and 16.3 we are now in possession of the estimators of the population mean and population total and estimators of their precision, and they can be calculated from a single sample.

It can be shown that when the population contains a very large number of units, i.e., N is very large, then the estimator of the standard error of $\overline{X}$ takes the simplified form $s_{\overline{X}} = s_X/\sqrt{n}$ (which is the familiar formula for infinite populations).

The estimator of the coefficient of variation of $\overline{X}$ (or of $N\overline{X}$) is equal to $(s_{\overline{X}}/\overline{X}) \cdot 100\%$.

When an estimator $\hat{\theta}$ is a biased estimator of θ, then its expected value is not equal to θ. Suppose it is θ^*; then the *bias* is equal to $\theta^* - \theta = \beta_{\hat{\theta}}$, say. The *root mean square error* (RMSE) is then the proper measure of accuracy, and this is given by:

$$\text{RMSE}(\hat{\theta}) = \sqrt{\sigma^2_{\hat{\theta}} + \beta^2_{\hat{\theta}}}$$

For example, the sample variance s^2_X is a biased estimator of σ^2_X, because it can be shown that its expected value is equal to $[N/(N - 1)]\sigma^2_X$. Hence

$$\beta_{s^2_X} = \frac{N}{N - 1}\sigma^2_X - \sigma^2_X = \frac{1}{N - 1}\sigma^2_X$$

Therefore:

$$\text{RMSE}(s^2_X) = \sqrt{\sigma^2_{s^2_X} + \left[\left(\frac{1}{N - 1}\right)\sigma^2_X\right]^2}$$

which can be evaluated when $\sigma^2_{s^2_X}$ is derived. This, however, is beyond the scope of this book.

We now illustrate Theorems 16.1 and 16.3 with an example to show the ease of calculation.

Example 16.1 With a particular enumeration area comprising 150 households, a simple random sample of 11 households resulted in the following data with respect to number of children.

Household in sample	Number of children (below 18)
019	2
022	6
043	0
064	4
075	1
086	2
097	3
108	7
139	11
140	2
149	0

$\overline{X} = 3.45 \qquad N\overline{X} = (150)(3.45) = 518.18 \qquad s_X^2 = 11.273$

$s_X = 3.358 \qquad \frac{N - 1}{N} s_X^2 = \frac{149}{150}(11.273) = 11.198$

$\sqrt{\frac{N - 1}{N}} s_X = 3.347 \qquad s_{\overline{X}} = \sqrt{\frac{150 - 11}{(11)(150)}} (3.358) = 0.975$

$s_{N\overline{X}} = 150(0.975) = 146.25$

Estimated CV of $\overline{X} = (s_{\overline{X}}/\overline{X}) \times 100\% = (0.975/3.45) \times 100\% = 28\%$

Hence the average number of children for the enumeration area is estimated to be 3.45 children with an estimated standard error of 0.98 children; i.e., the variability of the sample mean over all possible samples is estimated to be 0.98 children. The total number of children for the area is estimated to be 518.18 children with an estimated standard error of 146.3 children. The estimated coefficient of variation of the sample mean or estimated total is equal to 28%.

16.6 ESTIMATION FORMULAS WHEN THE SAMPLE DATA ARE ARRANGED IN A FREQUENCY DISTRIBUTION

Often sample data are arranged in a frequency distribution. In such a case the formulas are modified in order to take midpoints and frequencies into account, as was done in Chapter 4. Doing this we obtain:

$$\overline{X} = \frac{1}{n}\sum_{i=1}^{h} X_i f_i = \text{sample mean}$$

where

$$h = \text{number of classes}$$

$$n = \sum_{i=1}^{h} f_i = \text{sample size}$$

$$X_i = \text{midpoint of ith class}$$

$$f_i = \text{frequency of the ith class}$$

The population total is estimated, as before, by $N\overline{X}$, and the estimators of the standard error of $\overline{X}$ and $N\overline{X}$ are, respectively:

$$s_{\overline{X}} = \sqrt{\frac{(N - n)}{nN}}\, s_X \qquad s_{N\overline{X}} = N\sqrt{\frac{(N - n)}{nN}}\, s_X$$

where

$$s_X = \sqrt{\frac{\sum_{i=1}^{h} (X_i - \overline{X})^2 f_i}{n - 1}}$$

$$= \sqrt{\frac{\sum_{i=1}^{h} X_i^2 f_i - (1/n)\left(\sum_{i=1}^{h} X_i f_i\right)^2}{n - 1}}$$

Since we are familiar with calculating s_X, the estimates of the standard errors are trivial. An expression for the estimator of the coefficient of variation is obtained by taking the appropriate ratio.

16.7 SIMPLE RANDOM SAMPLING WITHOUT REPLACEMENT TO ESTIMATE PROPORTIONS AND PERCENTAGES

In this case each X_i assumes two values, namely, 0 and 1, by letting X_i be equal to 1 if the unit possesses the characteristic, and letting X_i be equal to 0 if the unit does not have the characteristic. The estimating formulas are:

Estimated proportion $= \hat{p} = \frac{1}{n} \sum_{i=1}^{n} X_i = \bar{X}$

Estimated total $= N\hat{p}$

Estimated sampling variance of $\hat{p} = s_{\hat{p}}^2 = \frac{N - n}{N(n - 1)} \hat{p}\hat{q}$,

where $\hat{q} = 1 - \hat{p}$

Estimated variance of the total $= s_{N\hat{p}}^2 = N^2 \left[\frac{(N - n)}{N(n - 1)}\right] \hat{p}\hat{q}$

From these expressions one immediately obtains the standard errors $s_{\hat{p}}$ and $s_{N\hat{p}}$. See Exercise 16.5 for an example. Also, for large N and n close to n - 1 the formula for $s_{\hat{p}}^2$ becomes simply $\hat{p}\hat{q}/n$.

16.8 SIZE OF A SIMPLE RANDOM SAMPLE WITHOUT REPLACEMENT

In every sample survey we are confronted with the question of how many units to include in the sample. We have dealt with this problem in Chapter 10 for the case of infinite populations. The development for the finite population case is similar, except that we have added one more method based on budget considerations. Let us deal with this one first.

1. Sample size for fixed budget. Often accuracy considerations do not enter into the picture when determining the sample size. This comes about because a fixed budget is allotted for the sample survey. A simple cost equation which explains the total cost of a sample survey is given by:

$$B = C_0 + C_1 \cdot n$$

where C_0 is the total overhead cost for the whole sample survey and C_1 is the cost per sampling unit. Thus when the budget allotted for the survey is equal to B_0 dollars, then the number of sampling units which can be accommodated is equal to

$$n^* = \frac{B_0 - C_0}{C_1}$$

The precision of the sample mean for this n* is then equal to

$$\sigma_{\overline{X}} = \sqrt{\frac{N - n^*}{n^*(N - 1)}}\,\sigma_X$$

so that when σ_X is known from previous surveys or has been estimated by a pilot survey, the precision is estimated ahead of the survey.

As an example, consider a projected mail survey regarding the attitudes of parents in "crime on TV" in a city having 50,000 households with children under 18. Assume that this is done by a polling company. Let $10,000 be the allotted budget and suppose that the fixed cost, i.e., the cost which is independent of the sample size, is equal to $5000. If the cost of one questionnaire, envelope, and stamp comes to $2.50, then the sample size is equal to:

$$n^* = \frac{10{,}000\text{-}5000}{2.50} = 2000$$

Hence 2000 parents can be reached by the mail survey. If the standard deviation of one of the variables in the questionnaire, say "number of hours of watching TV" is known from previous studies to be equal to 3.7 hr, then the sample size of 2000 leads to an estimated precision of the mean of:

$$\hat{\sigma}_{\overline{X}} = \sqrt{\frac{50{,}000 - 2000}{2000(49{,}999)}}\,(3.7) \doteq 0.1 \text{ hr}$$

If no advance estimate of σ_X is available, then the precision is estimated after the survey by using the unbiased estimator of σ_X.

2. *Sample size for fixed coefficients of variation.* One of the drawbacks of the fixed budget approach is that it prevents us from obtaining estimates of a prescribed precision. With this in mind, it is quite logical to make the sample size depend on the precision to be attained, and then calculate the resulting total cost. This means that we fix the precision in advance; e.g., $CV(\overline{X}) = k\%$, and then from this obtain a solution for the sample size. Using the CV of $\overline{X}$, we have:

$$CV(\overline{X}) = \frac{\sigma_{\overline{X}}}{\mu_X} = \frac{k}{100}$$

i.e.,

$$\frac{\sigma^2_{\overline{X}}}{\mu^2_X} = \left(\frac{k}{100}\right)^2$$

i.e.,

$$\frac{[(N - n)/n(N - 1)]\sigma^2_X}{\mu^2_X} = \left(\frac{k}{100}\right)^2$$

i.e.,

$$n^* = \frac{N(\sigma^2_X/\mu^2_X)}{(k/100)^2(N - 1) + \sigma^2_X/\mu^2_X}$$

$$= \frac{NV^2_X}{(k/100)^2(N - 1) + V^2_X}$$

where V^2_X = relative variance of $X = \sigma^2_X/\mu^2_X$.

Notice that an advance estimate of V^2_X is required in order to calculate the sample size. Using the simple cost equation, one then obtains the total cost of the sample survey as $B^* = C_0 + C_1n^*$.

For very large N the formula for n* reduces to the one developed in Chapter 10; i.e.,

$$n^* = \frac{\sigma^2_X/\mu^2_X}{(k/100)^2} = \frac{V^2_X}{(k/100)^2}$$

For example, when N = 1000 and the precision or CV of $\overline{X}$ is set at 5%, and for a certain variable X it is known from a previous survey that V_X is 20%, then the required sample size is equal to:

$$n^* = \frac{(1000)(0.20)^2}{(0.05)^2(999) + (0.20)^2} = 15.8$$

i.e., a sample of size 16.

For proportions or percentages the sample size is derived by the same formulas and by noting that X is a dichotomous variable, so that $\mu_X = p$ and $\sigma_X^2 = pq$.

3. *Sample size for fixed confidence tolerance.* When the frequency distribution of the variable X can be reasonably approximated by the normal distribution, the problem of sample size can be formulated using the confidence interval approach. Suppose that k is the error bound or tolerance on the sample mean, i.e., $|\overline{X} - \mu_X| = k$, and that we want to be 95% confident that the absolute difference $|\overline{X} - \mu_X|$ will be less than k. This means that

$$P[|\overline{X} - \mu_X| < k] = 0.95$$

i.e.,

$$P\left[|Z| < \frac{k}{\sigma_{\overline{X}}}\right] = 0.95$$

Hence:

$$k = 1.96\sigma_{\overline{X}}$$

$$k = 1.96\sqrt{\frac{N - n}{n(N - 1)}}\,\sigma_X$$

$$k^2 = (1.96)^2\,\frac{N - n}{n(N - 1)}\,\sigma_X^2$$

$$n^* = \frac{N\sigma_X^2}{[(N - 1)k^2/(1.96)^2] + \sigma_X^2}$$

Since $(1.96)^2 \doteq 4.0$, this formula for large populations reduces to

$$n^* = \frac{4\sigma_X^2}{k^2}$$

which has already been derived in Chapter 10.

For proportions or percentages, we substitute $\sigma_X^2 = pq$ in the above formula. Of course the budget can be obtained by calculating $B = C_0 + n^*C_1$. The reader is invited to provide an illustration.

It should be clear that probabilities other than 95% can be used in calculating the sample size. The choice will depend on what kind of chances one is willing to take to remain within the specified error bound.

Remark. The formulas given above also hold for sample sizes to estimate the population total, since all statements regarding $N\overline{X}$ can be reduced to a statement about $\overline{X}$. Also, the formulas can be used if σ_X^2 is known from a previous study or estimated from a pilot study.

16.9 STRATIFIED SIMPLE RANDOM SAMPLING

The idea of stratification or grouping of a population into homogeneous subpopulations is closely related to blocking of experimental units in experimental design. When this idea is carried through, and sampling is done in each subpopulation, then one would expect a higher precision of estimates of population parameters, because an interfering variable has now been controlled. We formally define stratified simple random sampling as follows:

DEFINITION 16.9 *When, on the basis of a variable, a population is divided into disjoint groups such that the units in a group are more alike with respect to some variable of interest than are the units in the population as a whole and a simple random sample without replacement is drawn from each group, then this procedure is called stratified simple random sampling.*

The groups are referred to as *strata* and the process of dividing the population is called *stratification*. The variable used for forming the groups is known as the *stratification variable*.

For example, when a sociologist is interested in studying the practices of manufacturing industries in hiring blue collar workers of minority groups, then stratification of these industries according to size of blue collar labor force will result in more homogeneous subpopulations for carrying out the study.

In many surveys *geographical stratification* is applied, when the assumption that the units tend to be more alike in a geographical region than in the overall population is valid.

Stratification is especially effective when the units in a population possess extreme values. When such a population is divided into strata it is clear that the units within each stratum will show considerably less variability than the units in the overall population.

16.10 POPULATION PARAMETERS IN STRATIFIED SIMPLE RANDOM SAMPLING

Suppose there are h strata with N_i units in the ith stratum, i = 1, 2, ..., h. Let X_{ij} be the value of the variable X for the jth unit in the ith stratum; then

$$N = N_1 + N_2 + \cdots + N_h = \sum_{i=1}^{h} N_i$$

= total number of units in the population

$$\tau_i = X_{i1} + X_{i2} + \cdots + X_{iN_i} = \sum_{j=1}^{N_i} X_{ij}$$

= total of X for the ith stratum

$$\mu_i = \frac{1}{N_i} \sum_{j=1}^{N_i} X_{ij} = \frac{\tau_i}{N_i}$$

= mean of X for the ith stratum

$$\tau = \tau_1 + \tau_2 + \cdots + \tau_h = \sum_{i=1}^{h} \tau_i = \sum_{i=1}^{h} \sum_{j=1}^{N_i} X_{ij}$$

= population total of X

$$\mu = \frac{\tau}{N} = \frac{1}{N}\sum_{i=1}^{h} \tau_i = \frac{1}{N}\sum_{i=1}^{h}\sum_{j=1}^{N_i} X_{ij} = \text{population mean of X}$$

These expressions look horrendous, but when they are organized in a table such as Table 16.1, they become intelligible.

Having defined the totals and means, we are now ready to define the variances:

$$\sigma_i^2 = \frac{1}{N_i}\sum_{j=1}^{N_i} (X_{ij} - \mu_i)^2$$

$$= \text{variance of X in the ith stratum}$$

$$\sigma^2 = \frac{1}{N}\sum_{i=1}^{h}\sum_{j=1}^{N_i} (X_{ij} - \mu)^2$$

$$= \text{variance of X in the overall population}$$

If we set the relative size of the ith stratum equal to p_i, i.e., $p_i = N_i/N$, then we may rewrite the population mean μ and the population variance σ^2 as:

$$\mu = \sum_{i=1}^{h} p_i\mu_i$$

and

$$\sigma^2 = \sum_{i=1}^{h} p_i\sigma_i^2 + \sum_{i=1}^{h} p_i(\mu_i - \mu)^2$$

Hence we see that the variance in the population can be written as the sum of the weighted variances of the strata plus the weighted variability of strata means, the weights being the p_i's.

All the above expressions are just generalizations of the expressions for the simple random sampling case, because this is stratified sampling with one stratum. Note that we have dropped the subscript X in the parameters to keep the notation simple.

Table 16.1 Population Structure in Stratified Sampling

Strata				
1	2	$\cdots$	h	
X_{11}	X_{21}	$\cdots$	X_{h1}	
X_{12}	X_{22}	$\cdots$	X_{h2}	
$\vdots$	$\vdots$		$\vdots$	
X_{1N_1}	X_{2N_2}	$\cdots$	X_{hN_h}	
$\sum_{j=1}^{N_1} X_{1j}$	$\sum_{j=1}^{N_2} X_{2j}$	$\cdots$	$\sum_{j=1}^{N_j} X_{hj}$	$\sum_{i=1}^{h} \sum_{j=1}^{N_i} X_{ij} = \tau$
$= \tau_1$	$= \tau_2$	$\cdots$	$= \tau_h$	$= \tau_1 + \tau_2 + \cdots + \tau_h$
$\mu_1 = \tau_1/N_1$	$\mu_2 = \tau_2/N_2$	$\cdots$	$\mu_h = \tau_h/N_h$	$\mu = \tau/N$

16.11 ESTIMATION OF PARAMETERS IN STRATIFIED SIMPLE RANDOM SAMPLING

Assume that n_1 units are drawn from the first stratum, n_2 units from the second stratum, ..., n_h units from the hth stratum. The sample size $n = n_1 + n_2 + \cdots + n_h = \sum_{i=1}^{h} n_i$.

We may now form the following estimators:

$$\overline{X}_i = \frac{1}{n_i} \sum_{j=1}^{n_i} X_{ij}$$

= sample mean for the ith stratum

= estimator of μ_i

$N_i\overline{X}_i$ = estimator of the total τ_i

$$\overline{X} = \sum_{i=1}^{h} p_i\overline{X}_i = \text{estimator of } \mu, \text{ where } p_i = N_i/N$$

$$N\overline{X} = \sum_{i=1}^{h} N_i\overline{X}_i = \text{estimator of } \tau$$

$$\frac{N_i - 1}{N_i} s_i^2 = \frac{(N_i - 1)}{N_i(n_i - 1)} \sum_{j=1}^{n_i} (X_{ij} - \overline{X}_i)^2$$

$$= \text{sample variance for the ith stratum}$$

$$= \text{estimator of } \sigma_i^2$$

This leads us to the following theorem, which is similar in nature to Theorem 16.1.

THEOREM 16.4 *The estimators* $\overline{X}_i$, $N_i\overline{X}_i$, $[(N_i - 1)/N_i]s_i^2$, $\overline{X}$, *and* $N\overline{X}$ *are, respectively, unbiased estimators of the stratum mean* μ_i, *the stratum total* τ_i, *the stratum variance* σ_i^2, *the population mean* μ, *and the population total* τ.

The next theorem lays the groundwork for obtaining the standard errors of the estimators of the stratum means, stratum totals, the overall population mean, and the overall population total.

THEOREM 16.5 *The sampling variances of* $\overline{X}_i$, $N_i\overline{X}_i$, $\overline{X}$, *and* $N\overline{X}$ *are, respectively, equal to*

$$\sigma_{\overline{X}_i}^2 = \frac{(N_i - n_i)}{n_i(N_i - 1)} \sigma_i^2$$

$$\sigma_{N_i\overline{X}_i}^2 = \frac{N_i^2(N_i - n_i)}{n_i(N_i - 1)} \sigma_i^2$$

$$\sigma_{\overline{X}}^2 = \sum_{i=1}^{h} p_i^2 \frac{(N_i - n_i)}{n_i(N_i - 1)} \sigma_i^2$$

and

$$\sigma_{N\overline{X}}^2 = N^2 \sum_{i=1}^{h} p_i^2 \frac{(N_i - n_i)}{n_i(N_i - 1)} \sigma_i^2$$

Hence we obtain the standard errors of $\overline{X}_i$, $N_i\overline{X}_i$, $\overline{X}$, and $N\overline{X}$, which have to be estimated to provide measures of precision:

$$\sigma_{\overline{X}_i} = \sqrt{\frac{(N_i - n_i)}{n_i(N_i - 1)}}\,\sigma_i$$

$$\sigma_{N_i\overline{X}_i} = \sqrt{\frac{N_i^2(N_i - n_i)}{n_i(N_i - 1)}}\,\sigma_i$$

$$\sigma_{\overline{X}} = \sqrt{\sum_{i=1}^{h} p_i^2 \frac{(N_i - n_i)}{n_i(N_i - 1)}\sigma_i^2}$$

and

$$\sigma_{N\overline{X}} = N\sqrt{\sum_{i=1}^{h} p_i^2 \frac{(N_i - n_i)}{n_i(N_i - 1)}\sigma_i^2}$$

which can be simplified to:

$$= \sqrt{\sum_{i=1}^{h} N_i^2 \frac{(N_i - n_i)}{n_i(N_i - 1)}\sigma_i^2}$$

The following theorem provides the estimation formulas for the variances, and hence the standard errors.

THEOREM 16.6 *Unbiased estimators of the sampling variances* $\sigma^2_{\overline{X}_i}$, $\sigma^2_{N_i\overline{X}_i}$, $\sigma^2_{\overline{X}}$; *and* $\sigma^2_{N\overline{X}}$ *are given by*

$$s^2_{\overline{X}_i} = \frac{N_i - n_i}{n_i N_i} s_i^2$$

$$s^2_{N_i\overline{X}_i} = \frac{N_i^2(N_i - n_i)}{n_i N_i} s_i^2$$

$$s^2_{\overline{X}} = \sum_{i=1}^{h} p_i^2 \frac{(N_i - n_i)}{n_i N_i} s_i^2$$

and

$$s^2_{N\overline{X}} = N^2 \sum_{i=1}^{h} p_i^2 \frac{(N_i - n_i)}{n_i N_i} s_i^2$$

Hence the estimators of the standard errors of $\overline{X}_i$, $N_i\overline{X}_i$, $\overline{X}$, and $N\overline{X}$ are:

$$s_{\overline{X}_i} = \sqrt{\frac{N_i - n_i}{n_i N_i}}\, s_i$$

$$s_{N_i\overline{X}_i} = \sqrt{\frac{N_i^2(N_i - n_i)}{n_i N_i}}\, s_i = N_i\sqrt{\frac{N_i - n_i}{n_i N_i}}\, s_i$$

$$s_{\overline{X}} = \sqrt{\sum_{i=1}^{h} p_i^2 \frac{(N_i - n_i)}{n_i N_i} s_i^2}$$

and

$$s_{N\overline{X}} = N\sqrt{\sum_{i=1}^{h} p_i^2 \frac{(N_i - n_i)}{n_i N_i} s_i^2}$$

From these expressions one obtains the estimators of the CVs, i.e,.

$$CV(\overline{X}_i) = CV(N\overline{X}_i) = \frac{s_{\overline{X}_i}}{\overline{X}_i} \cdot 100\%$$

and

$$CV(\overline{X}) = CV(N\overline{X}) = \frac{s_{\overline{X}}}{\overline{X}} \cdot 100\%$$

Example 16.2 We set up the following small example to illustrate the calculations in all its details.

Stratum	N_i	n_i	X_{ij}
1	12	4	2,1,1,0
2	9	4	3,2,3,1
3	4	3	4,4,6

Using Theorems 16.4 and 16.6, we obtain the following results:

$$\overline{X}_1 = \frac{1}{4}(2 + 1 + 1 + 0) = 1 \qquad N_1\overline{X}_1 = (12)(1) = 12$$

$$\overline{X}_2 = \frac{1}{4}(3 + 2 + 3 + 1) = 2.25 \qquad N_2\overline{X}_2 = (9)(2.25) = 20.25$$

$$\overline{X}_3 = \frac{1}{3}(4 + 4 + 6) = 4.67 \qquad N_3\overline{X}_3 = (4)\left(\frac{14}{3}\right) = 18.67$$

$$\overline{X} = \frac{12}{25}(1) + \frac{9}{25}(2.25) + \frac{4}{25}(4.67) = 2.04$$

$$N\overline{X} = 25(2.04) = 51.00$$

$$s_1^2 = \frac{1}{3}[(2^2 + 1^2 + 1^2 + 0^2) - \frac{1}{4}(2 + 1 + 1 + 0)^2] = 0.67$$

$$s_2^2 = \frac{1}{3}[(3^2 + 2^2 + 3^2 + 1^2) - \frac{1}{4}(3 + 2 + 3 + 1)^2] = 0.92$$

$$s_3^2 = \frac{1}{2}[(4^2 + 4^2 + 6^2) - \frac{1}{3}(4 + 4 + 6)^2] = 1.34$$

$$s_{\overline{X}_1} = \sqrt{\frac{(12 - 4)}{(12)(4)}(0.67)} = \sqrt{0.11} = 0.33$$

$$s_{\overline{X}_2} = \sqrt{\frac{(9 - 4)}{(9)(4)}(0.92)} = \sqrt{0.13} = 0.36$$

$$s_{\overline{X}_3} = \sqrt{\frac{(4 - 3)}{(4)(3)}(1.34)} = \sqrt{0.11} = 0.33$$

$$s_{N_1\overline{X}_1} = 12(0.33) = 3.96$$

$$s_{N_2\overline{X}_2} = 9(0.36) = 3.24$$

$$s_{N_3\overline{X}_3} = 4(0.33) = 1.32$$

$$s_{\overline{X}} = \sqrt{\left(\frac{12}{25}\right)^2(0.11) + \left(\frac{9}{25}\right)^2(0.13) + \left(\frac{4}{25}\right)^2(0.11)} = 0.22$$

$$s_{N\overline{X}} = (25)(0.22) = 5.59$$

Hence, estimates of the parameters μ_1, μ_2, μ_3, τ_1, τ_2, τ_3, μ, τ, $\sigma_{\overline{X}_1}$, $\sigma_{\overline{X}_2}$, $\sigma_{\overline{X}_3}$, $\sigma_{N_1\overline{X}_1}$, $\sigma_{N_2\overline{X}_2}$, $\sigma_{N_3\overline{X}_3}$, $\sigma_{\overline{X}}$, and $\sigma_{N\overline{X}}$ are given, respectively, by 1, 2.25, 4.67, 12, 20.25, 18.67, 2.04, 51.00, 0.33, 0.36, 0.33, 3.96, 3.24, 1.32, 0.22, and 5.59.

16.12 SAMPLE SIZE ALLOCATION AMONG THE STRATA FOR GIVEN OVERALL SAMPLE SIZE n

A problem of considerable interest is how to allocate the sample sizes to the various strata. Suppose that the overall sample size for the stratified sample survey is given, i.e., we are told that n units are to be drawn from a specified population. We can do the following:

1. *Equal sample size for each stratum.* That is,

$$n_i = \frac{n}{h} \qquad i = 1, 2, \ldots, h$$

This procedure does not take into account the sizes of strata nor the variability in the strata.

2. *Proportional allocation.* In this case

$$n_i = \frac{N_i}{N} \cdot n = p_i \cdot n$$

i.e., we take account of the size of the stratum. For this type of allocation we get a nice simplification in the formulas as an added bonus. Thus:

$$\overline{X} = \frac{1}{n} \sum_{i=1}^{h} \sum_{j=1}^{n_i} X_{ij}$$

$$= \text{ordinary mean of the sample}$$

and

$$s_{\overline{X}}^2 = \frac{N - n}{nN} \sum_{i=1}^{h} p_i s_i^2$$

3. Arbitrary allocation. In this case the n_i's are allocated without taking any criteria into account. This is the most general approach in developing the theory, which we have done already.

4. Optimum allocation. The allocation of the n_i's to the strata such that the resulting variance of $\overline{X}$ will be a minimum is given by:

$$n_i = \frac{p_i\sqrt{N_i/(N_i - 1)}\sigma_i}{\sum_{j=1}^{h} p_j\sqrt{N_j/(N_j - 1)}\sigma_j} n$$

where advance estimates of $\sigma_1, \sigma_2, \ldots, \sigma_h$ are needed in order to apply the formula. This type of allocation is the best type.

Example 16.3 Suppose that a population containing the following number of units and advance estimates of the σ_i's has been stratified into five strata:

Stratum	Number of units	Advance estimates
1	$N_1 = 3{,}000$	$\sigma_1 = 5$
2	$N_2 = 4{,}000$	$\sigma_2 = 7$
3	$N_3 = 1{,}200$	$\sigma_3 = 8$
4	$N_4 = 1{,}800$	$\sigma_4 = 6$
5	$N_5 = 2{,}000$	$\sigma_5 = 6$
	$N = 12{,}000$	

The p_i's are as follows: $p_1 = 1/4$, $p_2 = 1/3$, $p_3 = 1/10$, $p_4 = 3/20$, $p_5 = 1/6$. Suppose that a 10% sample is desired, i.e., $n = 1200$; then for *equal allocation* $n_1 = n_2 = n_3 = n_4 = n_5 = 1200/5 = 240$, and for *proportional allocation* $n_1 = (1/4)(1200) = 300$, $n_2 = (1/3)(1200) = 400$, $n_3 = (1/10)(1200) = 120$, $n_4 = (3/20)(1200) = 180$, and $n_5 = (1/6)(1200) = 200$.

Since $\sqrt{N_i/(N_i - 1)}$ is close to 1 for all strata, we ignore this factor, and the following calculation results for the *optimum allocation:*

Stratum	$p_j\sigma_j$	n_i
1	$\frac{5}{4} = 1.25$	$n_1 = \frac{1.25}{6.28}(1200) = 239$
2	$\frac{7}{3} = 2.33$	$n_2 = \frac{2.33}{6.28}(1200) = 445$
3	$\frac{8}{10} = 0.80$	$n_3 = \frac{0.80}{6.28}(1200) = 153$
4	$\frac{18}{20} = 0.90$	$n_4 = \frac{0.90}{6.28}(1200) = 172$
5	$1 = 1.00$	$n_5 = \frac{1}{6.28}(1200) = 191$
	$\Sigma p_j\sigma_j = 6.28$	

Remark. There are many more probability sampling techniques, such as cluster sampling, two-stage sampling, systematic sampling, and double sampling. The reader is advised to read about these elsewhere, since these topics fall beyond the scope of this book. A good reference is Deming (1960). At the beginning of this chapter we listed many of the practical problems associated with sample surveys. An excellent example of these problems is encountered in the World Fertility Survey of the International Statistical Institute. This project was directed by Kendall (1975).

EXERCISES

16.1 Construct a small example to show that the sample variance $s_X^2 = \frac{1}{n-1}\sum_{i=1}^{n}(X_i - \overline{X})^2$ is a biased estimator of σ_X^2 in simple random sampling without replacement, for finite populations.

16.2 A simple random sample without replacement of 2072 small businesses in Ontario with 75,308 businesses resulted in the following frequency distribution of income in dollars:

Income	Number of businesses
0-10,000	263
10,000-20,000	140
20,000-30,000	188
30,000-40,000	440
40,000-50,000	350
50,000-60,000	263
60,000-70,000	120
70,000-80,000	120
80,000-90,000	99
90,000-100,000	46
100,000-120,000	43

(a) Estimate the mean income of businesses for Ontario.

(b) Estimate the total income of businesses for Ontario.

(c) Estimate the percentage CV of the sample mean income for Ontario.

16.3 Give an example of nonprobability sampling that is not mentioned in this chapter, and state why it should not be used.

16.4 A random sample without replacement of 10 undergraduate B.Sc. Honors students at a university was asked the question: How many hours do you study per week outside of lecture periods and laboratory hours? The answers were: 19, 15, 12, 18, 10, 21, 14, 11, 10, 14. If the enrollment of B.Sc. Honors students from which the sample was selected was equal to 250, then:

(a) Using random numbers describe how the students were selected.

(b) Estimate the mean number of hours for all B.Sc. Honors students.

(c) Estimate the standard error of the sample mean and give a repeated sampling interpretation.

16.5 Suppose that a sample without replacement of 315 households in a community was drawn in a village having 15,762 households Each household was asked whether it owned or rented the home and also whether it had TV or not. The results were as follows:

Television	Owned	Rented
With TV	153	121
Without TV	10	31

(a) For households who rent, estimate the proportion in the community without TV and also estimate the standard error.

(b) Estimate the total number of renting households and provide an estimate for the standard error.

16.6 Using the CV in Exercise 16.2 as an advance estimate for a forthcoming survey, find the sample size if a CV of 5% is desired.

16.7 Illustrate, with your own example, the sample size calculation for fixed confidence tolerance.

16.8 Consider the following small population of N = 5 units and the corresponding values of a variable X:

Population	X values
Unit 1	1000
Unit 2	500
Unit 3	250
Unit 4	10
Unit 5	5

Procedure (a): Select simple random samples without replacement of 3 units and obtain the expected value and the variance of the estimator of the total.

Procedure (b): Divide the population into two strata by putting unit 1 all by itself into a stratum and the rest of them into a second stratum. Obtain the expected value and the variance of the estimator of the total for $n_1 = 1$ and $n_2 = 2$. Which procedure is better?

16.9 In a socio-economic survey the metropolitan area under study was divided into five strata, with the following numbers of households in each stratum and in each sample:

	Number of households in	
Stratum	Stratum	Sample
1	3136	373
2	451	56
3	1873	258
4	8908	874
5	2542	383

A variable X was observed and the sample results were:

Stratum	Sample mean $\overline{X}_i$	Sample variance s_i^2
1	4.70	2.72
2	3.85	2.40
3	2.97	1.21
4	4.16	3.13
5	1.40	1.12

Find estimates of μ_1, μ_2, μ_3, μ_4, μ_5, τ_1, τ_2, τ_3, τ_4, τ_5, μ, τ, $\sigma_{\overline{X}_1}$, $\sigma_{\overline{X}_2}$, $\sigma_{\overline{X}_3}$, $\sigma_{\overline{X}_4}$, $\sigma_{\overline{X}_5}$, $\sigma_{N_1\overline{X}_1}$, $\sigma_{N_2\overline{X}_2}$, $\sigma_{N_3\overline{X}_3}$, $\sigma_{N_4\overline{X}_4}$, $\sigma_{N_5\overline{X}_5}$, $\sigma_{\overline{X}}$, and $\sigma_{N\overline{X}}$ and describe what you have found.

16.10 Illustrate equal and proportional allocation of a 10% sample for a forthcoming survey for the population in Exercise 16.9. Using your estimates of the standard deviations in Exercise 16.9, give the optimum allocation.

16.11 Which statement is most nearly correct?

(a) Regardless of cost, stratification should be used whenever it can be accomplished.

(b) For stratified designs the sample estimates of the variances are unfortunately biased.

(c) Proportional allocation is superior to optimum allocation in stratified sampling.

16.12 What is the essential difference between:

(a) A census and a sample survey?

(b) An experiment and a survey?

(c) Probability and nonprobability sampling?

(d) Sampling with and without replacement?

(e) Unbiased and biased estimators?

(f) Simple random sampling (SRS) and stratified SRS?

APPENDIX

Statistical Tables

Table A.1 Binomial Probabilities*

n	x	.05	.10	.15	.20	p .25	.30	.35	.40	.45	.50†
1	0	.9500	.9000	.8500	.8000	.7500	.7000	.6500	.6000	.5500	.5000
	1	.0500	.1000	.1500	.2000	.2500	.3000	.3500	.4000	.4500	.5000
2	0	.9025	.8100	.7225	.6400	.5625	.4900	.4225	.3600	.3025	.2500
	1	.0950	.1800	.2550	.3200	.3750	.4200	.4550	.4800	.4950	.5000
	2	.0025	.0100	.0225	.0400	.0625	.0900	.1225	.1600	.2025	.2500
3	0	.8574	.7290	.6141	.5120	.4219	.3430	.2746	.2160	.1664	.1250
	1	.1354	.2430	.3251	.3840	.4219	.4410	.4436	.4320	.4084	.3750
	2	.0071	.0270	.0574	.0960	.1406	.1890	.2389	.2880	.3341	.3750
	3	.0001	.0010	.0034	.0080	.0156	.0270	.0429	.0640	.0911	.1250
4	0	.8145	.6561	.5220	.4096	.3164	.2401	.1785	.1296	.0915	.0625
	1	.1715	.2916	.3685	.4096	.4219	.4116	.3845	.3456	.2995	.2500
	2	.0135	.0486	.0975	.1536	.2109	.2646	.3105	.3456	.3675	.3750
	3	.0005	.0036	.0115	.0256	.0469	.0756	.1115	.1536	.2005	.2500
	4	.0000	.0001	.0005	.0016	.0039	.0081	.0150	.0256	.0410	.0625
5	0	.7738	.5905	.4437	.3277	.2373	.1681	.1160	.0778	.0503	.0312
	1	.2036	.3280	.3915	.4096	.3955	.3602	.3124	.2592	.2059	.1562
	2	.0214	.0729	.1382	.2048	.2637	.3087	.3364	.3456	.3369	.3125
	3	.0011	.0081	.0244	.0512	.0879	.1323	.1811	.2304	.2757	.3125
	4	.0000	.0004	.0022	.0064	.0146	.0284	.0488	.0768	.1128	.1562
	5	.0000	.0000	.0001	.0003	.0010	.0024	.0053	.0102	.0185	.0312
6	0	.7351	.5314	.3771	.2621	.1780	.1176	.0754	.0467	.0277	.0156
	1	.2321	.3543	.3993	.3932	.3560	.3025	.2437	.1866	.1359	.0938
	2	.0305	.0984	.1762	.2458	.2966	.3241	.3280	.3110	.2780	.2344
	3	.0021	.0146	.0415	.0819	.1318	.1852	.2355	.2765	.3032	.3125
	4	.0001	.0012	.0055	.0154	.0330	.0595	.0951	.1382	.1861	.2344
	5	.0000	.0001	.0004	.0015	.0044	.0102	.0205	.0369	.0609	.0938
	6	.0000	.0000	.0000	.0001	.0002	.0007	.0018	.0041	.0083	.0156

* Entries in this table are values of $C^n_x p^x (1 - p)^{n-x}$ for the indicated values of n, x, and p. Reproduced with permission from Burington and May, *Handbook of Probability and Statistics with Tables*, McGraw-Hill, New York, 1970.

† For $p > 1/2$, use the relation $P[X = x; n, p] = P[X = n - x; n, 1 - p]$; e.g., $P[X = 4; n = 10, p = 3/4] = P[X = 6; 10, 1/4] = 0.0162$.

Table A.1 (Continued)

n	x	p .05	.10	.15	.20	.25	.30	.35	.40	.45	.50
7	0	.6983	.4783	.3206	.2097	.1335	.0824	.0490	.0280	.0152	.0078
	1	.2573	.3720	.3960	.3670	.3115	.2471	.1848	.1306	.0872	.0547
	2	.0406	.1240	.2097	.2753	.3115	.3177	.2985	.2613	.2140	.1641
	3	.0036	.0230	.0617	.1147	.1730	.2269	.2679	.2903	.2918	.2734
	4	.0002	.0026	.0109	.0287	.0577	.0972	.1442	.1935	.2388	.2734
	5	.0000	.0002	.0012	.0043	.0115	.0250	.0466	.0774	.1172	.1641
	6	.0000	.0000	.0001	.0004	.0013	.0036	.0084	.0172	.0320	.0547
	7	.0000	.0000	.0000	.0000	.0001	.0002	.0006	.0016	.0037	.0078
8	0	.6634	.4305	.2725	.1678	.1001	.0576	.0319	.0168	.0084	.0039
	1	.2793	.3826	.3847	.3355	.2670	.1977	.1373	.0896	.0548	.0312
	2	.0515	.1488	.2376	.2936	.3115	.2965	.2587	.2090	.1569	.1094
	3	.0054	.0331	.0839	.1468	.2076	.2541	.2786	.2787	.2568	.2188
	4.	.0004	.0046	.0185	.0459	.0865	.1361	.1875	.2322	.2627	.2734
	5	.0000	.0004	.0026	.0092	.0231	.0467	.0808	.1239	.1719	.2188
	6	.0000	.0000	.0002	.0011	.0038	.0100	.0217	.0413	.0703	.1094
	7	.0000	.0000	.0000	.0001	.0004	.0012	.0033	.0079	.0164	.0312
	8	.0000	.0000	.0000	.0000	.0000	.0001	.0002	.0007	.0017	.0039
9	0	.6302	.3874	.2316	.1342	.0751	.0404	.0207	.0101	.0046	.0020
	1	.2985	.3874	.3679	.3020	.2253	.1556	.1004	.0605	.0339	.0176
	2	.0629	.1722	.2597	.3020	.3003	.2668	.2162	.1612	.1110	.0703
	3	.0077	.0446	.1069	.1762	.2336	.2668	.2716	.2508	.2119	.1641
	4	.0006	.0074	.0283	.0661	.1168	.1715	.2194	.2508	.2600	.2461
	5	.0000	.0008	.0050	.0165	.0389	.0735	.1181	.1672	.2128	.2461
	6	.0000	.0001	.0006	.0028	.0087	.0210	.0424	.0743	.1160	.1641
	7	.0000	.0000	.0000	.0003	.0012	.0039	.0098	.0212	.0407	.0703
	8	.0000	.0000	.0000	.0000	.0001	.0004	.0013	.0035	.0083	.0176
	9	.0000	.0000	.0000	.0000	.0000	.0000	.0001	.0003	.0008	.0020
10	0	.5987	.3487	.1969	.1074	.0563	.0282	.0135	.0060	.0025	.0010
	1	.3151	.3874	.3474	.2684	.1877	.1211	.0725	.0403	.0207	.0098
	2	.0746	.1937	.2759	.3020	.2816	.2335	.1757	.1209	.0763	.0439
	3	.0105	.0574	.1298	.2013	.2503	.2668	.2522	.2150	.1665	.1172
	4	.0010	.0112	.0401	.0881	.1460	.2001	.2377	.2508	.2384	.2051
	5	.0001	.0015	.0085	.0264	.0584	.1029	.1536	.2007	.2340	.2461
	6	.0000	.0001	.0012	.0055	.0162	.0368	.0689	.1115	.1596	.2051
	7	.0000	.0000	.0001	.0008	.0031	.0090	.0212	.0425	.0746	.1172
	8	.0000	.0000	.0000	.0001	.0004	.0014	.0043	.0106	.0229	.0439
	9	.0000	.0000	.0000	.0000	.0000	.0001	.0005	.0016	.0042	.0098
	10	.0000	.0000	.0000	.0000	.0000	.0000	.0000	.0001	.0003	.0010

Table A.1 (Continued)

						p					
n	x	.05	.10	.15	.20	.25	.30	.35	.40	.45	.50
11	0	.5688	.3138	.1673	.0859	.0422	.0198	.0088	.0036	.0014	.0005
	1	.3293	.3835	.3248	.2362	.1549	.0932	.0518	.0266	.0125	.0054
	2	.0867	.2131	.2866	.2953	.2581	.1998	.1395	.0887	.0513	.0269
	3	.0137	.0710	.1517	.2215	.2581	.2568	.2254	.1774	.1259	.0806
	4	.0014	.0158	.0536	.1107	.1721	.2201	.2428	.2365	.2060	.1611
	5	.0001	.0025	.0132	.0388	.0803	.1321	.1830	.2207	.2360	.2256
	6	.0000	.0003	.0023	.0097	.0268	.0566	.0985	.1471	.1931	.2256
	7	.0000	.0000	.0003	.0017	.0064	.0173	.0379	.0701	.1128	.1611
	8	.0000	.0000	.0000	.0002	.0011	.0037	.0102	.0234	.0462	.0806
	9	.0000	.0000	.0000	.0000	.0001	.0005	.0018	.0052	.0126	.0269
	10	.0000	.0000	.0000	.0000	.0000	.0000	.0002	.0007	.0021	.0054
	11	.0000	.0000	.0000	.0000	.0000	.0000	.0000	.0000	.0002	.0005
12	0	.5404	.2824	.1422	.0687	.0317	.0138	.0057	.0022	.0008	.0002
	1	.3413	.3766	.3012	.2062	.1267	.0712	.0368	.0174	.0075	.0029
	2	.0988	.2301	.2924	.2835	.2323	.1678	.1088	.0639	.0339	.0161
	3	.0173	.0852	.1720	.2362	.2581	.2397	.1954	.1419	.0923	.0537
	4	.0021	.0213	.0683	.1329	.1936	.2311	.2367	.2128	.1700	.1208
	5	.0002	.0038	.0193	.0532	.1032	.1585	.2039	.2270	.2225	.1934
	6	.0000	.0005	.0040	.0155	.0401	.0792	.1281	.1766	.2124	.2256
	7	.0000	.0000	.0006	.0033	.0115	.0291	.0591	.1009	.1489	.1934
	8	.0000	.0000	.0001	.0005	.0024	.0078	.0199	.0420	.0762	.1208
	9	.0000	.0000	.0000	.0001	.0004	.0015	.0048	.0125	.0277	.0537
	10	.0000	.0000	.0000	.0000	.0000	.0002	.0008	.0025	.0068	.0161
	11	.0000	.0000	.0000	.0000	.0000	.0000	.0001	.0003	.0010	.0029
	12	.0000	.0000	.0000	.0000	.0000	.0000	.0000	.0000	.0001	.0002
13	0	.5133	.2542	.1209	.0550	.0238	.0097	.0037	.0013	.0004	.0001
	1	.3512	.3672	.2774	.1787	.1029	.0540	.0259	.0113	.0045	.0016
	2	.1109	.2448	.2937	.2680	.2059	.1388	.0836	.0453	.0220	.0095
	3	.0214	.0997	.1900	.2457	.2517	.2181	.1651	.1107	.0660	.0349
	4	.0028	.0277	.0838	.1535	.2097	.2337	.2222	.1845	.1350	.0873
	5	.0003	.0055	.0266	.0691	.1258	.1803	.2154	.2214	.1989	.1571
	6	.0000	.0008	.0063	.0230	.0559	.1030	.1546	.1968	.2169	.2095
	7	.0000	.0001	.0011	.0058	.0186	.0442	.0833	.1312	.1775	.2095
	8	.0000	.0000	.0001	.0011	.0047	.0142	.0336	.0656	.1089	.1571
	9	.0000	.0000	.0000	.0001	.0009	.0034	.0101	.0243	.0495	.0873
	10	.0000	.0000	.0000	.0000	.0001	.0006	.0022	.0065	.0162	.0349
	11	.0000	.0000	.0000	.0000	.0000	.0001	.0003	.0012	.0036	.0095
	12	.0000	.0000	.0000	.0000	.0000	.0000	.0000	.0001	.0005	.0016
	13	.0000	.0000	.0000	.0000	.0000	.0000	.0000	.0000	.0000	.0001

Table A.1 (Continued)

n	x	.05	.10	.15	.20	p .25	.30	.35	.40	.45	.50
14	0	.4877	.2288	.1028	.0440	.0178	.0068	.0024	.0008	.0002	.0001
	1	.3593	.3559	.2539	.1539	.0832	.0407	.0181	.0073	.0027	.0009
	2	.1229	.2570	.2912	.2501	.1802	.1134	.0634	.0317	.0141	.0056
	3	.0259	.1142	.2056	.2501	.2402	.1943	.1366	.0845	.0462	.0222
	4	.0037	.0349	.0998	.1720	.2202	.2290	.2022	.1549	.1040	.0611
	5	.0004	.0078	.0352	.0860	.1468	.1963	.2178	.2066	.1701	.1222
	6	.0000	.0013	.0093	.0322	.0734	.1262	.1759	.2066	.2088	.1833
	7	.0000	.0002	.0019	.0092	.0280	.0618	.1082	.1574	.1952	.2095
	8	.0000	.0000	.0003	.0020	.0082	.0232	.0510	.0918	.1398	.1833
	9	.0000	.0000	.0000	.0003	.0018	.0066	.0183	.0408	.0762	.1222
	10	.0000	.0000	.0000	.0000	.0003	.0014	.0049	.0136	.0312	.0611
	11	.0000	.0000	.0000	.0000	.0000	.0002	.0010	.0033	.0093	.0222
	12	.0000	.0000	.0000	.0000	.0000	.0000	.0001	.0005	.0019	.0056
	13	.0000	.0000	.0000	.0000	.0000	.0000	.0000	.0001	.0002	.0009
	14	.0000	.0000	.0000	.0000	.0000	.0000	.0000	.0000	.0000	.0001
15	0	.4633	.2059	.0874	.0352	.0134	.0047	.0016	.0005	.0001	.0000
	1	.3658	.3432	.2312	.1319	.0668	.0305	.0126	.0047	.0016	.0005
	2	.1348	.2669	.2856	.2309	.1559	.0916	.0476	.0219	.0090	.0032
	3	.0307	.1285	.2184	.2501	.2252	.1700	.1110	.0634	.0318	.0139
	4	.0049	.0428	.1156	.1876	.2252	.2186	.1792	.1268	.0780	.0417
	5	.0006	.0105	.0449	.1032	.1651	.2061	.2123	.1859	.1404	.0916
	6	.0000	.0019	.0132	.0430	.0917	.1472	.1906	.2066	.1914	.1527
	7	.0000	.0003	.0030	.0138	.0393	.0811	.1319	.1771	.2013	.1964
	8	.0000	.0000	.0005	.0035	.0131	.0348	.0710	.1181	.1647	.1964
	9	.0000	.0000	.0001	.0007	.0034	.0116	.0298	.0612	.1048	.1527
	10	.0000	.0000	.0000	.0001	.0007	.0030	.0096	.0245	.0515	.0916
	11	.0000	.0000	.0000	.0000	.0001	.0006	.0024	.0074	.0191	.0417
	12	.0000	.0000	.0000	.0000	.0000	.0001	.0004	.0016	.0052	.0139
	13	.0000	.0000	.0000	.0000	.0000	.0000	.0001	.0003	.0010	.0032
	14	.0000	.0000	.0000	.0000	.0000	.0000	.0000	.0000	.0001	.0005
	15	.0000	.0000	.0000	.0000	.0000	.0000	.0000	.0000	.0000	.0000
16	0	.4401	.1853	.0743	.0281	.0100	.0033	.0010	.0003	.0001	.0000
	1	.3706	.3294	.2097	.1126	.0535	.0228	.0087	.0030	.0009	.0002
	2	.1463	.2745	.2775	.2111	.1336	.0732	.0353	.0150	.0056	.0018
	3	.0359	.1423	.2285	.2463	.2079	.1465	.0888	.0468	.0215	.0085
	4	.0016	.0514	.1311	.2001	.2252	.2040	.1553	.1014	.0572	.0278
	5	.0008	.0137	.0555	.1201	.1802	.2099	.2008	.1623	.1123	.0667
	6	.0001	.0028	.0180	.0550	.1101	.1649	.1982	.1983	.1684	.1222

Table A.1 (Continued)

n	x	.05	.10	.15	.20	p .25	.30	.35	.40	.45	.50
16	7	.0000	.0004	.0045	.0197	.0524	.1010	.1524	.1889	.1969	.1746
	8	.0000	.0001	.0009	.0055	.0197	.0487	.0923	.1417	.1812	.1964
	9	.0000	.0000	.0001	.0012	.0058	.0185	.0442	.0840	.1318	.1746
	10	.0000	.0000	.0000	.0002	.0014	.0056	.0167	.0392	.0755	.1222
	11	.0000	.0000	.0000	.0000	.0002	.0013	.0049	.0142	.0337	.0667
	12	.0000	.0000	.0000	.0000	.0000	.0002	.0011	.0040	.0115	.0278
	13	.0000	.0000	.0000	.0000	.0000	.0000	.0002	.0008	.0029	.0085
	14	.0000	.0000	.0000	.0000	.0000	.0000	.0000	.0001	.0005	.0018
	15	.0000	.0000	.0000	.0000	.0000	.0000	.0000	.0000	.0001	.0002
	16	.0000	.0000	.0000	.0000	.0000	.0000	.0000	.0000	.0000	.0000
17	0	.4181	.1668	.0631	.0225	.0075	.0023	.0007	.0002	.0000	.0000
	1	.3741	.3150	.1893	.0957	.0426	.0169	.0060	.0019	.0005	.0001
	2	.1575	.2800	.2673	.1914	.1136	.0581	.0260	.0102	.0035	.0010
	3	.0415	.1556	.2359	.2393	.1893	.1245	.0701	.0341	.0144	.0052
	4	.0076	.0605	.1457	.2093	.2209	.1868	.1320	.0796	.0411	.0182
	5	.0010	.0175	.0668	.1361	.1914	.2081	.1849	.1379	.0875	.0472
	6	.0001	.0039	.0236	.0680	.1276	.1784	.1991	.1839	.1432	.0944
	7	.0000	.0007	.0065	.0267	.0668	.1201	.1685	.1927	.1841	.1484
	8	.0000	.0001	.0014	.0084	.0279	.0644	.1134	.1606	.1883	.1855
	9	.0000	.0000	.0003	.0021	.0093	.0276	.0611	.1070	.1540	.1855
	10	.0000	.0000	.0000	.0004	.0025	.0095	.0263	.0571	.1008	.1484
	11	.0000	.0000	.0000	.0001	.0005	.0026	.0090	.0242	.0525	.0944
	12	.0000	.0000	.0000	.0000	.0001	.0006	.0024	.0081	.0215	.0472
	13	.0000	.0000	.0000	.0000	.0000	.0001	.0005	.0021	.0068	.0182
	14	.0000	.0000	.0000	.0000	.0000	.0000	.0001	.0004	.0016	.0052
	15	.0000	.0000	.0000	.0000	.0000	.0000	.0000	.0001	.0003	.0010
	16	.0000	.0000	.0000	.0000	.0000	.0000	.0000	.0000	.0000	.0001
	17	.0000	.0000	.0000	.0000	.0000	.0000	.0000	.0000	.0000	.0000
18	0	.3972	.1501	.0536	.0180	.0056	.0016	.0004	.0001	.0000	.0000
	1	.3763	.3002	.1704	.0811	.0338	.0126	.0042	.0012	.0003	.0001
	2	.1683	.2835	.2556	.1723	.0958	.0458	.0190	.0069	.0022	.0006
	3	.0473	.1680	.2406	.2297	.1704	.1046	.0547	.0246	.0095	.0031
	4	.0093	.0700	.1592	.2153	.2130	.1681	.1104	.0614	.0291	.0117
	5	.0014	.0218	.0787	.1507	.1988	.2017	.1664	.1146	.0666	.0327
	6	.0002	.0052	.0301	.0816	.1436	.1873	.1941	.1655	.1181	.0708
	7	.0000	.0010	.0091	.0350	.0820	.1376	.1792	.1892	.1657	.1214
	8	.0000	.0002	.0022	.0120	.0376	.0811	.1327	.1734	.1864	.1669
	9	.0000	.0000	.0004	.0033	.0139	.0386	.0794	.1284	.1694	.1855

Table A.1 (Continued)

n	x	.05	.10	.15	.20	p .25	.30	.35	.40	.45	.50
18	10	.0000	.0000	.0001	.0008	.0042	.0149	.0385	.0771	.1248	.1669
	11	.0000	.0000	.0000	.0001	.0010	.0046	.0151	.0374	.0742	.1214
	12	.0000	.0000	.0000	.0000	.0002	.0012	.0047	.0145	.0354	.0708
	13	.0000	.0000	.0000	.0000	.0000	.0002	.0012	.0045	.0134	.0327
	14	.0000	.0000	.0000	.0000	.0000	.0000	.0002	.0011	.0039	.0117
	15	.0000	.0000	.0000	.0000	.0000	.0000	.0000	.0002	.0009	.0031
	16	.0000	.0000	.0000	.0000	.0000	.0000	.0000	.0000	.0001	.0006
	17	.0000	.0000	.0000	.0000	.0000	.0000	.0000	.0000	.0000	.0001
	18	.0000	.0000	.0000	.0000	.0000	.0000	.0000	.0000	.0000	.0000
19	0	.3774	.1351	.0456	.0144	.0042	.0011	.0003	.0001	.0000	.0000
	1	.3774	.2852	.1529	.0685	.0268	.0093	.0029	.0008	.0002	.0000
	2	.1787	.2852	.2428	.1540	.0803	.0358	.0138	.0046	.0013	.0003
	3	.0533	.1796	.2428	.2182	.1517	.0869	.0422	.0175	.0062	.0018
	4	.0112	.0798	.1714	.2182	.2023	.1491	.0909	.0467	.0203	.0074
	5	.0018	.0266	.0907	.1636	.2023	.1916	.1468	.0933	.0497	.0222
	6	.0002	.0069	.0374	.0955	.1574	.1916	.1844	.1451	.0949	.0518
	7	.0000	.0014	.0122	.0443	.0974	.1525	.1844	.1797	.1443	.0961
	8	.0000	.0002	.0032	.0166	.0487	.0981	.1489	.1797	.1771	.1442
	9	.0000	.0000	.0007	.0051	.0198	.0514	.0980	.1464	.1771	.1762
	10	.0000	.0000	.0001	.0013	.0066	.0220	.0528	.0976	.1449	.1762
	11	.0000	.0000	.0000	.0003	.0018	.0077	.0233	.0532	.0970	.1442
	12	.0000	.0000	.0000	.0000	.0004	.0022	.0083	.0237	.0529	.0961
	13	.0000	.0000	.0000	.0000	.0001	.0005	.0024	.0085	.0233	.0518
	14	.0000	.0000	.0000	.0000	.0000	.0001	.0006	.0024	.0082	.0222
	15	.0000	.0000	.0000	.0000	.0000	.0000	.0001	.0005	.0022	.0074
	16	.0000	.0000	.0000	.0000	.0000	.0000	.0000	.0001	.0005	.0018
	17	.0000	.0000	.0000	.0000	.0000	.0000	.0000	.0000	.0001	.0003
	18	.0000	.0000	.0000	.0000	.0000	.0000	.0000	.0000	.0000	.0000
	19	.0000	.0000	.0000	.0000	.0000	.0000	.0000	.0000	.0000	.0000
20	0	.3585	.1216	.0388	.0115	.0032	.0008	.0002	.0000	.0000	.0000
	1	.3774	.2702	.1368	.0576	.0211	.0068	.0020	.0005	.0001	.0000
	2	.1887	.2852	.2293	.1369	.0669	.0278	.0100	.0031	.0008	.0002
	3	.0596	.1901	.2428	.2054	.1339	.0716	.0323	.0123	.0040	.0011
	4	.0133	.0898	.1821	.2182	.1897	.1304	.0738	.0350	.0139	.0046
	5	.0022	.0319	.1028	.1746	.2023	.1789	.1272	.0746	.0365	.0148
	6	.0003	.0089	.0454	.1091	.1686	.1916	.1712	.1244	.0746	.0370
	7	.0000	.0020	.0160	.0545	.1124	.1643	.1844	.1659	.1221	.0739
	8	.0000	.0004	.0046	.0222	.0609	.1144	.1614	.1797	.1623	.1201
	9	.0000	.0001	.0011	.0074	.0271	.0654	.1158	.1597	.1771	.1602

Table A.1 (Continued)

						p					
n	x	.05	.10	.15	.20	.25	.30	.35	.40	.45	.50
20	10	.0000	.0000	.0002	.0020	.0099	.0308	.0686	.1171	.1593	.1762
	11	.0000	.0000	.0000	.0005	.0030	.0120	.0336	.0710	.1185	.1602
	12	.0000	.0000	.0000	.0001	.0008	.0039	.0136	.0355	.0727	.1201
	13	.0000	.0000	.0000	.0000	.0002	.0010	.0045	.0146	.0366	.0739
	14	.0000	.0000	.0000	.0000	.0000	.0002	.0012	.0049	.0150	.0370
	15	.0000	.0000	.0000	.0000	.0000	.0000	.0003	.0013	.0049	.0148
	16	.0000	.0000	.0000	.0000	.0000	.0000	.0000	.0003	.0013	.0046
	17	.0000	.0000	.0000	.0000	.0000	.0000	.0000	.0000	.0002	.0011
	18	.0000	.0000	.0000	.0000	.0000	.0000	.0000	.0000	.0000	.0002
	19	.0000	.0000	.0000	.0000	.0000	.0000	.0000	.0000	.0000	.0000
	20	.0000	.0000	.0000	.0000	.0000	.0000	.0000	.0000	.0000	.0000

Table A.2 Areas under the Normal Curve

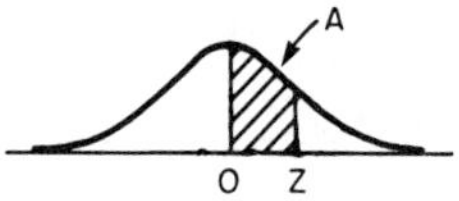

Z	A	Z	A	Z	A	Z	A
0.00	0.0000	0.30	0.1179	0.60	0.2258	0.90	0.3159
.01	.0040	.31	.1217	.61	.2291	.91	.3186
.02	.0080	.32	.1255	.62	.2324	.92	.3212
.03	.0120	.33	.1293	.63	.2357	.93	.3238
.04	.0160	.34	.1331	.64	.2389	.94	.3264
.05	.0199	.35	.1368	.65	.2422	.95	.3289
.06	.0239	.36	.1406	.66	.2454	.96	.3315
.07	.0279	.37	.1443	.67	.2486	.97	.3340
.08	.0319	.38	.1480	.68	.2518	.98	.3365
.09	.0359	.39	.1517	.69	.2549	.99	.3389
.10	.0398	.40	.1554	.70	.2580	1.00	.3413
.11	.0438	.41	.1591	.71	.2612	1.01	.3438
.12	.0478	.42	.1628	.72	.2642	1.02	.3461
.13	.0517	.43	.1664	.73	.2673	1.03	.3485
.14	.0557	.44	.1700	.74	.2704	1.04	.3508
.15	.0596	.45	.1736	.75	.2734	1.05	.3531
.16	.0636	.46	.1772	.76	.2764	1.06	.3554
.17	.0675	.47	.1808	.77	.2794	1.07	.3577
.18	.0714	.48	.1844	.78	.2823	1.08	.3599
.19	.0754	.49	.1879	.79	.2852	1.09	.3621
.20	.0793	.50	.1915	.80	.2881	1.10	.3643
.21	.0832	.51	.1950	.81	.2910	1.11	.3665
.22	.0871	.52	.1985	.82	.2939	1.12	.3686
.23	.0910	.53	.2019	.83	.2967	1.13	.3708
.24	.0948	.54	.2054	.84	.2996	1.14	.3729
.25	.0987	.55	.2088	.85	.3023	1.15	.3749
.26	.1026	.56	.2123	.86	.3051	1.16	.3770
.27	.1064	.57	.2157	.87	.3079	1.17	.3790
.28	.1103	.58	.2190	.88	.3106	1.18	.3810
.29	.1141	.59	.2224	.89	.3133	1.19	.3830

Table A.2 (Continued)

Z	A	Z	A	Z	A	Z	A
1.20	0.3849	1.50	0.4332	1.80	0.4641	2.10	0.4821
1.21	.3869	1.51	.4345	1.81	.4649	2.11	.4826
1.22	.3888	1.52	.4357	1.82	.4656	2.12	.4830
1.23	.3907	1.53	.4370	1.83	.4664	2.13	.4834
1.24	.3925	1.54	.4382	1.84	.4671	2.14	.4838
1.25	.3944	1.55	.4394	1.85	.4678	2.15	.4842
1.26	.3962	1.56	.4406	1.86	.4686	2.16	.4846
1.27	.3980	1.57	.4418	1.87	.4693	2.17	.4850
1.28	.3997	1.58	.4430	1.88	.4700	2.18	.4854
1.29	.4015	1.59	.4441	1.89	.4706	2.19	.4857
1.30	.4032	1.60	.4452	1.90	.4713	2.20	.4861
1.31	.4049	1.61	.4463	1.91	.4719	2.21	.4865
1.32	.4066	1.62	.4474	1.92	.4726	2.22	.4868
1.33	.4082	1.63	.4485	1.93	.4732	2.23	.4871
1.34	.4099	1.64	.4495	1.94	.4738	2.24	.4875
		1.645	.4500				
1.35	.4115	1.65	.4505	1.95	.4744	2.25	.4878
1.36	.4131	1.66	.4515	1.96	.4750	2.26	.4881
1.37	.4147	1.67	.4525	1.97	.4756	2.27	.4884
1.38	.4162	1.68	.4535	1.98	.4762	2.28	.4887
1.39	.4177	1.69	.4545	1.99	.4767	2.29	.4890
1.40	.4192	1.70	.4554	2.00	.4773	2.30	.4893
1.41	.4207	1.71	.4564	2.01	.4778	2.31	.4896
1.42	.4222	1.72	.4573	2.02	.4783	2.32	.4898
1.43	.4236	1.73	.4582	2.03	.4788	2.33	.4901
1.44	.4251	1.74	.4591	2.04	.4793	2.34	.4904
1.45	.4265	1.75	.4599	2.05	.4798	2.35	.4906
				2.055	.4800		
1.46	.4279	1.76	.4608	2.06	.4803	2.36	.4909
1.47	.4292	1.77	.4616	2.07	.4808	2.37	.4911
1.48	.4306	1.78	.4625	2.08	.4812	2.38	.4913
1.49	.4319	1.79	.4633	2.09	.4817	2.39	.4916

Table A.2 (Continued)

Z	A	Z	A	Z	A	Z	A
2.40	0.4918	2.70	0.4965	3.00	0.4987	3.30	0.4995
2.41	.4920	2.71	.4966	3.01	.4987	3.31	.4995
2.42	.4922	2.72	.4967	3.02	.4987	3.32	.4996
2.43	.4925	2.73	.4968	3.03	.4988	3.33	.4996
2.44	.4927	2.74	.4969	3.04	.4988	3.34	.4996
2.45	.4929	2.75	.4970	3.05	.4989	3.35	.4996
2.46	.4931	2.76	.4971	3.06	.4989	3.36	.4996
2.47	.4932	2.77	.4972	3.07	.4989	3.37	.4996
2.48	.4934	2.78	.4973	3.08	.4990	3.38	.4996
2.49	.4936	2.79	.4974	3.09	.4990	3.39	.4997
2.50	.4938	2.80	.4974	3.10	.4990	3.40	.4997
2.51	.4940	2.81	.4975	3.11	.4991	3.41	.4997
2.52	.4941	2.82	.4976	3.12	.4991	3.42	.4997
2.53	.4943	2.83	.4977	3.13	.4991	3.43	.4997
2.54	.4945	2.84	.4977	3.14	.4992	3.44	.4997
2.55	.4946	2.85	.4978	3.15	.4992	3.45	.4997
2.56	.4948	2.86	.4979	3.16	.4992	3.46	.4997
2.57	.4949	2.87	.4980	3.17	.4992	3.47	.4997
2.575	.4950						
2.58	.4951	2.88	.4980	3.18	.4993	3.48	.4998
2.59	.4952	2.89	.4981	3.19	.4993	3.49	.4998
2.60	.4953	2.90	.4981	3.20	.4993	3.50	.4998
2.61	.4955	2.91	.4982	3.21	.4993	3.51	.4998
2.62	.4956	2.92	.4983	3.22	.4994	3.52	.4998
2.63	.4957	2.93	.4983	3.23	.4994	3.53	.4998
2.64	.4959	2.94	.4984	3.24	.4994	3.54	.4998
2.65	.4960	2.95	.4984	3.25	.4994	3.55	.4998
2.66	.4961	2.96	.4985	3.26	.4994	3.56	.4998
2.67	.4962	2.97	.4985	3.27	.4995	3.57	.4998
2.68	.4963	2.98	.4986	3.28	.4995	3.58	.4998
2.69	.4964	2.99	.4986	3.29	.4995	3.59	.4998

Table A.2 (Continued)

Z	A	Z	A	Z	A
3.60	0.4998	3.70	0.4999	3.80	0.4999
3.61	.4999	3.71	.4999	3.81	.4999
3.62	.4999	3.72	.4999	3.82	.4999
3.63	.4999	3.73	.4999	3.83	.4999
3.64	.4999	3.74	.4999	3.84	.4999
3.65	.4999	3.75	.4999	3.85	.4999
3.66	.4999	3.76	.4999	3.86	.4999
3.67	.4999	3.77	.4999	3.87	.5000
3.68	.4999	3.78	.4999	3.88	.5000
3.69	.4999	3.79	.4999	3.89	.5000

Table A.3 Ordinates (Y Values) for the Normal Curve

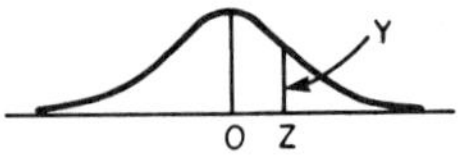

Z	Y	Z	Y	Z	Y	Z	Y
0.00	0.3989	0.25	0.3867	0.50	0.3521	0.75	0.3011
.01	.3989	.26	.3857	.51	.3503	.76	.2989
.02	.3989	.27	.3847	.52	.3485	.77	.2966
.03	.3988	.28	.3836	.53	.3467	.78	.2943
.04	.3986	.29	.3825	.54	.3448	.79	.2920
.05	.3984	.30	.3814	.55	.3429	.80	.2897
.06	.3982	.31	.3802	.56	.3410	.81	.2874
.07	.3980	.32	.3790	.57	.3391	.82	.2850
.08	.3977	.33	.3778	.58	.3372	.83	.2827
.09	.3973	.34	.3765	.59	.3352	.84	.2803
.10	.3969	.35	.3752	.60	.3332	.85	.2780
.11	.3965	.36	.3739	.61	.3312	.86	.2756
.12	.3961	.37	.3725	.62	.3292	.87	.2732
.13	.3956	.38	.3711	.63	.3271	.88	.2709
.14	.3950	.39	.3697	.64	.3251	.89	.2685
.15	.3945	.40	.3683	.65	.3230	.90	.2661
.16	.3939	.41	.3668	.66	.3209	.91	.2637
.17	.3932	.42	.3653	.67	.3187	.92	.2613
.18	.3925	.43	.3637	.68	.3166	.93	.2589
.19	.3918	.44	.3621	.69	.3144	.94	.2565
.20	.3910	.45	.3605	.70	.3122	.95	.2541
.21	.3902	.46	.3589	.71	.3101	.96	.2516
.22	.3894	.47	.3572	.72	.3078	.97	.2492
.23	.3885	.48	.3555	.73	.3056	.98	.2468
.24	.3876	.49	.3538	.74	.3034	.99	.2444

Table A.3 (Continued)

Z	Y	Z	Y	Z	Y	Z	Y
1.00	0.2420	1.30	0.1714	1.60	0.1109	1.90	0.0656
1.01	.2395	1.31	.1691	1.61	.1091	1.91	.0644
1.02	.2371	1.32	.1669	1.62	.1074	1.92	.0632
1.03	.2347	1.33	.1647	1.63	.1057	1.93	.0619
1.04	.2323	1.34	.1626	1.64	.1040	1.94	.0608
1.05	.2299	1.35	.1604	1.65	.1023	1.95	.0596
1.06	.2275	1.36	.1582	1.66	.1006	1.96	.0584
1.07	.2251	1.37	.1561	1.67	.0989	1.97	.0573
1.08	.2226	1.38	.1539	1.68	.0973	1.98	.0562
1.09	.2202	1.39	.1518	1.69	.0957	1.99	.0551
1.10	.2178	1.40	.1497	1.70	.0940	2.00	.0540
1.11	.2155	1.41	.1476	1.71	.0925	2.01	.0529
1.12	.2131	1.42	.1456	1.72	.0909	2.02	.0519
1.13	.2107	1.43	.1435	1.73	.0893	2.03	.0508
1.14	.2083	1.44	.1415	1.74	.0878	2.04	.0498
1.15	.2059	1.45	.1394	1.75	.0863	2.05	.0488
1.16	.2036	1.46	.1374	1.76	.0848	2.06	.0478
1.17	.2012	1.47	.1354	1.77	.0833	2.07	.0468
1.18	.1989	1.48	.1334	1.78	.0818	2.08	.0459
1.19	.1965	1.49	.1315	1.79	.0804	2.09	.0449
1.20	.1942	1.50	.1295	1.80	.0789	2.10	.0440
1.21	.1919	1.51	.1276	1.81	.0775	2.11	.0431
1.22	.1895	1.52	.1257	1.82	.0761	2.12	.0422
1.23	.1872	1.53	.1238	1.83	.0748	2.13	.0413
1.24	.1849	1.54	.1219	1.84	.0734	2.14	.0404
1.25	.1826	1.55	.1200	1.85	.0721	2.15	.0395
1.26	.1804	1.56	.1182	1.86	.0707	2.16	.0387
1.27	.1781	1.57	.1163	1.87	.0694	2.17	.0379
1.28	.1758	1.58	.1145	1.88	.0681	2.18	.0371
1.29	.1736	1.59	.1127	1.89	.0669	2.19	.0363

Table A.3 (Continued)

Z	Y	Z	Y	Z	Y	Z	Y
2.20	0.0355	2.50	0.0175	2.80	0.0079	3.10	0.0033
2.21	.0347	2.51	.0171	2.81	.0077	3.11	.0032
2.22	.0339	2.52	.0167	2.82	.0075	3.12	.0031
2.23	.0332	2.53	.0162	2.83	.0073	3.13	.0030
2.24	.0325	2.54	.0158	2.84	.0071	3.14	.0029
2.25	.0317	2.55	.0154	2.85	.0069	3.15	.0028
2.26	.0310	2.56	.0151	2.86	.0067	3.16	.0027
2.27	.0303	2.57	.0147	2.87	.0065	3.17	.0026
2.28	.0296	2.58	.0143	2.88	.0063	3.18	.0025
2.29	.0290	2.59	.0139	2.89	.0061	3.19	.0025
2.30	.0283	2.60	.0136	2.90	.0059	3.20	.0024
2.31	.0277	2.61	.0132	2.91	.0058	3.21	.0023
2.32	.0270	2.62	.0129	2.92	.0056	3.22	.0022
2.33	.0264	2.63	.0126	2.93	.0054	3.23	.0022
2.34	.0258	2.64	.0122	2.94	.0053	3.24	.0021
2.35	.0252	2.65	.0119	2.95	.0051	3.25	.0020
2.36	.0246	2.66	.0116	2.96	.0050	3.26	.0020
2.37	.0241	2.67	.0113	2.97	.0048	3.27	.0019
2.38	.0235	2.68	.0110	2.98	.0047	3.28	.0018
2.39	.0229	2.69	.0107	2.99	.0046	3.29	.0018
2.40	.0224	2.70	.0104	3.00	.0044	3.30	.0017
2.41	.0219	2.71	.0101	3.01	.0043	3.31	.0017
2.42	.0213	2.72	.0099	3.02	.0042	3.32	.0016
2.43	.0208	2.73	.0096	3.03	.0040	3.33	.0016
2.44	.0203	2.74	.0093	3.04	.0039	3.34	.0015
2.45	.0198	2.75	.0091	3.05	.0038	3.35	.0015
2.46	.0194	2.76	.0088	3.06	.0037	3.36	.0014
2.47	.0189	2.77	.0086	3.07	.0036	3.37	.0014
2.48	.0184	2.78	.0084	3.08	.0035	3.38	.0013
2.49	.0180	2.79	.0081	3.09	.0034	3.39	.0013

Table A.3 (Continued)

Z	Y	Z	Y	Z	Y	Z	Y
3.40	0.0012	3.55	0.0007	3.70	0.0004	3.85	00.0002
3.41	.0012	3.56	.0007	3.71	.0004	3.86	.0002
3.42	.0011	3.57	.0007	3.72	.0004	3.87	.0002
3.43	.0011	3.58	.0007	3.73	.0004	3.88	.0002
3.44	.0011	3.59	.0006	3.74	.0004	3.89	.0002
3.45	.0010	3.60	.0006	3.75	.0003	3.90	.0002
3.46	.0010	3.61	.0006	3.76	.0003	3.91	.0002
3.47	.0010	3.62	.0006	3.77	.0003	3.92	.0002
3.48	.0009	3.63	.0005	3.78	.0003	3.93	.0002
3.49	.0009	3.64	.0005	3.79	.0003	3.94	.0002
3.50	.0009	3.65	.0005	3.80	.0003	3.95	.0002
3.51	.0008	3.66	.0005	3.81	.0003	3.96	.0002
3.52	.0008	3.67	.0005	3.82	.0003	3.97	.0001
3.53	.0008	3.68	.0005	3.83	.0003	3.98	.0001
3.54	.0008	3.69	.0004	3.84	.0002	3.99	.0001

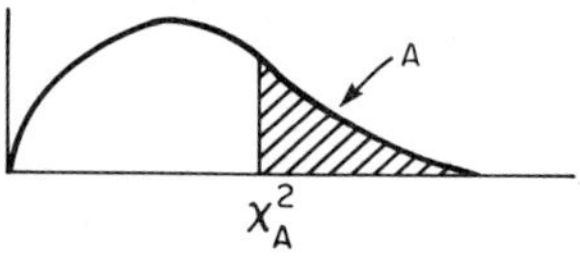

Table A.4 Chi-Square Distribution

ν or d.f.	A = 0.99	A = 0.98	A = 0.95	A = 0.90	A = 0.80	A = 0.70	A = 0.50
1	.00016	.00063	.0039	.016	.064	.15	.46
2	.02	.04	.10	.21	.45	.71	1.39
3	.12	.18	.35	.58	1.00	1.42	2.37
4	.30	.43	.71	1.06	1.65	2.20	3.36
5	.55	.75	1.14	1.61	2.34	3.00	4.35
6	.87	1.13	1.64	2.20	3.07	3.83	5.35
7	1.24	1.56	2.17	2.83	3.82	4.67	6.35
8	1.65	2.03	2.73	3.49	4.59	5.53	7.34
9	2.09	2.53	3.32	4.17	5.38	6.39	8.34
10	2.56	3.06	3.94	4.86	6.18	7.27	9.34
11	3.05	3.61	4.58	5.58	6.99	8.15	10.34
12	3.57	4.18	5.23	6.30	7.81	9.03	11.34
13	4.11	4.76	5.89	7.04	8.63	9.93	12.34
14	4.66	5.37	6.57	7.79	9.47	10.82	13.34
15	5.23	5.98	7.26	8.55	10.31	11.72	14.34
16	5.81	6.61	7.96	9.31	11.15	12.62	15.34
17	6.41	7.26	8.67	10.08	12.00	13.53	16.34
18	7.02	7.91	9.39	10.86	12.86	14.44	17.34
19	7.63	8.57	10.12	11.65	13.72	15.35	18.34
20	8.26	9.24	10.85	12.44	14.58	16.27	19.34
21	8.90	9.92	11.59	13.24	15.44	17.18	20.34
22	9.54	10.60	12.34	14.04	16.31	18.10	21.34
23	10.20	11.29	13.09	14.85	17.19	19.02	22.34
24	10.86	11.99	13.85	15.66	18.06	19.94	23.34
25	11.52	12.70	14.61	16.47	18.94	20.87	24.34
26	12.20	13.41	15.38	17.29	19.82	21.79	25.34
27	12.88	14.12	16.15	18.11	20.70	22.72	26.34
28	13.56	14.85	16.93	18.94	21.59	23.65	27.34
29	14.26	15.57	17.71	19.77	22.48	24.58	28.34
30	14.95	16.31	18.49	20.60	23.36	25.51	29.34

This is abridged from Table IV of Fisher and Yates, *Statistical Tables for Biological, Agricultural, and Medical Research*, Oliver and Boyd, Edinburgh, by permission of the authors and publishers.

Table A.4 (Continued)

A = 0.30	A = 0.20	A = 0.10	A = 0.05	A = 0.02	A = 0.01	A = 0.001
1.07	1.64	2.71	3.84	5.41	6.64	10.83
2.41	3.22	4.60	5.99	7.82	9.21	13.82
3.66	4.64	6.25	7.82	9.84	11.34	16.27
4.88	5.99	7.78	9.49	11.67	13.28	18.46
6.06	7.29	9.24	11.07	13.39	15.09	20.52
7.23	8.56	10.64	12.59	15.03	16.81	22.46
8.38	9.80	12.02	14.07	16.62	18.48	24.32
9.52	11.03	13.36	15.51	18.17	20.09	26.12
10.66	12.24	14.68	16.92	19.68	21.67	27.88
11.78	13.44	15.99	18.31	21.16	23.21	29.59
12.90	14.63	17.28	19.68	22.62	24.72	31.26
14.01	15.81	18.55	21.03	24.05	26.22	32.91
15.12	16.98	19.81	22.36	25.47	27.69	34.53
16.22	18.15	21.06	23.68	26.87	29.14	36.12
17.32	19.31	22.31	25.00	28.26	30.58	37.70
18.42	20.46	23.54	26.30	29.63	32.00	39.25
19.51	21.62	24.77	27.59	31.00	33.41	40.79
20.60	22.76	25.99	28.87	32.35	34.80	42.31
21.69	23.90	27.20	30.14	33.69	36.19	43.82
22.78	25.04	28.41	31.41	35.02	37.57	45.32
23.86	26.17	29.62	32.67	36.34	38.93	46.80
24.94	27.30	30.81	33.92	37.66	40.29	48.27
26.02	28.43	32.01	35.17	38.97	41.64	49.73
27.10	29.55	33.20	36.42	40.27	42.98	51.18
28.17	30.68	34.38	37.65	41.57	44.31	52.62
29.25	31.80	35.56	38.88	42.86	45.64	54.05
30.32	32.91	36.74	40.11	44.14	46.96	55.48
31.39	34.03	37.92	41.34	45.42	48.28	56.89
32.46	35.14	39.09	42.56	46.69	49.59	58.30
33.53	36.25	40.26	43.77	47.96	50.89	59.70

Table A.5 Percentage Points of the t Distribution

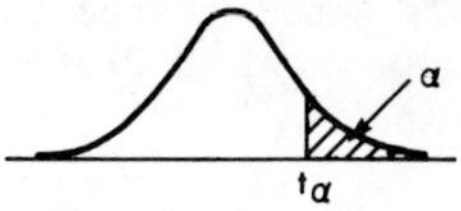

d.f.	α = 0.10	α = 0.05	α = 0.025	α = 0.010	α = 0.005
1	3.078	6.314	12.706	31.821	65.657
2	1.886	2.920	4.303	6.965	9.925
3	1.638	2.353	3.182	4.541	5.841
4	1.533	2.132	2.776	3.747	4.604
5	1.476	2.015	2.571	3.365	4.032
6	1.440	1.943	2.447	3.143	3.707
7	1.415	1.895	2.365	2.998	3.499
8	1.397	1.860	2.306	2.896	3.355
9	1.383	1.833	2.262	2.821	3.250
10	1.372	1.812	2.228	2.764	3.169
11	1.363	1.796	2.201	2.718	3.106
12	1.356	1.782	2.179	2.681	3.055
13	1.350	1.771	2.160	2.650	3.012
14	1.345	1.761	2.145	2.624	2.977
15	1.341	1.753	2.131	2.602	2.947
16	1.337	1.746	2.120	2.583	2.921
17	1.333	1.740	2.110	2.567	2.898
18	1.330	1.734	2.101	2.552	2.878
19	1.328	1.729	2.093	2.539	2.861
20	1.325	1.725	2.086	2.528	2.845
21	1.323	1.721	2.080	2.518	2.831
22	1.321	1.717	2.074	2.508	2.819
23	1.319	1.714	2.069	2.500	2.807
24	1.318	1.711	2.064	2.492	2.797
25	1.316	1.708	2.060	2.485	2.787
26	1.315	1.706	2.056	2.479	2.779
27	1.314	1.703	2.052	2.473	2.771
28	1.313	1.701	2.048	2.467	2.763
29	1.311	1.699	2.045	2.462	2.756
Inf	1.282	1.645	1.960	2.326	2.576

From "Table of Percentage points of the t Distribution," Computed by Maxine Merrington, *Biometrika* 32(1941), p.300. Reproduced by permission of the *Biometrika* Trustees.

Table A.6.1 Percentage Points of the F Distribution, $\alpha = 0.05$

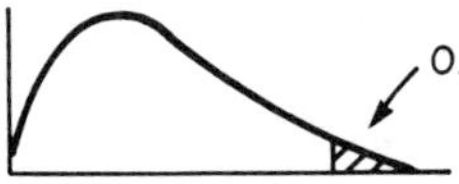

d.f.$_1$ / d.f.$_2$	1	2	3	4	5	6	7	8	9
1	161.4	199.5	215.7	224.6	230.2	234.0	236.8	238.9	240.5
2	18.51	19.00	19.16	19.25	19.30	19.33	19.35	19.37	19.38
3	10.13	9.55	9.28	9.12	9.01	8.94	8.89	8.85	8.81
4	7.71	6.94	6.59	6.39	6.26	6.16	6.09	6.04	6.00
5	6.61	5.79	5.41	5.19	5.05	4.95	4.88	4.82	4.77
6	5.99	5.14	4.76	4.53	4.39	4.28	4.21	4.15	4.10
7	5.59	4.74	4.35	4.12	3.97	3.87	3.79	3.73	3.68
8	5.32	4.46	4.07	3.84	3.69	3.58	3.50	3.44	3.39
9	5.12	4.26	3.86	3.63	3.48	3.37	3.29	3.23	3.18
10	4.96	4.10	3.71	3.48	3.33	3.22	3.14	3.07	3.02
11	4.84	3.98	3.59	3.36	3.20	3.09	3.01	2.95	2.90
12	4.75	3.89	3.49	3.26	3.11	3.00	2.91	2.85	2.80
13	4.67	3.81	3.41	3.18	3.03	2.92	2.83	2.77	2.71
14	4.60	3.74	3.34	3.11	2.96	2.85	2.76	2.70	2.65
15	4.54	3.68	3.29	3.06	2.90	2.79	2.71	2.64	2.59
16	4.49	3.63	3.24	3.01	2.85	2.74	2.66	2.59	2.54
17	4.45	3.59	3.20	2.96	2.81	2.70	2.61	2.55	2.49
18	4.41	3.55	3.16	2.93	2.77	2.66	2.58	2.51	2.46
19	4.38	3.52	3.13	2.90	2.74	2.63	2.54	2.48	2.42
20	4.35	3.49	3.10	2.87	2.71	2.60	2.51	2.45	2.39
21	4.32	3.47	3.07	2.84	2.68	2.57	2.49	2.42	2.37
22	4.30	3.44	3.05	2.82	2.66	2.55	2.46	2.40	2.34
23	4.28	3.42	3.03	2.80	2.64	2.53	2.44	2.37	2.32
24	4.26	3.40	3.01	2.78	2.62	2.51	2.42	2.36	2.30
25	4.24	3.39	2.99	2.76	2.60	2.49	2.40	2.34	2.28
26	4.23	3.37	2.98	2.74	2.59	2.47	2.39	2.32	2.27
27	4.21	3.35	2.96	2.73	2.57	2.46	2.37	2.31	2.25
28	4.20	3.34	2.95	2.71	2.56	2.45	2.36	2.29	2.24
29	4.18	3.33	2.93	2.70	2.55	2.43	2.35	2.28	2.22
30	4.17	3.32	2.92	2.69	2.53	2.42	2.33	2.27	2.21
40	4.08	3.23	2.84	2.61	2.45	2.34	2.25	2.18	2.12
60	4.00	3.15	2.76	2.53	2.37	2.25	2.17	2.10	2.04
120	3.92	3.07	2.68	2.45	2.29	2.17	2.09	2.02	1.96
∞	3.84	3.00	2.60	2.37	2.21	2.10	2.01	1.94	1.88

From "Tables of Percentage Points of the Inverted Beta (F)-Distribution," *Biometrika* 33(1943), pp. 73-78, by Maxine Merrington and Catherine M. Thompson. Reproduced by permission of the *Biometrika* Trustees.

Table A.6.1 (Continued)

10	12	15	20	24	30	40	60	120	∞
241.9	243.9	245.9	248.0	249.1	250.1	251.1	252.2	253.3	254.3
19.40	19.41	19.43	19.45	19.45	19.46	19.47	19.48	19.49	19.50
8.79	8.74	8.70	8.66	8.64	8.62	8.59	8.57	8.55	8.53
5.96	5.91	5.86	5.80	5.77	5.75	5.72	5.69	5.66	5.63
4.74	4.68	4.62	4.56	4.53	4.50	4.46	4.43	4.40	4.36
4.06	4.00	3.94	3.87	3.84	3.81	3.77	3.74	3.70	3.67
3.64	3.57	3.51	3.44	3.41	3.38	3.34	3.30	3.27	3.23
3.35	3.28	3.22	3.15	3.12	3.08	3.04	3.01	2.97	2.93
3.14	3.07	3.01	2.94	2.90	2.86	2.83	2.79	2.75	2.71
2.98	2.91	2.85	2.77	2.74	2.70	2.66	2.62	2.58	2.54
2.85	2.79	2.72	2.65	2.61	2.57	2.53	2.49	2.45	2.40
2.75	2.69	2.62	2.54	2.51	2.47	2.43	2.38	2.34	2.30
2.67	2.60	2.53	2.46	2.42	2.38	2.34	2.30	2.25	2.21
2.60	2.53	2.46	2.39	2.35	2.31	2.27	2.22	2.18	2.13
2.54	2.48	2.40	2.33	2.29	2.25	2.20	2.16	2.11	2.07
2.49	2.42	2.35	2.28	2.24	2.19	2.15	2.11	2.06	2.01
2.45	2.38	2.31	2.23	2.19	2.15	2.10	2.06	2.01	1.96
2.41	2.34	2.27	2.19	2.15	2.11	2.06	2.02	1.97	1.92
2.38	2.31	2.23	2.16	2.11	2.07	2.03	1.98	1.93	1.88
2.35	2.28	2.20	2.12	2.08	2.04	1.99	1.95	1.90	1.84
2.32	2.25	2.18	2.10	2.05	2.01	1.96	1.92	1.87	1.81
2.30	2.23	2.15	2.07	2.03	1.98	1.94	1.89	1.84	1.78
2.27	2.20	2.13	2.05	2.01	1.96	1.91	1.86	1.81	1.76
2.25	2.18	2.11	2.03	1.98	1.94	1.89	1.84	1.79	1.73
2.24	2.16	2.09	2.01	1.96	1.92	1.87	1.82	1.77	1.71
2.22	2.15	2.07	1.99	1.95	1.90	1.85	1.80	1.75	1.69
2.20	2.13	2.06	1.97	1.93	1.88	1.84	1.79	1.73	1.67
2.19	2.12	2.04	1.96	1.91	1.87	1.82	1.77	1.71	1.65
2.18	2.10	2.03	1.94	1.90	1.85	1.81	1.75	1.70	1.64
2.16	2.09	2.01	1.93	1.89	1.84	1.79	1.74	1.68	1.62
2.08	2.00	1.92	1.84	1.79	1.74	1.69	1.64	1.58	1.51
1.99	1.92	1.84	1.75	1.70	1.65	1.59	1.53	1.47	1.39
1.91	1.83	1.75	1.66	1.61	1.55	1.50	1.43	1.35	1.25
1.83	1.75	1.67	1.57	1.52	1.46	1.39	1.32	1.22	1.00

Table A.6.2 Percentage Points of the F Distribution, α = 0.01

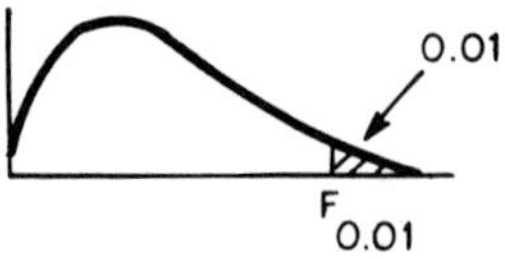

d.f.₂ \ d.f.₁	1	2	3	4	5	6	7	8	9
1	4052	4999.5	5403	5625	5764	5859	5928	5982	6022
2	98.50	99.00	99.17	99.25	99.30	99.33	99.36	99.37	99.39
3	34.12	30.82	29.46	28.71	28.24	27.91	27.67	27.49	27.35
4	21.20	18.00	16.69	15.98	15.52	15.21	14.98	14.80	14.66
5	16.26	13.27	12.06	11.39	10.97	10.67	10.46	10.29	10.16
6	13.75	10.92	9.78	9.15	8.75	8.47	8.26	8.10	7.98
7	12.25	9.55	8.45	7.85	7.46	7.19	6.99	6.84	6.72
8	11.26	8.65	7.59	7.01	6.63	6.37	6.18	6.03	5.91
9	10.56	8.02	6.99	6.42	6.06	5.80	5.61	5.47	5.35
10	10.04	7.56	6.55	5.99	5.64	5.39	5.20	5.06	4.94
11	9.65	7.21	6.22	5.67	5.32	5.07	4.89	4.74	4.63
12	9.33	6.93	5.95	5.41	5.06	4.82	4.64	4.50	4.39
13	9.07	6.70	5.74	5.21	4.86	4.62	4.44	4.30	4.19
14	8.86	6.51	5.56	5.04	4.69	4.46	4.28	4.14	4.03
15	8.68	6.36	5.42	4.89	4.56	4.32	4.14	4.00	3.89
16	8.53	6.23	5.29	4.77	4.44	4.20	4.03	3.89	3.78
17	8.40	6.11	5.18	4.67	4.34	4.10	3.93	3.79	3.68
18	8.29	6.01	5.09	4.58	4.25	4.01	3.84	3.71	3.60
19	8.18	5.93	5.01	4.50	4.17	3.94	3.77	3.63	3.52
20	8.10	5.85	4.94	4.43	4.10	3.87	3.70	3.56	3.46
21	8.02	5.78	4.87	4.37	4.04	3.81	3.64	3.51	3.40
22	7.95	5.72	4.82	4.31	3.99	3.76	3.59	3.45	3.35
23	7.88	5.66	4.76	4.26	3.94	3.71	3.54	3.41	3.30
24	7.82	5.61	4.72	4.22	3.90	3.67	3.50	3.36	3.26
25	7.77	5.57	4.68	4.18	3.85	3.63	3.46	3.32	3.22
26	7.72	5.53	4.64	4.14	3.82	3.59	3.42	3.29	3.18
27	7.68	5.49	4.60	4.11	3.78	3.56	3.39	3.26	3.15
28	7.64	5.45	4.57	4.07	3.75	3.53	3.36	3.23	3.12
29	7.60	5.42	4.54	4.04	3.73	3.50	3.33	3.20	3.09
30	7.56	5.39	4.51	4.02	3.70	3.47	3.30	3.17	3.07
40	7.31	5.18	4.31	3.83	3.51	3.29	3.12	2.99	2.89
60	7.08	4.98	4.13	3.65	3.34	3.12	2.95	2.82	2.72
120	6.85	4.79	3.95	3.48	3.17	2.96	2.79	2.66	2.56
∞	6.63	4.61	3.78	3.32	3.02	2.80	2.64	2.51	2.41

Table A.6.2 (Continued)

10	12	15	20	24	30	40	60	120	∞
6056	6106	6157	6209	6235	6261	6287	6313	6339	6366
99.40	99.42	99.43	99.45	99.46	99.47	99.47	99.48	99.49	99.50
27.23	27.05	26.87	26.69	26.60	26.50	26.41	26.32	26.22	26.13
14.55	14.37	14.20	14.02	13.93	13.84	31.75	13.65	13.56	13.46
10.05	9.89	9.72	9.55	9.47	9.38	9.29	9.20	9.11	9.02
7.87	7.72	7.56	7.40	7.31	7.23	7.14	7.06	6.97	6.88
6.62	6.47	6.31	6.16	6.07	5.99	5.91	5.82	5.74	5.65
5.81	5.67	5.52	5.36	5.28	5.20	5.12	5.03	4.95	4.86
5.26	5.11	4.96	4.81	4.73	4.65	4.57	4.48	4.40	4.31
4.85	4.71	4.56	4.41	4.33	4.25	4.17	4.08	4.00	3.91
4.54	4.40	4.25	4.10	4.02	3.94	3.86	3.78	3.69	3.60
4.30	4.16	4.01	3.86	3.78	3.70	3.62	3.54	3.45	3.36
4.10	3.96	3.82	3.66	3.59	3.51	3.43	3.34	3.25	3.17
3.94	3.80	3.66	3.51	3.43	3.35	3.27	3.18	3.09	3.00
3.80	3.67	3.52	3.37	3.29	3.21	3.13	3.05	2.96	2.87
3.69	3.55	3.41	3.26	3.18	3.10	3.02	2.93	2.84	2.75
3.59	3.46	3.31	3.16	3.08	3.00	2.92	2.83	2.75	2.65
3.51	3.37	3.23	3.08	3.00	2.92	2.84	2.75	2.66	2.57
3.43	3.30	3.15	3.00	2.92	2.84	2.76	2.67	2.58	2.49
3.37	3.23	3.09	2.94	2.86	2.78	2.69	2.61	2.52	2.42
3.31	3.17	3.03	2.88	2.80	2.72	2.64	2.55	2.46	2.36
3.26	3.12	2.98	2.83	2.75	2.67	2.58	2.50	2.40	2.31
3.21	3.07	2.93	2.78	2.70	2.62	2.54	2.45	2.35	2.26
3.17	3.03	2.89	2.74	2.66	2.58	2.49	2.40	2.31	2.21
3.13	2.99	2.85	2.70	2.62	2.54	2.45	2.36	2.27	2.17
3.09	2.96	2.81	2.66	2.58	2.50	2.42	2.33	2.23	2.13
3.06	2.93	2.78	2.63	2.55	2.47	2.38	2.29	2.20	2.10
3.03	2.90	2.75	2.60	2.52	2.44	2.35	2.26	2.17	2.06
3.00	2.87	2.73	2.57	2.49	2.41	2.33	2.23	2.14	2.03
2.98	2.84	2.70	2.55	2.47	2.39	2.30	2.21	2.11	2.01
2.80	2.66	2.52	2.37	2.29	2.20	2.11	2.02	1.92	1.80
2.63	2.50	2.35	2.20	2.12	2.03	1.94	1.84	1.73	1.60
2.47	2.34	2.19	2.03	1.95	1.86	1.76	1.66	1.63	1.38
2.32	2.18	2.04	1.88	1.79	1.70	1.59	1.47	1.32	1.00

Table A.7 Transformation of r to Z

$$\left(Z = \frac{1}{2} \log \frac{1 + r}{1 - r}\right)$$

r	Z	r	Z	r	Z
.00	.000				
.01	.010	.36	.377	.71	.887
.02	.020	.37	.388	.72	.908
.03	.030	.38	.400	.73	.929
.04	.040	.39	.412	.74	.950
.05	.050	.40	.424	.75	.973
.06	.060	.41	.436	.76	.996
.07	.070	.42	.448	.77	1.020
.08	.080	.43	.460	.78	1.045
.09	.090	.44	.472	.79	1.071
.10	.100	.45	.485	.80	1.099
.11	.110	.46	.497	.81	1.127
.12	.121	.47	.510	.82	1.157
.13	.131	.48	.523	.83	1.188
.14	.141	.49	.536	.84	1.221
.15	.151	.50	.549	.85	1.256
.16	.161	.51	.563	.86	1.293
.17	.172	.52	.576	.87	1.333
.18	.182	.53	.590	.88	1.376
.19	.192	.54	.604	.89	1.422
.20	.203	.55	.618	.90	1.472
.21	.213	.56	.633	.91	1.528
.22	.224	.57	.648	.92	1.589
.23	.234	.58	.662	.93	1.658
.24	.245	.59	.678	.94	1.738
.25	.255	.60	.693	.95	1.832
.26	.266	.61	.709	.96	1.946
.27	.277	.62	.725	.97	2.092
.28	.288	.63	.741	.98	2.298
.29	.299	.64	.758	.99	2.647
.30	.310	.65	.775		
.31	.321	.66	.793		
.32	.332	.67	.811		
.33	.343	.68	.829		
.34	.354	.69	.848		
.35	.365	.70	.867		

Abridged from Table VII of Fisher and Yates, *Statistical Tables for Biological, Agricultural, and Medical Research*, Oliver and Boyd, Edinburgh, by permission of the authors and publishers.

Table A.8 Random Numbers

Column / Line	(1)	(2)	(3)	(4)	(5)	(6)	(7)	(8)	(9)	(10)	(11)	(12)	(13)	(14)
1	10480	15011	01536	02011	81647	91646	69179	14194	62590	36207	20969	99570	91291	90700
2	22368	46573	25595	85393	30995	89198	27982	53402	93965	34095	52666	19174	39615	99505
3	24130	48360	22527	97265	76393	64809	15179	24830	49340	32081	30680	19655	63348	58629
4	42167	93093	06243	61680	07856	16376	39440	53537	71341	57004	00849	74917	97758	16379
5	37570	39975	81837	16656	06121	91782	60468	81305	49684	60672	14110	06927	01263	54613
6	77921	06907	11008	42751	27756	53498	18602	70659	90655	15053	21916	81825	44394	42880
7	99562	72905	56420	69994	98872	31016	71194	18738	44013	48830	63213	21069	10634	12952
8	96301	91977	05463	07972	18876	20922	94595	56869	69014	60045	18425	84903	42508	32307
9	89579	14342	63661	10281	17453	18103	57740	84378	25331	12566	58678	44947	05585	56941
10	85475	36857	53342	53988	53060	59533	38867	62300	08158	17983	16439	11458	18593	64952
11	28918	69578	88231	33276	70997	79936	56865	05859	90106	31595	01547	85590	91610	78188
12	63553	40961	48235	03427	49626	69445	18663	72695	52180	20847	12234	90511	33703	90322
13	09429	93969	52636	92737	88974	33488	36320	17617	30015	08272	84115	27156	30613	74952
14	10365	61129	87529	85689	48237	52267	67689	93394	01511	26358	85104	20285	29975	89868
15	07119	97336	71048	08178	77233	13916	47564	81056	97735	85977	29372	74461	28551	90707
16	51085	12765	51821	51259	77452	16308	60756	92144	49442	53900	70960	63990	75601	40719
17	02368	21382	52404	60268	89368	19885	55322	44819	01188	65255	64835	44919	05944	55157
18	01011	54092	33362	94904	31273	04146	18594	29852	71585	85030	51132	01915	92747	64951
19	52162	53916	46369	58586	23216	14513	83149	98736	23495	64350	94738	17752	35156	35749
20	07056	97628	33787	09998	42698	06691	76988	13602	51851	46104	88916	19509	25625	58104
21	48663	91245	85828	14346	09172	30168	90229	04734	59193	22178	30421	61666	99904	32812
22	54164	58492	22421	74103	47070	25306	76468	26384	58151	06646	21524	15227	96909	44592
23	32639	32363	05597	24200	13363	38005	94342	28728	35806	06912	17012	64161	18296	22851
24	29334	27001	87637	87308	58731	00256	45834	15398	46557	41135	10367	07684	36188	18510
25	02488	33062	28834	07351	19731	92420	60952	61280	50001	67658	32586	86679	50720	94953

Abridged from *Handbook of Tables for Probability and Statistics*, Second Edition, edited by William H. Beyer, © The Chemical Rubber Co., 1968. Used by permission of the Chemical Rubber Co.

ANSWERS TO SELECTED EXERCISES

Chapter 1

1.1 (a) F (b) T (c) T (d) F (e) T (f) F (g) F (h) T (i) T (j) T (k) F (l) T (m) F

1.2 (a) Nominal (b) Ratio (continuous) (c) Ratio (continuous) (d) Nominal (e) Ordinal (discrete) (f) Ratio (continuous) (g) Ratio (discrete) (h) Ratio (discrete)

1.9 Ordinal

Chapter 2

2.1 (a) Pie chart (b) Sources of the fiscal yearly income for the University of Guelph for 1971 and 1974 (c) 7.7% and 10.5% (d) Actual dollar amounts from the various sources (e) $6,198,060

2.3 (a) T (b) T (c) F (d) F (e) T

2.7

C.B.	f
38-48	6
49-59	3
60-70	6
71-81	7
82-92	3

C.B.	f
38-55	7
56-73	9
74-91	9

C.B.	f	F
38-46	4	4
47-55	3	7
56-64	7	14
65-73	2	16
74-82	7	23
83-91	2	25

Chapter 3

3.1 (a) 18 (b) 78 (c) -5 (d) 19 (e) 54 (f) 3.6 (g) 3.3 (h) 0 (i) 8.16 (j) 3.3 (k) 73 (l) 60 (m) 210

3.3 (a) $\sum_{i=1}^{k} X_i$ (b) $\sum_{i=1}^{k-1} X_i$ (c) $\sum_{i=1}^{k} X_i^2$ (d) $\sum_{i=1}^{k} (X_i - 10)^2$

(e) $\frac{1}{k} \sum_{i=1}^{k} X_i$ (f) $\frac{1}{k - 1} \sum_{i=1}^{k} (X_i - \overline{X})^2$

3.4 (a) $\sum_{j=1}^{6} Z_j$ (b) $\sum_{j=1}^{m} Y_j^2$ (c) $\sum_{j=1}^{k} X_j Y_j$ (d) $\sum_{i=1}^{n} X_i f_i$

(e) $\sum_{i=1}^{k} X_i^2 Y_i$ (f) $\sum_{i=1}^{k} i X_i Y_i^3$ (g) $\sum_{j=1}^{5} (X_j - \overline{X})$

(h) $\frac{1}{k}\left(\sum_{i=1}^{k} X_i\right)^2$ (i) $\sum_{i=0}^{3} X_i Y_i$ (j) $\sum_{i=1}^{k} (X_i - \overline{X})^2 f_i$

(k) $\sum_{j=0}^{n} a_j X^j$ (l) $\sum_{n=1}^{6} n^2$

3.9 (a) X_{ij} is the number of responses to the ith rank type and jth answer type, i = 1, 2, 3, 4, j = 1, 2.

(b) Two-way (c) 2, 23, 6, 19, 11, 14, 13, 12.

(d) $\sum_{i=1}^{4} X_{i1}$, $\sum_{i=1}^{4} X_{i2}$, $\sum_{j=1}^{2} X_{1j}$, $\sum_{j=1}^{2} X_{2j}$, $\sum_{j=1}^{2} X_{3j}$, $\sum_{j=1}^{2} X_{4j}$

3.10 1560; sum of squares of the numbers of responses

3.11 10,000; square of the sum of the numbers of responses

3.12
$$\sum_{i=1}^{k} (X_i - \overline{X}) = X_1 - \overline{X} + X_2 - \overline{X} + \cdots + X_k - \overline{X}$$
$$= X_1 + X_2 + \cdots + X_k - k\overline{X}$$
$$= \sum_{i=1}^{k} X_i - k\left(\frac{1}{k} \sum_{i=1}^{k} X_i\right) = \Sigma X_i - \Sigma X_i = 0$$

3.13
$$\sum_{i=1}^{c} (X_i - \overline{X}) f_i = \Sigma (X_i f_i - \overline{X} f_i)$$
$$= \Sigma X_i f_i - \overline{X} \Sigma f_i = k\overline{X} - \overline{X}k = 0$$

3.14 $$\sum_{i=1}^{c} (X_i + 1)^2 f_i = \Sigma(X_i^2 + 2X_i + 1)f_i$$
$$= \Sigma X_i^2 f_i + \Sigma 2X_i f_i + \Sigma f_i$$
$$= \Sigma X_i^2 f_i + 2\Sigma X_i f_i + k$$

3.16 $$\sum_{i=1}^{c} (X_i - \overline{X})^2 = \Sigma(X_i^2 - 2\overline{X}X_i + \overline{X}^2) = \Sigma X_i^2 - 2\overline{X}\Sigma X_i + c\overline{X}^2$$
$$= \Sigma X_i^2 - 2\overline{X}(c\overline{X}) + c\overline{X}^2$$
$$= \Sigma X_i^2 - c\overline{X}^2 = \Sigma X_i^2 - \frac{1}{c}(\Sigma X_i)^2$$

Chapter 4

4.1 (a) 43.5, 43, 41 and 43 (bimodal), 43.35
(c) 8, 2.2, 2.7, 6.25% (d) 70% and 100%

4.2 57.9 mph

4.3 (b) 5.03, 5.06, 5, 7.55 (c) $\sqrt{3.97} \doteq 2$, 39.8% (indicative that these data are "rather" variable)

4.6 (a) F (b) F (c) T (d) T (e) F (f) T

4.9 (b) Mean = 114.2

4.13 (a) $s_X = 2.16$ (b) $s_X = 2.16$. Adding a constant to each observation does not alter the standard deviation.

Chapter 5

5.1 Let A_i (i = 1, 2, ..., 6) be the six individuals. The sample space consists of 15 outcomes: (A_1, A_2), (A_1, A_3), (A_1, A_4), (A_1, A_5), (A_1, A_6), (A_2, A_3), (A_2, A_4), (A_2, A_5), (A_3, A_4), (A_3, A_5), (A_3, A_6), (A_4, A_5), (A_4, A_6), (A_5, A_6), (A_2, A_6)

5.2 S = {(H, T), (T, H), (H, H), (T, T)}

5.5 1326

5.6 A = {HHT, THH, HTH} = B; yes

5.9 No; since, for example, $P(A) \neq P(A|B)$

5.11 (a) 4 (b) 16 (c) 12

5.15 $25 \times 26 \times 26 \times 10^3 = 16{,}900{,}000$

5.16 8!

5.17 12!/4!

5.19 13^4

5.20 $C_1^3 C_1^5 C_2^4 C_1^7 C_1^3 = 1890$ lunches

5.21 $10!/3!3!2! = 50{,}400$

5.22 26^4

5.23 (a) $r = 0$ or 1 (b) 15 (c) 20

5.24 If "conditions" were repeatable an infinite number of times, the proportion of times it will rain tomorrow is 80/100.

5.25 (a) 30/36 (b) 30/36 (c) 10/36 (d) 2/36 (e) 34/36 (f) 26/36 (g) 34/36 (h) 26/36 (i) 6/36

5.26 0.53

5.27 0.94

5.28 (a)

i	P[X = i]
0	1/32
1	5/32
2	10/32
3	10/32
4	5/32
5	1/32

(c) 31/32, 26/32, 26/32

5.29 0.64

5.31 0.36

5.33 0.60

5.34 0.76

5.35 27

5.37 Besides the trivial cases ($B = S$ and $C = A_i$, for any i), let $B = A_1 \cup A_2$, $C = A_1 \cup A_3$

5.38 $C_0^{10} C_3^{40} / C_3^{50}$

5.41 (a) 1/2 (b) 1/3

5.43 (a) 0.185 (b) 0.145

5.44 (a) 0.63 (b) 0.625

5.45 0.4

5.48 Without replacement: $C_0^{10} C_{10}^{40} / C_{10}^{50}$

5.50
$$\begin{aligned} P(\bar{A} \cup \bar{B}) &= P(\overline{A \cup B}) = 1 - P(A \cup B) = 1 - P(A) - P(B) + P(A \cap B) \\ &= P(\bar{A}) - P(B)\{1 - P(A)\} = P(\bar{A}) - P(B)P(\bar{A}) \\ &= P(\bar{A})\{1 - P(B)\} = P(\bar{A})P(\bar{B}) \end{aligned}$$

5.51

X	P(X)
3	3/36
4	8/36
5	14/36
6	8/36
7	3/36

5.52 1/9

5.54 1/3

5.55 53/90

5.56 4/10; 4/10

Chapter 6

6.1 $\mu_X = 0.9$, $\sigma_X = 0.3$

6.2 $\mu_X = 3.5$, $\sigma_X = 1.32$

(a) 99/128 (b) 120/128 (c) 120/128

6.4

X	P(X)
0	24/91
1	45/91
2	20/91
3	2/91

$\mu_X = 1$

6.5 $P[1 \le X \le 4] = 0.76$

6.7 0.38

6.9 (a) 0.27 (b) 0.66

6.13 (a) 0.12 (b) 0.14

6.14 (a) 40 (b) 6

6.15 0.68

6.17 $\overline{X} = 14.54$, $s_X = 2.235$; 68.4% and 94.9%

6.18 $P[X > 180] = P[Z > \frac{180 - 170}{5} = 2] = 0.0227$

6.19 0.2866

6.21 140.7

6.23 (a) T (b) F (c) T (d) T (e) T (f) T (g) T (h) F (i) T (j) T

6.24 (a) 0.4332 (b) 0.2291 (c) 0.9909 (d) 0.1469 (e) 0.1814 (f) 0.0668 (g) 0.0638 (h) 0.7333 (i) 0.8254 (j) 0.95

6.26 0.736

6.27 (a) 0.0227 (b) 0.4773

6.28 (a) 3.564 (b) -2.683

6.29 (a) 294 (b) 6 (c) 32 (d) 32

6.30 170.23

Chapter 7

7.1 X = number of yes votes out of 7

$P[X = k] = C_k^7 (1/2)^7$, k = 0, 1, 2, ..., 7

f_B = number of cities, under binomial model; i.e.,

$f_B(k) = 128 \cdot P[X_B = k]$, $\mu_{X_B} = 3.5$, $\sigma_{X_B} = 1.323$

f_N = number of cities, under normal model; i.e.,

$f_N(k) = 128 \cdot P[k - 1/2 \le X_N \le k + 1/2]$

k	$P[X_B = k]$	f_B	$P[k - 1/2 \le X_N \le k + 1/2]$	f_N	$\frac{\lvert f_N - f_B \rvert}{f_B}$
0	0.0078	1	0.0104	1.3	0.30
1	0.0547	7	0.0536	6.9	0.01
2	0.1641	21	0.1595	20.4	0.03
3	0.2734	35	0.2752	35.2	0.01
4	0.2734	35	0.2752	35.2	0.01
5	0.1641	21	0.1595	20.4	0.03
6	0.0547	7	0.0536	6.9	0.01
7	0.0078	1	0.0104	1.3	0.30

Note: $P[X_B = 4] = P[3.5 < X_N < 4.5] = P\left[\frac{3.5 - 3.5}{1.323} < Z < \frac{4.5 - 3.5}{1.323}\right]$

$= P[0 < Z < 0.756] = 0.2752$

7.2 Both (a) and (b) are very close to zero.

7.5 The normal approximation underestimates at small values of X and overestimates at large values of X.

7.6 H = 2.6 (a) $H \ge 0$ (b) H = 0 if $o_i = e_i$ for each i; that is, when there is perfect agreement between observed and expected frequencies. (c) $\Sigma(o - e) = \Sigma 128(P_B - P_N) = 128[\Sigma P_B - \Sigma P_N] \doteq 0$, where P_B and P_N are the binomial and normal probabilities. This is also true in other cases.

7.7 (a) 25% (b) 13% (c) 5%

7.8 (a) 108 (b) 97.2 (c) 0.995 (d) "Events" (shipwrecked) are independent

7.10

Data X	Data f	Fitting Classes	Z	Expected P	Expected f_e	PRE
8	8	6.5-9.5	-2.08	0.0188	10.0	0.24
11	86	9.5-12.5	-0.83	0.1845	97.8	0.14
14	259	12.5-15.5	0.42	0.4595	243.5	0.06
17	157	15.5-18.5	1.67	0.2897	153.5	0.02
20	20	18.5-21.5	2.92	0.0458	24.3	0.21
Sum	530			0.9983	529.1	

Note: $\hat{\mu}_X = \overline{X} = 14.5$, $\hat{\sigma}_X = s_X = 2.4$, $Z = (X - 14.5)/2.4$,
Y (see Table A.3), $f^* = \frac{3(530)}{2.4} Y$, $f_e = 530 \cdot P$,

$$P = P[12.5 < X < 15.5] = P\left[\frac{12.5 - 14.5}{2.4} < Z < \frac{15.5 - 14.5}{2.4}\right]$$

$$= P[-0.83 < Z < 0.42] = 0.4595,$$

$$\text{PRE} = |f - f_e|/f$$

$$H = \sum \frac{(f - f_e)^2}{f_e} = 3.6$$

Chapter 8

8.1 $P[\overline{X} > 25{,}000] = P\left[Z > \frac{25{,}000 - 14{,}725}{6219/5}\right] = P[Z > 8.26] < 0.0001$

8.2 97

8.3 0.5

8.4 In repeated sampling from a normal population the sample mean will exceed the value k, $100 \cdot \alpha\%$ of the time.

8.6 0.1056

8.7 (a) T (b) F (c) F (d) F (e) T (f) T (g) T (h) F
(i) T (j) F (k) T

8.8 This problem deals with sampling from a *finite* population.

(a) $\mu_X = 10/4 = 2.5$, $\sigma_X^2 = [1/(N - 1)]\Sigma(X_i - \overline{X})^2 = 5/3 = 1.67$

(b) There are $C_2^4 = 6$ samples.

(c)

Units	$\overline{X}$	s^2
1, 2	1.5	0.5
1, 3	2.0	2.0
1, 4	2.5	4.5
2, 3	2.5	0.5
2, 4	3.0	2.0
3, 4	3.5	0.5

(d) $\mu_{\overline{X}} = \frac{1.5 + \cdots + 3.5}{6} = \frac{15}{6} = 2.5$

so $\mu_X = \mu_{\overline{X}}$

(e) $\sigma_{\overline{X}}^2 = \frac{(1.5 - 2.5)^2 + \cdots + (3.5 - 2.5)^2}{5} = 0.50$

and $\sigma_X^2/n = 1.67/2 = 0.83$

so $\sigma_{\overline{X}}^2 \neq \sigma_X^2/n$

Remark . Had we used $\sigma_X^2 = (1/N)\Sigma(X_i - \overline{X})^2$, where N is the size of the population, then $\sigma_X^2 = 5/4 = 1.25$. Then using this same idea, the variance among the six sample means is $\sigma_{\overline{X}}^2 =$

$(1/6)\Sigma(\bar{X}_i - \bar{X})^2 = (1/6)(2.5) = 0.417$. Now, for *finite* populations, it is shown later (see Theorem 16.2) that $\sigma^2_{\bar{X}} = \sigma^2_X[(N - n)/n(N - 1)]$. In this example we have $1.25 \cdot (4 - 2)/[2(4 - 1)] = 0.417$. Therefore the relationship $\sigma^2_{\bar{X}} = \sigma^2_X/n$ holds only for infinite populations, since then $[(N - n)/n(N - 1)] = 1/n$.

Chapter 9

9.1 (a) Increase (b) 0.90 (c) Decrease

9.2 4; -3.182 and 3.182

9.3 $P[t_4 > 2.132] = 0.05$ (i.e., P[that in repeated sampling of size 5 from a normal population with unknown mean and unknown standard deviation the t value will be greater than 2.132] = 0.05); $P[-3.182 < t_3 < 3.182] = 0.95$.

9.5 $P[F_{7,19} > 3.77] = 0.01$; i.e., P[that in repeated sampling of sizes 8 and 20 from two independent normal populations with unknown means and given standard deviations the F value will be greater than 3.77] = 0.01.

9.7 t from the sample $= (\bar{X} - \mu_X)/s_{\bar{X}} = (6 - 5)/0.4 = 2.5$. Hence $P[t_9 > 2.5]$ = between 1% and 2.5% (from the t table).

9.9 $n_A = n_B = 6$; $\bar{X}_A = 29.3$, $\bar{X}_B = 18.0$; assume $\mu_A = \mu_B$, $\sigma^2_A = \sigma^2_B$
A: $\Sigma(X_i - \bar{X})^2 = 207.33$, B: $\Sigma(X_i - \bar{X})^2 = 4$; $s_{\bar{X}_A - \bar{X}_B} = 2.654$; the t value obtained is: $[(\bar{X}_A - \bar{X}_B) - (\mu_A - \mu_B)]/s_{\bar{X}_A - \bar{X}_B} = [(29.3 - 18.0) - 0]/2.654 = 4.26$. Hence $P[t_{10} > 4.26]$ = less than 0.005 (from the t table).

9.11 (a) 0.7540 (b) 0.99 (c) 0.95 (d) 3.22 (e) 2.55 (f) -3.012 and 3.012 (g) 2.447

9.17 For n = 2, $\bar{X} = (1/2)(X_1 + X_2)$ and $(n - 1)s^2_X = s^2_X = (X_1 - \bar{X})^2 + (X_2 - \bar{X})^2 = X_1^2 + X_2^2 - 2\bar{X}(X_1 + X_2) + 2\bar{X}^2 = X_1^2 + X_2^2 - X_1^2 - 2X_1X_2 - X_2^2 + (1/2)(X_1^2 + 2X_1X_2 + X_2^2) = (1/2)(X_1^2 - 2X_1X_2 + X_2^2) = (1/2)(X_1 - X_2)^2$. So $s_X = (1/\sqrt{2})|X_1 - X_2|$. Therefore $t = (\bar{X} - \mu_X)\sqrt{n}/s_X = [(1/2)(X_1 + X_2) - \mu_X]\sqrt{2}/(1/\sqrt{2})|X_1 - X_2| = (X_1 + X_2 - 2\mu_X)/|X_1 - X_2|$.

Chapter 10

10.1 (a) Bernoulli (b) 0.22, 0.44, 0.147

10.3 203.8, 28.7, 12.8 lb.

10.5 An estimator is said to be *precise* if its variance (or standard deviation) is small, while an *accurate* estimator has a small variance (or standard deviation) *and* small bias.

10.6 A point estimator yields a single value as an estimate of a parameter, whereas an interval estimator yields *two* values, a lower value L and an upper value U, such that one may assert that in repeated sampling the parameter will be in the interval, from L to U, with a given probability.

10.7 RMSE = 3.7

10.9 In repeated sampling the mean of the sampling distribution of $\overline{X}^2$ is not equal to σ_X^2, which is easily shown by example.

10.10 (6.34, 7.66). In repeated sampling of size 10, 90% of such intervals will contain the mean of the population and 10% will not contain them. We are 90% confident that (6.34, 7.66) covers the mean.

10.11 (a) (-0.21, 2.69)

10.12 We do not need the central limit theorem, when the large sample comes from a normal population, because we already know from Theorem 8.1 that $\overline{X}$ is also normally distributed.

10.14 (1757.6, 1842.4); 84.9

10.16 (16.6, 19.8). Since the value $\mu_X = 20$ does not belong to this interval, this sample is unlikely to have come from a population with mean 20 ($\alpha = 0.05$).

10.17 Not maintaining standard, at $\alpha = 0.05$. (Yes, at $\alpha = 0.01$.)

10.18 Yes, at $\alpha = 0.05$.

10.19 Yes, at $\alpha = 0.05$.

10.20 Buy Brand X (at $\alpha = 0.05$).

Chapter 11

11.1 (13.73, 16.27)

11.3 (95.3, 114.1)

11.4 (1.5, 3.1)

11.6 (a) (52.8, 97.8) (b) (1.3, 4.8)

11.7 (a) F (b) F (c) T (d) T (e) F

11.8 (a) As γ increases the t value increases for fixed d.f.

(b) The χ^2 distribution is useful in constructing confidence intervals for σ_X and σ_X^2.

(c) The t distribution for large sample size is approximately equal to the Z distribution, so that in confidence interval estimation the t value can be replaced by the Z value.

(d) Sampling from two independent normal populations with unknown means and standard deviations.

(e) The sample size is proportional to the square of the C.V.

11.11 Yes

11.12 (29.5, 37.3) if $\alpha = 0.05$

11.13 (137.8, 238.2) if $\alpha = 0.05$

11.14 The 90% C.I. for σ_X^2/σ_Y^2 is (0.23, 2.36) and the 90% C.I. for $\mu_X - \mu_Y$ is (-0.63, 0.25).

11.15 The 98% C.I. for σ_X^2/σ_Y^2 is (0.17, 8.45).

11.16 (-1669.2, 3669.2)

11.17 No, since the 95% confidence interval is (-2.9, 10.9).

Chapter 12

12.2 $\overline{X}$ = mean number of heads obtained from a sample.

12.4 Let H_0: $\mu_X = 2.3$ (given $\sigma_X = 0.41$); $\alpha = 0.05$. The test statistic is

$$Z = \frac{\overline{X} - \mu_X}{\sigma_X/\sqrt{n}}$$

Case 1. H_A: $\mu_X < 2.3$; $Z_{0.05} = -1.645$; decision rule: reject H_0 if $Z < -1.645$; $Z = -4.9$.

Case 2. H_A: $\mu_X \neq 2.3$; $Z_{0.025} = 1.96$; decision rule: reject H_0 if $|Z| > 1.96$; $|Z| = 4.9$.

Case 3. H_A: $\mu_X > 2.3$; $Z_{0.05} = 1.645$; decision rule: reject H_0 if $Z > 1.645$; $Z = 4.9$.

12.5 Power = P = 1 - P[accepting H_0 when in fact H_A is true] = probability of rejecting H_0 when in fact H_A is true. Suppose H_0: $\mu_X = 2.3$, H_A: $\mu_X = \mu_A$ where $\mu_A \neq 2.3$.

$$P = 1 - P[|Z| < 1.96, \text{ given } \mu_X = \mu_A \text{ (and } \sigma_X = 0.41, n = 16)]$$

$$= 1 - P\left[2.3 - 1.96\frac{\sigma_X}{\sqrt{n}} < \overline{X} < 2.3 + 1.96\frac{\sigma_X}{\sqrt{n}}, \text{ given } \mu_X = \mu_A\right]$$

$$= 1 - P[2.1 < \overline{X} < 2.5, \text{ given } \mu_X = \mu_A]$$

$$= 1 - P[20.48 - 9.76\mu_A < Z < 24.40 - 9.76\mu_A]$$

$$= 1 - P[C_1 < Z < C_2]$$

TABLE OF VALUES OF P FOR SELECTED VALUES OF μ_A

μ_A	C_1	C_2	$P[C_1 < Z < C_2]$	P
0	20.5	24.4	0	1
1.5	5.8	9.8	0	1
1.75	3.4	7.3	0	1
2	1.0	4.9	0.17	0.83
2.3	-2.0	2.0	0.95	0.05
2.6	-4.9	-1.0	0.17	0.83
3	-8.8	-4.9	0	1

Sketch of Power Function

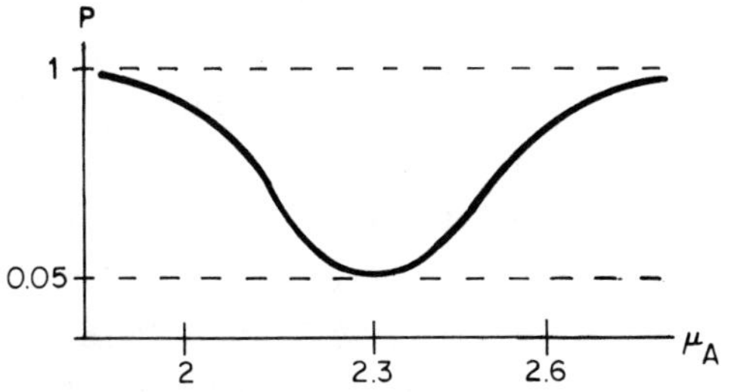

12.7 Yes; $Z = -3.3$

12.8 Type I error; type II error

12.9 Type II error; Type I error

12.10 (a) H_0: song is not effective; (b) H_0: song is effective

12.13 Yes; $Z = -3.02$

12.14 Yes; $Z = -1.74$

12.15 (a) F (α is a probability) (b) T (see Definition 12.8) (c) T (large samples, hence CLT) (d) F (we accept H_0 if 0 is in the confidence interval) (e) T (see Definition 12.3) (f) F (value of test statistic also depends on sample) (g) T (see Definition 12.7) (h) T (If H_0 is true only Type I error is possible.) (i) F (In fact no error has been made.) (j) T (Only type I error is possible.) (k) T ($\sigma_{\bar{X}} \to 0$ as n becomes larger and larger.) (l) T (see Definition 12.9) (m) T (Since, for example, $Z_{0.05} < Z_{0.01}$.)

12.16 The 99% C.I. for $p_1 - p_2$ is (-0.28, -0.02).

12.18 Examples are the following:

(a) Test statistics are used in testing null hypotheses about the mean of a normal population.

(b) The one-tail Z test is a statistical test.

(c) A simple null hypothesis may compete against a simple alternative hypothesis.

(d) The right-hand Z test at the 5% level has as critical region all Z values falling to the right of 1.645.

(e) Tabled Z values used in testing a null hypothesis depend on the choice of the significance level.

(f) A one-tail Z test for $H_0: \mu_X = 10$ cm against $H_A: \mu_X < 10$ cm consists in rejecting H_0 if the calculated Z is less than -1.645 when testing at the 5% level.

(g) The Z test is appropriate when testing null hypotheses about the mean of a population when the CLT is valid.

(h) The power of the two-tail Z test depends on values of the mean to the right and to the left of hypothesized value.

(i) Type II error can be expressed as one minus the power.

(j) The Z test for a difference between two means can be a one-tail Z test.

12.19 Reject H_0: $\mu_X - \mu_Y = 0$ at $\alpha = 0.05$ ($Z = -2.7$).

12.20 Reject H_0: $\mu_X - \mu_Y = 0$ at $\alpha = 0.01$ ($Z = -4.7$).

12.21 Reject H_0: $\mu_1 = \mu_2$ at $\alpha = 0.05$ ($Z = -1.8$).

12.23 Yes, at $\alpha = 0.05$ ($Z = 1.8$).

12.24 There is no significant difference in mortality rate. (In H_A: $p_1 \neq p_2$ we are assuming no prior belief on the effectiveness of the drug!) $Z = -0.90$.

12.26 Reject H_0: $p_1 = p_2$ at $\alpha = 0.01$ ($Z = 3.2$).

12.27 There is significant evidence that men (admitted to hospitals) have a higher rate than women. ($Z = 3.4$)

12.28 Do not reject H_0: $p_1 = p_2$ (versus H_A: $p_1 > p_2$) for $\alpha = 0.05$ ($Z = 1.5$). Thus the evidence is against Mr. Davis' claim.

12.29 Do not reject H_0: $p_1 = p_2$ for $\alpha = 0.05$ ($Z = 1.7$). So, there is n.s.d. in performance of the two machines.

12.30 (1)(a) 0.02; (b)(i) 0.025; (ii) 0.77; (iii) The first error because in this case he will play and lose his money in the long run.

(2)(a) 0.05; (b) If the value of the parameter increases, the probability of wrong decision increases.

(3)(a) 0.011; (b) If the value of the parameter decreases, the probability of wrong decision decreases.

(4)(a) 0.034; (b) If n increases, the probability of error increases.

(5)(a) 0.025; (b) 0.09; (c) Decide on p_2 if X = 12, 13, 14, 15 because $P[X = 12, 13, 14, 15 \text{ when } p = 0.5] = 0.0352$ but $P[X = 11, 12, 13, 14, 15] = 0.0769$ when $n = 15$, $p = 1/2$ (see Table A.1).

Chapter 13

13.1 (a) F (requires normality); (b) F (χ^2 value is always ≥ 0); (c) T (two sided alternative leads to two-tail test); (d) F

(We accept H_0!); (e) T (Does depend on σ_X^2/σ_Y^2, although this is 1 when H_0: $\sigma_X^2 = \sigma_Y^2$ is true.)

13.2 Let H_0: $\mu_X = 50$, H_A: $\mu_X < 50$ (insufficient amount!). Assume X is normal and sample is random. Choose $\alpha = 5\%$. The decision rule is: reject H_0 if $t < -t_{5,0.05} = -2.015$. The value of the test statistic (under H_0) is $t = (\overline{X} - 50)/(s_X/\sqrt{6}) = -1.3$. Thus do not reject H_0 and conclude that the manufacturer is not cheating his customers.

13.3 Reject H_0: $\mu_X = 3$ (H_A: $\mu_X \neq 3$) at $\alpha = 0.05$ ($t = 2.43$). The student has a tendency to overestimate the midpoint--a systematic error, on the average.

13.4 Reject H_0 ($t = 4.4$)

13.5 Remove the machine!

13.6 $t = 0.9$; n.s.d. between groups (assuming $\sigma_X = \sigma_Y$).

13.7 $\chi^2 = 4.5$; accept H_0: $\sigma_X = 2$.

13.8 $\chi^2 = 2.56$; accept H_0; sexes are equally represented.

13.9 $\chi^2 = 72$. Response is dependent on sex; in fact, females disagree less (18% versus 27%).

13.10 $F = 1.7$, $F_{4,5,0.05} = 5.19$; accept H_0: $\sigma_B^2 = \sigma_A^2$ (versus H_A: $\sigma_B^2 > \sigma_A^2$).

13.11 $\chi^2 = 27$, $\chi^2_{8,0.01} = 20.1$; distribution is not binomial.

13.12 (a) One should use the critical value $t_{n-1,\alpha}$. (b) The test is a two-tail χ^2 test. (c) χ^2 test for goodness-of-fit is valid when n is small. (d) d.f. = $r \cdot c - 1$.

(e) This test is valid for any two populations.

13.14 $\nu_1 \cdot F_{\nu_1,\infty,\alpha} = \chi^2_{\nu_1,\alpha}$ (For example, consider $\nu_1 = 4$, $\alpha = 0.01$.)

13.15 $F_{1,\nu,\alpha} = (t_{\nu,\alpha/2})^2$

13.19 Yates' correction for continuity is used in 2 by 2 contingency tables when $n < 200$. For example,

a	b
c	d

, $a + b + c + d = n$

Rule: if $ad > bc$, replace d by $d - 1/2$ and adjust cell frequencies so that marginal totals are the same, i.e.,

$a - \frac{1}{2}$	$b + \frac{1}{2}$
$c + \frac{1}{2}$	$d - \frac{1}{2}$

; if ad < bc, replace d by d + 1/2, and adjust

again, i.e.,

$a + \frac{1}{2}$	$b - \frac{1}{2}$
$c - \frac{1}{2}$	$d + \frac{1}{2}$

. Then

$$\chi^2 = \frac{(|ad - bd| - n/2)^2 n}{(a + b)(a + c)(b + d)(c + d)}$$

13.20 In a one-tailed F test the decision rule (if $\alpha = 0.10$) is: reject H_0 if $F > F_{\nu_1,\nu_2,0.10}$ (or $F < F_{\nu_1,\nu_2,0.90}$); and in a two-tailed F test we use: reject H_0 if F is outside the interval $(F_{\nu_1,\nu_2,0.95}, F_{\nu_1,\nu_2,0.05})$.

13.22 $\sigma_X^2 = \sigma_Y^2$ because $F = 1.5$ $(F_{6,11,0.05} = 3.09)$
$\mu_X = \mu_Y$ because $t = 1.2$ $(t_{17,0.025} = 2.11)$
Thus the two samples can be regarded as coming from the same population.

13.23 Let H_0: cancer is independent of smoking; H_A: cancer depends on smoking; choose $\alpha = 5\%$.

Data:			
	32(37.3)	60(54.7)	92
	11(5.7)	3(8.3)	14
	43	63	106

$\chi^2 = 9.6$; $\chi^2_{1,0.05} = 3.84$; reject H_0. There is a dependency. The data supports H_A. In fact 60/92 = 65% of smokers have lung cancer while 3/14 = 21% of nonsmokers have cancer.

13.24 $\chi^2 = 5.2$; $\chi^2_{4,0.05} = 9.49$; do not reject H_0.

13.25 $\chi^2 = 13$; $\chi^2_{1,0.05} = 3.84$; reject H_0. There is a greater conviction rate (77%) among identical twins than among nonidentical twins (12%).

13.26 $\chi^2 = 19.1$

13.27 $\chi^2 = 7.1$. More men prefer A (35%) than women (14%) and more women prefer B (86%) than men (65%).

13.29 $\chi^2 = 9.25$

13.30 $\chi^2 = 1.46$; fit is good

13.31 $\chi^2 = 9.6$

Chapter 14

14.2 (a) F (b) T (c) F (d) T (e) T (f) F (g) F (h) T
(i) F (j) T (k) T (l) T

14.3 (b) $\hat{\mu}_{Y|X} = -9.04 + 8.51X$ (d) (5.5, 11.5) (e) Yes
(f) 95% C.I. for $\mu_{Y|10}$ is (66.9, 85.3).

14.5 (a) r = +0.92 (b) Yes (c) reject H_0 (Z = 4)
(d) (0.77, 0.98)

14.7 r = +0.87

14.8 Yes, in fact in every case!

14.9 r^2 is the coefficient of determination. It measures the amount of variation in Y which can be attributed to the variation in X. So r^2 "determines" how important X is in relation to Y. $1 - r^2$ is the coefficient of alienation. It measures the amount of variation in Y which can be attributable to other sources than the variation of X. So $1 - r^2$ measures the "foreign" or "alien" portion of the variation in Y. For example, in Exercise 14.3, r = 0.922, $r^2 = 0.85$, $1 - r^2 = 0.15$. So 85% of the variation in Y can be attributed to the variation in X and 15% can be attributed to all other sources.

14.12 1. (e); 2. (d); 3. (d) (n > 2); 4. (a); 5. (d); 6. (e);
7. (b); 8. (b)

14.13 (a) Spearman (b) There is a weak (i.e., -0.39) inverse linear relationship between proficiency and neatness. (Sloppiness is positively correlated with proficiency!?)

14.15 a = 7.57, b = 1.28, r = 0.97; (a) 97.5 mph (b) Case A.

14.16 (a) No (b) Yes (c) If β were equal to zero then there is no regression! (d) None (e) r = +1 indicates positive (or direct) whereas r = -1 implies negative (or inverse) strong linear relationship.

14.17 (a) A freehand line is a subjective line. (b) Y intercept of a line through the origin is zero. (c) In repeated sampling of n pairs 90% of the calculated confidence intervals contain β and you have found one particular interval, namely, (0.26, 0.80). (d) The t values for testing H_0: $\alpha = 0$ and H_0: $\beta = 0$ depend, respectively, on s_a and s_b.

14.18 (a) Case B; (b) X = per capita income in province, Y = provincial expenditure per student; (c) Take one year and get 9 pairs of data. Perform a regression and correlation analysis. Repeat for several years.

14.19 Table 1: a = 191.9, b = -0.025, r = -0.87

Table 2: a = 99.96, b = -0.0075, r = -0.65

14.20 a = -18.01, b = 0.816, r = 0.92, $s_{Y|X} = 8.81$, $s_a = 4.12$, $s_b = 0.08$

Chapter 15

15.1 A nonrandomized experimental design is a design in which the treatments are allocated to the units in a systematic fashion rather than a random fashion. An example of such a design with two treatments is to allocate treatment A to four selected units, say the first four, and treatment B to the remaining four units. Schematically this systematic design looks like

A	A	A	A	B	B	B	B

A disadvantage of a nonrandomized design is the introduction of subjective bias in the experiment, so that the treatments cannot be compared on a fair and square basis.

15.2 Two levels of a drug A are to be compared in combination with three levels of a drug B. If the levels are denoted by a_0, a_1 and b_0, b_1, and b_2, respectively, then the six treatments combinations a_0b_0, a_0b_1, a_0b_2, a_1b_0, a_1b_1, and a_1b_2 constitute the treatment design.

15.4 Aptitude tests were given to 10 randomly selected female and 10 randomly selected male students in order to assess the difference of the sexes. Clearly sex is the qualitative factor in this experiment.

15.5 In order to compare the effects of two diets D_1 and D_2 relative to loss of weight of women diet D_1 was given over a period of time to highly overweight women and diet D_2 was given to moderately overweight women. In this experiment the difference in loss of weight due to the two diets is confounded with the difference in loss of weight due to the two weight groups.

15.6 Replication is a set of homogeneous experimental units to which each treatment has been allocated at least once. Using the treatment design of Exercise 15.2, if six mice of a litter are assigned randomly to the treatments (or vice versa) then the six mice comprise a replicate.

15.7 In an absolute experiment a single treatment is under study. Experimental error is the variability of the measurements due to the inherent variability among the units and variability due to other uncontrolled sources. It is not due to the treatment, since all units were treated alike. The experimental error is estimated by the variance or standard deviation of the measurements in an absolute experiment.

15.8 Randomization serves the purpose to obtain unbiased estimates of differences due to the treatments and a valid estimate of the experimental error.

15.9 (a) Randomly allocate the 5 diets to the 20 subjects such that each diet is applied 4 times. (For particulars, see the paragraph after Definition 15.10.)

(b)

B	D	A	C	A
D	E	B	D	C
E	D	C	A	B
A	C	E	B	E

(c)

Source	d.f.	SS	MS	F	$F_{4,15,0.05}$
Treatments	4	270.8	67.7	9.9	3.06
Error	15	103.0	6.9	--	--
Total	19	373.8	--	--	--

We are testing H_0: $\mu_A = \mu_B = \cdots = \mu_E$ versus H_A: there is some difference. Assuming the five populations are independent, homoscedastic normal populations, we reject H_0 since 9.9 > 3.06.

(d) $LSD = t_{15,0.025}\sqrt{6.9(1/4 + 1/4)} = 3.95$. $\mu_A > \mu_D = \mu_B = \mu_C > \mu_E$.

(e) The 95% C.I. for $\mu_A - \mu_B$ is 5 ± 3.95 or (1.05, 8.95). (Since this does not contain zero, and in fact it is to the right of zero, $\mu_A > \mu_B$.)

15.11 (c)

Source	d.f.	SS	MS	F	F_{table}
Blocks	3	335.35	--		--
Treatments	4	2782.70	695.7	4.8	$F_{4,12,0.05} = 3.26$
Error	12	1728.90	144.1	--	--
Total	19	4846.95	--	--	--

(d) (-20.2, 16.7)

15.12 crd: $H_0 = \mu_1 = \mu_2 = \cdots = \mu_v$

H_A: at least two μ_i's are not the same.

rcbd: $\bar{\mu}_{1.} = \bar{\mu}_{2.} = \cdots = \bar{\mu}_{v.}$

H_A: at least two $\bar{\mu}_i$'s are not the same.

15.13 A crd is proper when the experimental units are relatively homogeneous, so that the v treatments can be randomly allocated to them. If the units are not homogeneous and on some basis they may be grouped to form homogeneous groups and the treatments are randomly allocated to units within each group, then a rcbd is proper.

15.14 t = -1.2

15.15 The effect of unequal repetitions on the standard error of a difference between two means, is that for each difference between two means there is a different standard error. If repetitions are the same then any difference between two means has the same standard error.

15.16 (a) F (b) F (c) F (d) T (e) F (f) F (g) T (h) F
(i) F (j) F (k) T (1) F (m) F (n) T (o) T

15.17

Source	d.f.	SS	MS	F	F_{table}
Treatments	2	3194.52	1597.3	11.3	3.55
Error	18	2542.43	141.2	--	--
Total	20	5736.95	--	--	--

15.18

Source	d.f.	SS	MS	F	F_{table}
Blocks	2	81.50	--	--	--
Treatments	3	1267.58	422.53	6.0	4.76
Error	6	423.17	70.53	--	--
Total	11	1772.25	--	--	--

We reject H_0. There are significant differences between the four treatments. It is now important to discover the nature of these differences; e.g., is it displays or prices which are the important "factors"?; there are methods available to solve such problems.

15.19

Source	d.f.	SS	MS	F	F_{table}
Blocks	2	58.173	29.086	--	--
Treatments	1	0.081	0.081	6.9	5.32
Error	8	0.094	0.012	--	--
Total	11	58.348	--	--	--

There is a significant difference (at $\alpha = 5\%$) in achievement between bilingual and monolingual children in this experiment. There is a suggestion that on the average the monolingual achieve a slightly better score (over grades). The difference can be described by a confidence interval estimate.

15.20

Source	d.f.	SS	MS	F	F_{table}
Blocks	6	13.70	2.23	--	--
Treatments	4	38.36	9.59	15.6	2.78
Error	24	14.73	0.61	--	--
Total	34	66.79	--	--	--

There is a difference in the five treatments. Since LSD = 0.82 it can be shown that $\mu_1 > \mu_2 > \mu_3 = \mu_4 > \mu_5$. Thus there is no difference in cued 1-7 and cued 1-4; but constrained > cued. Also cued and constrained is > control.

Chapter 16

16.2 (a) $42,492.76 (b) 3.2 billion dollars (c) 1.3%

16.4 (b) 14.4 hr (c) 1.2 hr

16.5 $\hat{p} \pm SE = 0.20 \pm 0.03$; 7605 ± 440

16.9

Stratum i	Estimates of μ_i	τ_i	$\sigma_{\bar{X}_i}$	$\sigma_{N_i\bar{X}_i}$
1	4.70	14739	0.0802	251.4
2	3.85	1736	0.1937	87.4
3	2.97	5563	0.0636	119.1
4	4.16	37057	0.0568	506.3
5	1.40	3559	0.0498	126.7

$N = 16{,}910$, $\hat{\mu} = \bar{X} = 3.705$, $\hat{\tau} = 62{,}654$, $\hat{\sigma}_{\bar{X}} = s_{\bar{X}} = 0.035$,

$\hat{\sigma}_{N\bar{X}} = s_{N\bar{X}} = 591.8$

16.12 (a) A census is a complete enumeration of all the units in a population, while a sample survey is an incomplete enumeration.

(b) In an experiment a phenomenon is studied such that the conditions are under the control of the experimenter and they may be changed, while in a survey the phenomenon is studied under existing conditions.

(c) In probability sampling the probability that a unit will be included in the sample can be calculated, while in a nonprobability sample this cannot be achieved.

(d) In sampling with replacement the unit is replaced after each drawing, while in sampling without replacement the unit is not replaced.

(e) Unbiased estimators have the property that their averages in repeated sampling are equal to the parameters being estimated, while biased estimators do not have this property.

(f) Simple random sampling is applied to a population which has not been grouped into relatively homogeneous groups, while in stratified simple random sampling the sampling is done from such groups.

REFERENCES

1 Ansari, J. M. A. (1975). A study of 65 impotent males. *British Journal of Psychiatry* 127: 337-341.

2 Armore, S. J. (1967). *Fundamental Statistical Analysis and Inferences*, Fourth Edition. Toronto: Wiley.

3 Bain, B. (1973). Toward a theory of perception: Participation as a function of body-flexibility. *Journal of General Psychology* 89: 157-296.

4 Bernstein, A. L. (1964). *A Handbook of Statistics Solutions for the Behavioral Sciences*. Toronto: Holt, Rinehart and Winston.

5 Black. F. W. (1974). Self-concept as related to achievement and age in learning-disabled children. *Child Development* 45: 1137-1140.

6 Bolster, R., and R. L. Schuster (1973). Magnitude and duration of cocaine reinforcement. *Journal of Experimental Analysis of Behavior* 20: 119-129.

7 Bull, H. C., and P. H. Venables (1974). Speech perception in schizophrenia. *British Journal of Psychiatry* 125: 350-354.

8 Chun, K., and J. B. Campbell (1974). Dimensionality of the Rotter Interpersonal Trust Scale. *Psychological Reports* 35: 1059-1070.

9 Clark, D. F., and A. W. Johnston (1974). XYZ individuals in a special school. *British Journal of Psychiatry* 125: 390-396.

10 Cornfield, J. (1956). A statistical problem arising from retrospective studies. *Proceedings of the 3rd Berkeley Symposium* 4: 135-148.

11 Cox, D. R. (1969). *Analysis of Binary Data*. London: Methuen.

12 Deming, W. E. (1960). *Sample Design and Business Research*. New York: Wiley.

13 Deutscher, I. (1964). The quality of post parental life: Definitions of the situation. *Journal of Marriage and the Family* 26 (No. 1): 53.

14 Dougherty, J., and R. Pickens (1973). Fixed-interval schedules of intravenous cocaine presentation in rats. *Journal of the Experimental Analysis of Behavior* 20 (No. 1): 111-118.

15 Dufrenoy, J. (1938). The publishing behavior of biologists. *Quarterly Review of Biology* 13: 207-210.

16 Ecker, J., J. Levine, and E. Zigler (1973). Impaired sex-role identification in schizophrenia expressed in the comprehension of humor stimuli. *Journal of Psychology* 83: 67-77.

17 Edwards, W. (1953). Probability preferences in gambling. *American Journal of Psychology* 66: 349-364.

18 Ellis, K. J., and S. H. Cohn (1975). Correlation between skeletal calcium mass and muscle mass in man. *Journal of Applied Physiology* 38:(No. 3, March): 455-460.

19 Emerson, R. A., and E. M. East (1913). The inheritance of quantitative characters in maize. *Bulletin of the Agricultural Experiment Station of the University of Nebraska* No. 2, 5-120.

20 Fagot, B. I. (1973). Sex-related stereotyping of toddlers' behaviors. *Developmental Psychology* 9 (No. 3): 429.

21 Federer, W. T. (1973). *Statistics and Society*. New York: Marcel Dekker.

22 Federer, W. T., and L. N. Balaam (1972). *Bibliography on Experiment and Treatment Design, Pre-1968*. Edinburgh: Oliver and Boyd.

23 Feller, W. (1964). *An Introduction to Probability Theory and Its Applications*, Second Edition. New York: Wiley.

24 Ferguson, G. A. (1966). *Statistical Analysis in Psychology and Education*, Second Edition. New York: McGraw-Hill.

25 Finch, A. J., P. A. Deardorff, and L. E. Montgomery (1974). Reflection-impulsivity: Reliability of the matching familiar figures test with emotionally disturbed children. *Psychological Reports* 35: 1133-1134.

26 Fisher, R. A. (1935). *The Design of Experiments*, 8th ed., 1966. Edinburgh: Oliver and Boyd.

27 Gallup (1975). Gun control, from the *Gallup Opinion Index*, Report No. 123, Sept. 8-11.

28 Games, P. A., and G. R. Klare (1967). *Elementary Statistics: Data Analysis for the Beharioral Sciences*. Toronto: McGraw-Hill.

29 Gauss, K. F. (1809). *Theoria Motus Corporum Coelestium.* Hamburg: Perther and Bessen. [English translation by C. H. Davis (1957), Boston: Little, Brown.]

30 Goodwin, W. C., and M. T. Erickson (1973). Developmental problems and dental morphology. *American Journal of Mental Deficiency* 43: 199-204.

31 Hanley, J., W. R. Rickles, P. H. Crandall, and R. D. Walter (1972). Automatic recognition of EEG correlates of behavior in a chronic schizophrenic patient. *American Journal of Psychiatry* 128 (Part 3): 1424-1428.

32 Hays, W. L. (1963). *Statistics for Psychologists.* Toronto: Holt, Rinehart and Winston.

33 Hess, E. H., A. L. Seltzer, and J. M. Schlien (1965). Pupil response of hetero- and homosexual males to pictures of men and women: a pilot study. *Journal of Abnormal Psychology* 70: 165-168.

34 Hoel, P. G. (1960). *Introduction to Mathematical Statistics*, Second Edition. New York: Wiley.

35 Hubert, J. J. (1968). *Report on Compensation for Victims of Crime.* Institute of Law Research and Reform, University of Alberta.

36 Hubert, J. J. (1975). *A Rank-Frequency Model for the Scientific Productivity of Canadian Mathematicians.* Technical Report No. 1975-31, Dept. of Math. and Statist., University of Guelph, pp. 1-19.

37 Jenkins, S. (1963). *Comparative Recreation Needs and Services in New York Neighborhoods.* Research Dept., Community Council of Greater New York.

38 Jones, M. C., N. Bayley, J. W. Macfarlane, and M. P. Honzik (1971). *The Course of Human Development.* Toronto: Xerox College Publishing.

39 Kaufman, A. S., and N. L. Kaufman (1972). Tests built from Piaget's and Gesell's tasks as predictors of first-grade achievement. *Child Development* 43: 521-535.

40 Kendall, M. (1975). *The Fiji Fertility Survey: A Critical Commentary on Administration and Methodology.* No. 15 in the Occasional Papers Series of the International Statistical Institute, written by M. A. Sahib *et al.* (Director WFS: Sir M. Kendall).

41 Kenney, J. F., and E. S. Keeping (1954). *Mathematics of Statistics, Part One*, Third Edition. Toronto: Van Nostrand.

42 Lasaga, J. I., and A. M. Lasaga (1973). Sleep learning and progressive blurring or preception during sleep. *Perceptual and Motor Skills* 37: 51-62.

43 Lawton, M. S., and R. D. Seim (1973). Developmental investigation of tactual-visual integration and reading achievement. *Perceptual and Motor Skills* 36: 375-382.

44 Lazarsfeld, P. F. (1955). Interpretation of statistical relations as a research operation, in *The Language of Social Research* (P. F. Lazarsfeld and M. Rosenberg, Eds.). Glencoe, Ill.: The Free Press. Pp.115-125.

45 Liebert, A. M., and A. A. Baumeister (1973). Behavioral variability among retardates, children and college students. *Journal of Psychology* 83: 57-65.

46 Loether, H. J., and D. G. McTavish. *Inferential Statistics for Sociologists.* Boston: Allyn and Bacon.

47 Longhurst, T. M. (1974). Communication in retarded adolescents: Sex and intelligence level. *American Council of Mental Deficiency* 78 (No. 5): 607-618.

48 MacInnes, J. W., and L. L. Uphouse (1973). Effects of alcohol on acquisition and retention of passive-avoidance conditioning in different mouse strains. *Journal of Comparative and Physiological Psychology* 84: 398-402.

49 MacMillan, J. W. (1931). Linkage studies with the tomato. III. Fifteen factors in 6 groups. *Trans. R. Canadian Inst.* 18: 1-19.

50 Masters, J. C., and J. R. Mokros (1973). Effects of incentive magnitude upon discriminative learning and choice preference in young children. *Child Development* 44: 225-231.

51 Melamed, A. R., M. S. Silverman, and G. J. Lewis (1975). Personal orientation inventory. *Review of Religious Research* 16 (No. 2): 105-110.

52 Mendenhall, W., L. Ott, and R. F. Larson (1974). *Statistics: A Tool for the Social Sciences.* California: Wadsworth.

53 Michalos, A. C. (1974). *Crime in Canada and the United States of America.* Technical Report, University of Guelph. Pp.1-53.

54 Milicer, H. (1968). Age at menarche of girls in Wroclaw, Poland, in 1966. *Human Biology* 40: 249-259.

55 Mills, J. and Y. P. Seng (1954). The effect of age and parity of the mother on birth weight of the offspring. *Annals of Human Genetics* 19: 58-73.

56 Müller-Lyer, F. C. (1889). Optische Urteilstäuschungen. *Dubois-Reymonds-Archiv für Anatomie und Physiologie.* Supplement Volume: pp.263-270.

57 Parzen, E. (1960). *Modern Probability Theory and Its Applications.* New York: Wiley.

58 Parzen, E. (1962). On estimation of a probability density function and mode. *Annals of Mathematical Statistics* 33: (No. 3): 1065-1076.

59 Runyon, R. P., and A. Haber (1970). *Student Workbook to Accompany Fundamentals of Behavioral Statistics*, Second Edition. Don Mills, Ontario: Addison-Wesley.

60 Russet, B. M. (1964). *World Handbook of Political and Social Indicators*. New Haven: Yale Univ. Press.

61 Senter, R. J. (1969). *Analysis of Data: Introductory Statistics for the Behavioral Sciences*. Glenview, Ill.: Scott, Foresman.

62 Shaw, A., T. Fostvedt, and E. S. Keeping (1960). *Statistics Laboratory Manual*. University of Alberta, Dept. of Mathematics.

63 Silverman, B. I., and S. T. Battram (1975). Canadian's and American's mutual awareness. *Canadian Journal of Behavioral Science* 7 (No. 2): 104-117.

64 Slotnick, R. S., and J. Bleiberg (1974). Authoritarianism, occupational sex-typing, and attitudes toward work. *Psychological Reports* 35: 763-770.

65 Smith, F. L. (1939). A genetic analysis of red seed-coat colour in *Phaseolus vulgaris*. *Hilgardia* 12: 553-621.

66 Smith, R. M., and J. M. Hanna (1975). Skin folds and resting heat loss in cold air and water: temperature equivalence. *Journal of Applied Physiology* 39: (No. 1): 93-102.

67 Statistics Canada (1973a). *Family Expenditure in Canada (1969)*. Volume 1.

68 Statistics Canada (1973b). *Trusteed Pension Plans Financial Statistics*. Catalogue 74-201 Annual.

69 Statistics Canada (1973c). *Vital Statistics*, Vol. II, *Marriages and Divorces*. Catalogue 84-205 Annual.

70 Statistics Canada (1975). *Statistical Report on the Operation of Unemployment Insurance Act*.

71 Thorndike, R. L., and E. Hagen (1955). *Measurement and Evaluation in Psychology and Education*. New York: Wiley.

72 Tsushima, W. T., and T. P. Hogan (1975). Verbal ability and school achievement of bilingual and monolingual children of different ages. *Journal of Educational Research* 68 (No. 9): 349-353.

73 United Nations (1962). *Demographic Yearbook (1961)*. New York.

74 United Nations (1963). *Compendium of Social Statistics (1963)*. New York.

75 Webb, P., and M. Hiestand (1975). Sleep metabolism and age. *Journal of Applied Physiology* 38 (No. 2): 257-262.

76 Weiskott, G. N. (1974). Moon phases and telephone counseling calls. *Psychological Reports* 35: 752-754.

77 Wilkinson, B. (1968). World Series pools. *The American Statistician* 22 (No. 4): 35.

78 Williams, K. G., and L. R. Goulet (1975). The effects of cueing and constraint instructions on children's free recall performance. *Journal of Experimental Child Psychology* 19 (No. 3): 464-475.

79 Wyatt, R. J., R. Vaughan, M. Galan, J. Kaplan, and R. Green (1972). Behavioral changes in chronic schizophrenic patients given L-5-hydroxytryptophan. *Science* 177: 1124-1126.

80 Wynne, D. F., and T. F. Hartnagel (1975). Race and plea negotiations - an analysis of some Canadian data. *Canadian Journal of Sociology* 1 (No. 2): 147-155.

INDEX